Recent Advances in Applied Thermal Imaging for Industrial Applications

V. Santhi
VIT University, India

A volume in the Advances in Civil and Industrial Engineering (ACIE) Book Series

IGI GLOBAL
DISSEMINATOR OF KNOWLEDGE
www.igi-global.com

Published in the United States of America by
IGI Global
Engineering Science Reference (an imprint of IGI Global)
701 E. Chocolate Avenue
Hershey PA 17033
Tel: 717-533-8845
Fax: 717-533-8661
E-mail: cust@igi-global.com
Web site: http://www.igi-global.com

Library of Congress Cataloging-in-Publication Data

Names: Santhi, V., 1971- editor.
Title: Recent advances in applied thermal imaging for industrial applications / V. Santhi, editor.
Description: Hershey, PA : Engineering Science Reference, 2017. | Includes bibliographical references and index.
Identifiers: LCCN 2017002969| ISBN 9781522524236 (hardcover) | ISBN 9781522524243 (ebook)
Subjects: LCSH: Infrared imaging--Industrial applications.
Classification: LCC TA1570 .R4184 2017 | DDC 621.36/72--dc23 LC record available at https://lccn.loc.gov/2017002969

This book is published in the IGI Global book series Advances in Civil and Industrial Engineering (ACIE) (ISSN: 2326-6139; eISSN: 2326-6155)

British Cataloguing in Publication Data
A Cataloguing in Publication record for this book is available from the British Library.

Advances in Civil and Industrial Engineering (ACIE) Book Series

ISSN:2326-6139
EISSN:2326-6155

Mission

Private and public sector infrastructures begin to age, or require change in the face of developing technologies, the fields of civil and industrial engineering have become increasingly important as a method to mitigate and manage these changes. As governments and the public at large begin to grapple with climate change and growing populations, civil engineering has become more interdisciplinary and the need for publications that discuss the rapid changes and advancements in the field have become more in-demand. Additionally, private corporations and companies are facing similar changes and challenges, with the pressure for new and innovative methods being placed on those involved in industrial engineering.

The **Advances in Civil and Industrial Engineering (ACIE) Book Series** aims to present research and methodology that will provide solutions and discussions to meet such needs. The latest methodologies, applications, tools, and analysis will be published through the books included in **ACIE** in order to keep the available research in civil and industrial engineering as current and timely as possible.

Coverage

- Hydraulic Engineering
- Optimization Techniques
- Ergonomics
- Coastal Engineering
- Transportation Engineering
- Engineering Economics
- Operations Research
- Earthquake engineering
- Urban Engineering
- Quality Engineering

IGI Global is currently accepting manuscripts for publication within this series. To submit a proposal for a volume in this series, please contact our Acquisition Editors at Acquisitions@igi-global.com or visit: http://www.igi-global.com/publish/.

The Advances in Civil and Industrial Engineering (ACIE) Book Series (ISSN 2326-6139) is published by IGI Global, 701 E. Chocolate Avenue, Hershey, PA 17033-1240, USA, www.igi-global.com. This series is composed of titles available for purchase individually; each title is edited to be contextually exclusive from any other title within the series. For pricing and ordering information please visit http://www.igi-global.com/book-series/advances-civil-industrial-engineering/73673. Postmaster: Send all address changes to above address.

Titles in this Series

For a list of additional titles in this series, please visit: www.igi-global.com

Modeling and Simulation Techniques in Structural Engineering
Pijush Samui (National Institute of Technology Patna, India) Subrata Chakraborty (Indian Institute of Engineering Science and Technology (IIEST), Shibpur, India) and Dookie Kim (Kunsan National University, South Korea)
Engineering Science Reference • copyright 2017 • 524pp • H/C (ISBN: 9781522505884) • US $220.00 (our price)

Computational Modeling of Masonry Structures Using the Discrete Element Method
Vasilis Sarhosis (Newcastle University, UK) Katalin Bagi (Budapest University of Technology and Economics, Hungary) José V. Lemos (National Laboratory for Civil Engineering, Portugal) and Gabriele Milani (Technical University in Milan, Italy)
Engineering Science Reference • copyright 2016 • 505pp • H/C (ISBN: 9781522502319) • US $210.00 (our price)

Handbook of Research on Emerging Innovations in Rail Transportation Engineering
B. Umesh Rai (Chennai Metro Rail Limited, India)
Engineering Science Reference • copyright 2016 • 664pp • H/C (ISBN: 9781522500841) • US $235.00 (our price)

Advanced Research on Nanotechnology for Civil Engineering Applications
Anwar Khitab (Mirpur University of Science and Technology, Pakistan) and Waqas Anwar (Mirpur University of Science and Technology, Pakistan)
Engineering Science Reference • copyright 2016 • 339pp • H/C (ISBN: 9781522503446) • US $195.00 (our price)

Emerging Challenges and Opportunities of High Speed Rail Development on Business and Society
Raj Selladurai (Indiana University Northwest, USA) Peggy Daniels Lee (Kelley School of Business, Indiana University, USA) and George VandeWerken (Providence Bank, USA)
Engineering Science Reference • copyright 2016 • 289pp • H/C (ISBN: 9781522501022) • US $210.00 (our price)

Advanced Manufacturing Techniques Using Laser Material Processing
Esther Titilayo Akinlabi (Univeristy of Johannesburg, South Africa) Rasheedat Modupe Mahamood (University of Johannesburg, South Africa & University of Ilorin, Nigeria) and Stephen Akinwale Akinlabi (University of Johannesburg, South Africa)
Engineering Science Reference • copyright 2016 • 288pp • H/C (ISBN: 9781522503293) • US $175.00 (our price)

Editorial Advisory Board

Table of Contents

Detailed Table of Contents

Section 1
Applications of Computational Intelligence in Thermal Image Processing

Thermal images play a key role in medical, military and industry applications. It is also used to provide security for the nation through the detection of landmines and targets. Analyzing thermal images to get meaningful information is a tedious task. Therefore, digital thermal image processing is considered as very essential research areas in the current scenario to analyze and interpret the same. There are many traditional approaches exists to carry out thermal image processing but to ensure automaton in image processing the role computational intelligence is essential. Keeping this as objective, basics of computational intelligence and its advancement is presented in this section. In addition, thermal images are used for disease diagnosis in medical field. In this section, a novel approach using fuzzy logic is presented to detect flu using thermal images of human body. Similarly, fuzzy logic based approach is discussed in detail for tumor detection in brain. Thus, computational approaches can be used to carryout image processing, in particular thermal images with high accuracy for automation.

It is important to have a system for detecting infectious disease that spread very rapidly across the world via a community of human or animals or both, either through airborne particles or viruses. As it is a very big threat across the world, it is required to screen potential carriers for such disease at crucial locations like airports to safeguard human community. In this chapter, a fuzzy logic based approach is proposed for flu detection using thermal images. As a symptom of infection, the body temperature of a disease carrier is higher than normal people. The increased temperature in human body would be captured by thermal images and that would be helpful in detecting the flu. Authors have proposed system able to capture a thermogram of the human subject, detecting the eye region of the human subject, calculating the pixels values around the detected eye region, converted to temperature readings and further classified the subject's body temperature whether the subject satisfies a flu condition or not. It is proved that the proposed fuzzy logic based Viola Jones algorithm can trace out flu infectious personal from the input thermal images with high accuracy.

In general, MRI images play a major role in medical diagnosis. Imaging is referred as visual representation of an object form. These objects in images need to be segmented and classified with high accuracy as it plays a major role in making decision. In this chapter, authors have carried out a study on MRI tumor segmentation using Fuzzy C-Means methods and compared results other approaches too. Image segmentation is a technique which divides an image into its constituent regions or objects. Segmentation continues till area of interest or the specified object of target is reached. In order to measure and compare the accuracy, authors have used Dice coefficient and Jaccard's measure. It is observed that Fuzzy-C means approaches are performing better than traditional approaches.

Computational intelligence plays a major role in image processing in recent days. The basics of computational intelligence and its principle have been presented elaborately in this chapter. The author of this chapter explained intelligence components that are used predominantly in thermal image processing like Fuzzy logic, Fuzzy logic, Probabilistic reasoning, Neural networks, Genetic algorithms, Fuzzy association rules etc. In particular, neural networks and fuzzy logic concepts are most often used in thermal processing. In medical field, for disease diagnosis fuzzy logic approach and neural network approach has been well proved.

Section 2
Thermal Imaging in Human Health Care and Diagnosis System

Images in different modality play a major role for diagnosis in medical field. During imaging process, instruments come in contact to human body that may lead to severe pain, example MRI images. Similarly human body may also be exposed to radiation. Thus to avoid human contact and radiation, thermal imaging can be considered as complementary to other modality of images. These images are used to measure thermal energy radiated from human body. The measurement of heat helps us to diagnosis disease to greater extent and its cost is also very low compare to other images of modalities. Keeping this as objective, in this section the role thermal images in medical science is discussed elaborately. In addition, a novel approach using iris feature for human health diagnosis is also presented.

In recent days breast cancer is considered as a very serious disease that threats for women unless it is screened and treated at right time. Thus women need to undergo screening test regularly and periodically. But breast cancer screening is carried out on, in contact with human body which leads to pain and exposure to radiation too. It is required to have an alternative system for cancer screening; there comes the role of thermal imaging. Thermal imaging is an effective screening tool for breast cancer prediction. Thermal imaging is a non-destructive, non-contact system. In this chapter, the main concept and the different applications of thermal imaging is presented. In addition, elaborate studies in thermography have been carried out to justify that it can be considered as a complementary tool to detect breast diseases.

In general, biometric data is predominantly used in many real time applications for authentication, in particular iris features. It could also be used for medical diagnosis

In this chapter, optical analysis of solar concentrator with image processing is presented in order to improve thermal efficiency of the solar concentrator. The optical efficiency of solar concentrators has a high impact on its thermal performance. Thus a system determining the geometrical accuracy of a solar concentrator system is necessary. Thus the order of measurement techniques used to determine geometric errors of the solar concentrating system have been described in detail. As the optical efficiency of solar concentrators has a high impact on its thermal performance, a qualification method determines the geometrical accuracy of a solar concentrator system is necessary and it is presented in detail.

Thermal image quality enhancement is often required for its interpretation with high accuracy. It can be carried out through filtering approaches and contrast adjustment process. In this chapter, three recent methodologies for multidimensional processing on sequences of thermographic images are presented. All three methods for image enhancement techniques are presented in detail, starting with its mathematical foundations, algorithmic procedures, advantages and limitations.

In the recent world, providing security right from a home, marketing area, financial sector, etc. is a very challenging task. There are several ways to ensure security in real time; one among them is human identification with high accuracy through biometric data such as thumb impression and iris features. In this chapter, a new approach for personal authentication using finger knuckle geometric and texture feature measurement are used. In addition, the entire finger back region including proximal and distal phalanx of the finger knuckle surface also added for recognition. The geometric measures are derived through angular geometric and texture measures are derived through statistical-based texture analysis methods.

Preface

In general, one picture or one image conveys thousands of words. It would be easy to analyze and interpret image data for more information. Images are available in different modality based on the band of frequency used in spectrum for capturing it. Thermography is a technique in which thermal energy is used to measure heat emitted from an object. Thermography is also called as Infrared imaging in which surface temperature of objects are measured. This technology uses pseudo-color image processing to describe the temperature distribution of object surface with infrared temperature calibration. Infrared thermography cameras produce images of invisible infrared or heat radiation and provide precise non-contact temperature measurement capabilities. In real life, all the objects gets heated up before it fails to function, in these situations thermography based diagnostics tool could be used to avoid malfunctioning of devices or equipments. The measurement of heat helps us to carry out further analysis depending on the application for which it is applied.

Thermography technique could be used in many diverse applications because of its low cost. In industries it helps to improve product quality, work safety, efficiency by reducing failure rate of objects due to heat. In industrial environments thermal imaging is being used for finding hot-spots that can lead to failures in electrical and mechanical installations. Electrical cabinets and motor control centers are regularly scanned with a thermal imaging camera for temperature monitoring for early stage detection of anomalies to avoid disasters. On the other side, it plays a major role in military and defense department for detecting land mines, rescuing people from natural disasters. Thus, Infrared thermal imaging has been used extensively in military, industry and security applications. It is also used applications such as remote sensing, electrical, electronics, petrochemical, plant physiology, geology and medicine.

Thermal imaging devices could capture images with higher spatial resolution that leads to variety of engineering applications starting from the fundamental scientific researches to the development of advanced scientific researches. Similarly advancements in image processing techniques allow fast and accurate processing of thermal images. By combing advancements in both thermal imaging techniques and image processing techniques, many real world problems in military and industrial domains could be addressed with less effort. Keeping this in mind, the objective of this edited book is to provide the researchers of various fields like computer science, electrical, electronics and Information Technology the recent advances and approaches of infrared thermography and its applications in industries. To achieve these objectives, theoretical back ground, advances and its applications to real time problems are emphasized. I believe, this effort can make this collection interesting and highly attractive to research community.

Many of the researchers in different organizations across the globe have been doing research in thermal image processing. To keep abreast with this development, it is an effort to bring the recent advances in thermal image processing and its applications in a cohesive manner. The main objective is to bring most of the major developments in the above mentioned area in a precise manner, so that it can serve as a handbook for many researchers. I trust and hope that this book will help the researchers, who have interest in Computational Intelligence in thermal image processing and its recent developments in industrial applications, to keep insight into recent advances and their importance in real life applications.

While these are a few examples of issues, my intention in editing this book is to offer concepts of Infrared Thermography and its applications in a precise and clear manner to the research community. Editing this book is an attempt to provide frontier advances and applications of Infrared Thermography. The conceptual basis required achieving in depth knowledge in the field of computer science, electronics, electrical and information technology will be stressed upon. All these motivated me towards Infrared Thermography and its applications.

This book comprises of three sections. First section broadly covers foundation and principles of computational intelligence and its applications in thermal image processing. The second section of the edited book discusses about thermal imaging in human health care and diagnosis system. Third section discusses about various industrial applications of thermal images. In addition it also discusses about various approaches to enhance quality of thermal images.

Analyzing thermal images to get meaningful information is a tedious task. Therefore, digital thermal image processing is considered as very essential research areas in the current scenario to analyze and interpret the same. There are many traditional approaches exists to carry out thermal image processing but to ensure automaton in image processing the role computational intelligence is essential. Computational intelligence techniques have firmly established themselves as feasible alternate mathematical tools for more than a decade. They have been extensively employed in many systems and application domains, especially in signal processing, automatic control, industrial and consumer electronics, robotics, finance, manufacturing systems, electric power systems, and power electronics. In Chapter 1, a fuzzy logic based approach is proposed for flu detection using thermal images. As a symptom of infection, the body temperature of a disease carrier is higher than normal people. The increased temperature in human body would be captured by thermal images and that would be helpful in detecting the flu. Authors have proposed system able to capture a thermogram of the human subject, detecting the eye region of the human subject, calculating the pixels values around the detected eye region, converted to temperature readings and further classified the subject's body temperature whether the subject satisfies a flu condition or not. It is proved that the proposed fuzzy logic based Viola Jones algorithm can trace out flu infectious personal from the input thermal images with high accuracy. In Chapter 2, authors have carried out a study on the brain tumor segmentation using Fuzzy C-means methods and compared results other approaches too. Image segmentation is a technique which divides an image into its constituent regions or objects. Segmentation continues till area of interest or the specified object of target is reached. In order to measure and compare the accuracy, authors have used Dice coefficient and Jaccard's measure. It is observed that fuzzy-c means approaches are performing well than traditional approaches. Chapter 3 gives insight into basics of intelligence components that are used predominantly in thermal image processing like Fuzzy logic, Probabilistic reasoning, Neural networks, Genetic algorithms, Fuzzy association rules etc. In medical field, for disease diagnosis fuzzy logic approach and neural network approach have been well proved.

In general images of different modality play a major role in medical diagnosis. Imaging system capture images in contact with human body that may lead to severe pain also human body may be exposed to radiation also. Thus to avoid human contact and radiation thermal imaging can be considered as complementary to other modality images in medical applications. Thermal imaging is a non-destructive, non-contact system. In Chapter 4, the main concept and different applications of thermal imaging in medical application is discussed in detail. In addition, elaborate studies in thermography have been carried out to justify that it can be considered as a complementary tool to detect breast diseases. In Chapter 5, application of iris features such as texture and geometric features are deployed for human health diagnosis. The texture features present in the human iris are extracted using the mathematical statistical measure which is used to specify the characteristics of the texture of an image using gray-level co-occurrence matrix. The iris and pupil are extracted and correlated to the compactness features of the circle. Based on the comparison the results, prediction of abnormalities in the iris texture take place and helps to identify the affected person.

By combing advancements in both thermal imaging techniques and image processing techniques, many real world problems in military, and industrial domains could be addressed with less effort. In addition, usage of thermal images in security application and its enhancement methods are presented in detail. Moreover, optical analysis of solar concentrator with an image process is presented, to improve thermal efficiency of the solar concentrator. In Chapter 6, various components of thermal imaging system and its application is presented in detail. Infrared thermal imaging (IRTI) can provide the temperature mapping without a physical contact with the object of interest from a reasonable distance. In general, typical IRTI system comprises of a thermal camera equipped with infrared detector, a signal processing unit and an image acquisition system, usually in the form of an embedded system. Such cameras are utilized for applications like fault detection, irrigation management, motion detection, etc. This chapter briefly introduces use of thermal imaging in medicine, agriculture, environment, smart home/cities and security applications. In Chapter 7, optical analysis of solar concentrator with image processing is presented in order to improve thermal efficiency of the solar concentrator. The optical efficiency of solar concentrators has a high impact on its thermal performance. Thus a system determining the geometrical accuracy of a solar concentrator system is necessary. Thermal image quality enhancement is often required for its interpretation with high accuracy. It can be carried out by filtering approaches and contrast adjustment process. In Chapter 8, three recent methodologies for multidimensional processing on sequences of thermographic images are presented. All three methods for image enhancement techniques are presented in detail, starting with its mathematical foundations, algorithmic procedures, advantages and limitations. In recent world, providing security right from a home, marketing area, financial sector, etc ... is a very challenging task. There are several ways to ensure security in real time; one among them is human identification with high accuracy through biometric data such as thumb impression and iris features. In Chapter 9, a new approach for personal authentication using finger knuckle geometric and texture feature measurement are used. In addition, the entire finger back region including proximal and distal phalanx of the finger knuckle surface also added for recognition. The geometric measures are derived through angular geometric and texture measures are derived through statistical-based texture analysis methods.

Because of continued effort, it is ensured to keep the book reader-friendly. By a problem-solving approach, researchers learn the material through real life examples that provide the motivation behind the concepts and its relation to the real world problems. I trust and hope that the book will help the readers to further carryout their research in different directions.

While writing, contributors have referred several books and journals, I take this opportunity to thank all those authors and publishers. In addition, I would like to thank VIT University, India for providing facilities to complete this project. I am extremely thankful to the editorial board, reviewers for their support during the process of evaluation. I extend my sincere thanks to the production team of IGI Global, USA for their support, encouragement and extending their full cooperation in successful completion of this edited book.

V. Santhi
VIT University, India

Acknowledgment

It is with great pleasure and satisfaction, I would like to present my book titled *Recent Advances in Applied Thermal Imaging for Industrial Applications* and wish to express my gratitude to all those who helped me either directly or indirectly to complete this project successfully. I would like to sincerely thank all authors who have contributed to this edited book. I acknowledge, with sincere gratitude the kindness of the School of Computer Science and Engineering, VIT University, India to provide an opportunity to carry out this research work. Authors would have referred several books and journals while writing chapters for this edited book; I take this opportunity to thank all those authors and publishers. I also thank the editorial board members, reviewers for their support during the process of evaluation. I extend my sincere thanks to the production team of IGI Global, USA for continuous encouragement and extending their full cooperation in successful completion of this edited book.

V. Santhi
VIT University, India

Section 1
Applications of Computational Intelligence in Thermal Image Processing

Thermal images play a key role in medical, military and industry applications. It is also used to provide security for the nation through the detection of landmines and targets. Analyzing thermal images to get meaningful information is a tedious task. Therefore, digital thermal image processing is considered as very essential research areas in the current scenario to analyze and interpret the same. There are many traditional approaches exists to carry out thermal image processing but to ensure automaton in image processing the role computational intelligence is essential. Keeping this as objective, basics of computational intelligence and its advancement is presented in this section. In addition, thermal images are used for disease diagnosis in medical field. In this section, a novel approach using fuzzy logic is presented to detect flu using thermal images of human body. Similarly, fuzzy logic based approach is discussed in detail for tumor detection in brain. Thus, computational approaches can be used to carryout image processing, in particular thermal images with high accuracy for automation.

Chapter 1
An Intelligent Thermal Imaging System Adopting Fuzzy-Logic-Based Viola Jones Method in Flu Detection

Wai Kit Wong
Multimedia University of Melaka, Malaysia

Nur Izzati Nadiah Binti Ishak
Telekom Research and Development Sdn Bhd, Malaysia

Heng Siong Lim
Multimedia University, Malaysia

Jalil bin Md Desa
Telekom Research and Development Sdn Bhd, Malaysia

ABSTRACT

Some infectious diseases can spread rapidly via a community of human or animals or both, either through airborne particles or viruses. Such rapid spread diseases may become a local, national or international widespread and contagious threat. As a symptom of infection, the body temperature of a disease carrier is higher than normal people. In this chapter, flu detection system using thermal imaging tool and computer vision techniques are discussed. An automatic flu detection method adopting human object extraction algorithm and fuzzy logic based Viola Jones

DOI: 10.4018/978-1-5225-2423-6.ch001

algorithm are also discussed. The proposed system able to capture a thermogram of the human subject, detecting the eye region of the human subject, calculating the pixels values around the detected eye region, converted to temperature readings and further classified the subject's body temperature whether the subject satisfies a flu condition or not. Experimental results also shown that the proposed fuzzy logic based Viola Jones algorithm can trace out flu infectious personal from the input thermal images up to 80% of accuracy.

INTRODUCTION

Flu is a type of infectious disease induced by the outspread of influenza virus. The dominant symptoms on the flu infected personal comprise of: high fever, sore throat, running nose, headache, muscle pains, coughing, fatigue, congestion and muscle aches. From all these symptoms, a rise or drop in body temperature may be an early symptom of such disease, and as such it may be desirable to measure the body temperature of each person for flu detection. This may help to detect and quarantine infected individuals at certain locations such as airports or border, to prevent a disease entering a region, state or country. Conventional flu detection system using contact measurement system (clinical thermometer), whereby modern flu detection system applying contactless measurement system (infrared thermometer and infrared camera), both are to trace out whether there is an existence of infectious disease active carrier or not for release pass or quarantine purposes. The main merit of modern flu detection system as compare to conventional flu detection system is that in the contactless flu detection system, the security personal no need to have physical contact with the examinee. In a better version of contactless flu detection system, the image processing based flu detection system can help free up the long que faster, saving times, at the meantime offering the examinee a more relaxing body temperature measurement way.

In general, the contactless flu detection system can be divided into two main categories, one is by the infrared thermometer measurement and another one is by the infrared imaging capturement. Infrared thermometer measurement is by the use of medical infrared thermometer or the door type forehead thermometer. The infrared thermometer first invented by (Egawa, 1997) comprising an infrared sensor for receiving heat radiation from an object and an arithmetic circuit which calculates the temperature readings. The door type forehead thermometer manufactured by TECOS (Topteches 2015) is a walkthrough forehead thermometer. The examinee is allowed to enter the passageway through a door. An infrared optical recognition system above the door checks the temperature of the examinee. One problem encounter in infrared thermometer measurement is the requirement of long queuing, since only one person at a time enters the passageway to get the measurement. Hence it is better to apply the infrared imaging capturement tool that can help free up the long queue, since the photo of the people who pass through the camera will be captured in video rate, and it will be process with image processing technique to extract the human objects in the captured image and further identify flu infected personnel (human body temperature > 37.5degree Celcius). Infrared cameras make small temperature difference visible. They are currently applied widely in many new or existing security networks.

Conventional infrared imaging system normally employs human observers to analyze the captured video. In the long run this is vulnerable to error due to deterioration in attention of the human observer (Hampapur, 2003). It is a fact that a human's visual attention drops below acceptable levels when assigned to visual monitoring and this fact holds true even for a trained personnel (Hampapur, 2005; Green, 1999). The weakness in conventional infrared imaging system has raised the need for a smart flu detection system where it employs computer and pattern recognition techniques to analyze information from situated imaging tools and automatically detect flu infected personal. An automatic flu detection method is discussed in this chapter, this includes human object extraction algorithm and fuzzy logic based Viola Jones algorithm. With the algorithms proposed in this chapter, it can automatically detect the flu infected personal at certain location captured by the infrared imaging tool, without required long queue for one by one checking and not to disturb the moving personal proceed to their destination ahead. These monitoring and subsequent analyses of the images from the inspection can alert security personnel to take further action on either let pass the healthy personal or quarantine the flu infectious people effectively.

In this chapter, the fundamentals of human body temperature measurement system (contact or contactless sensor system), flu detection system, human operator versus computer vision technique in flu detection system are first discussed. The algorithm and implementation of some image processing techniques on human object detection and extraction will be discussed too, such as Background-foreground Segmentation (Radzi, 2011), Histograms of Oriented Gradients (Suard, 2006) and Head Curve Geometry technique (Wong, 2012). Automatic flu detection using fuzzy logic theory and Viola Jones method will be discussed next, and their design procedures in infrared image processing tasks will be included. Later, some experimental results to prove the algorithms proposed for the flu detection system are shown. In the last section of this chapter, the work is summarized and some future enhancement is shown.

FLU DETECTION SYSTEM

Flu is different from a cold and it usually comes on swiftly. People who suffer bad flu may have a high fever. Fever which also known as pyrexia is a common medical sign that indicated by an elevation of human body temperature above the normal range of 36.5-37.5 °C. There are two methods to measure the human body temperature which are contact and contactless method.

Contact Flu Detection System

Contact flu detection system is a direct measurement using thermometer on certain parts of the human body. There are five ways to take human body temperature, which are by rectal way (by the rectum), oral way (by the mouth), axillary way (under the armpit), temporal artery way (forehead) and tympanic method (in the ear). The commonly found temperature measurement tools are mercury thermometer, forehead thermometer, digital thermometer and ear thermometer.

The mercury thermometer is shown in Figure 1. It made up of of a bulb containing mercury sticked to a narrow diameter glass tube. The quantity of mercury in the bulb is much more than the quantity in the glass tube. The quantity of mercury alters moderately with temperature. The little change in quantity pushes the narrow mercury column comparatively long route up to the tube. The application procedure is first clean the thermometer with cool, soapy water and rinse. Then carefully place the tip of the thermometer under the tester's tongue (mouth closed)/rectum/armpit. Leave the thermometer in place for about 1 minute. After that remove the thermometer from tester and read the mercury level on the tube with scale reflecting the temperature reading. Finally, clean the thermometer with flowing water and antiseptic for hygiene issue, to be prepared for next use. Mercury thermometer is

Figure 1. Mercury thermometer

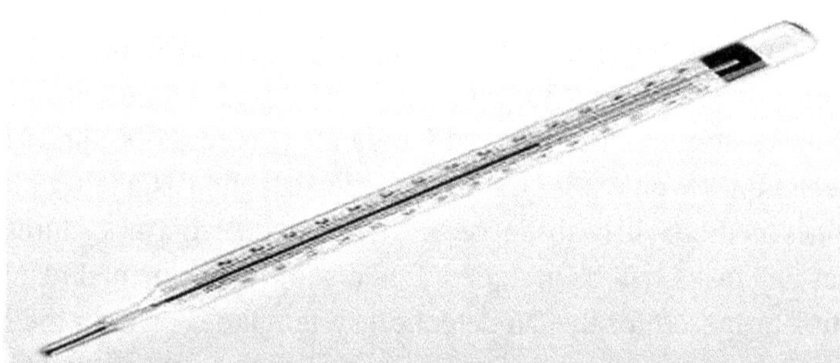

durable, accurate and can be easily calibrated. However, mercury thermometer is fragile since the tube is glass made and the mercury vapor is poisonous. Mercury thermometer is also slow responses and the display is harder to read.

The forehead thermometer strip is shown in Figure 2. It is a plastic strip printed with thermochromic chemicals (liquid crystals) that change color with temperature to reveal the tested part's temperature. The application procedure is first hold the forehead thermometer strip at each end and place firmly against dry forehead just above the eyebrows. Then hold in place until the colors stop changing in about 15 seconds. Green color exist on any of the reading bar indicates the correct temperature. If tester is tested alone, he/she may use a mirror to read the reversed lower scale. Forehead thermometer strip is inexpensive in price, more hygienic, safe (not easily break and contaminate human flesh), easy-to-use, fast reading (temperature shows in about 15 seconds) and reusable (upto 100's of times). However, forehead thermometer strip is not recommended by professionals because they do not give accurate readings. They show the temperature of the skin, not the body.

The digital thermometer is shown in Figure 3. It applies the electronic sensor such as resistance temperature detectors (RTDs), thermocouple or thermistor for measuring and displaying temperature reading. The electronic sensor is put in the body part where temperature is to be measured, and left long enough to reach equilibrium. The application procedure is very similar to mercury thermometer except

Figure 2. Forehead thermometer strip

Figure 3. Digital thermometer

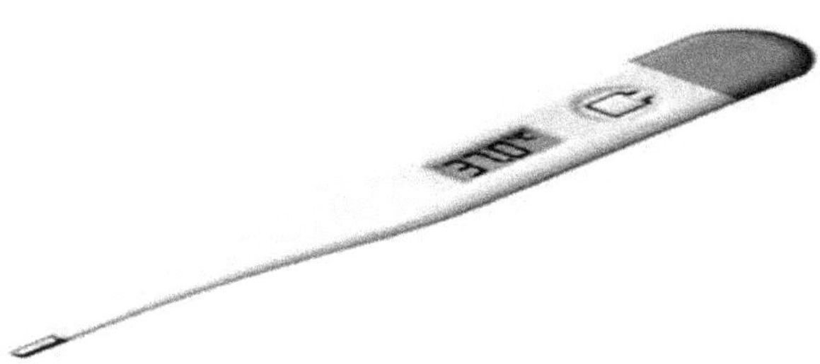

that the user has to place the thermometer in body part (tongue (mouth closed)/ rectum/armpit) for about 1 minute, until there is an indicator on it sound "beep", the user may then read the digital display shown on the mini LCD screen on it. Many of the digital thermometer show readings to great precision (0.2 °F, 0.1 °C or occasionally half of that), but the accuracy is un-guaranteed. The detailed accuracy should be reviewed in documentation and the thermometer should uphold calibration regularly. However, digital thermometer required physical contact with users and hence hygienic is of major concern.

The ear thermometer is shown in Figure 4. It measures the infrared energy emitted from the user's eardrum. A short tube with a protective sleeve is put into the ear, and a shutter is opened to allow the tympanic membrane emitted radiation fall on the infrared detector for duration of time (roughly 0.1 to 0.3 seconds). The device sound "Beep" when it completed the data collection and display the body temperature reading on a mini liquid crystal display. Ear thermometer is expensive and difficult to use. One major reason medical professionals point out that ear thermometer is inaccurate is due to the probe positioning into the ear canal is inconsistent, therefore causing inconsistent readings and constantly loss detecting fevers.

Furthermore, the user does not like foreign matter inserted into their ear, causing good positioning of the probe even more difficult.

In summary, the contact method is time consuming, had injury risk due to broken thermometer, non-comfortable if the temperature taken in sensitive area like anus and not suitable for public screening. On the other hand, a better body temperature measurement way using contactless method is proposed by scientists to overcome the above limitations.

Figure 4. Ear thermometer

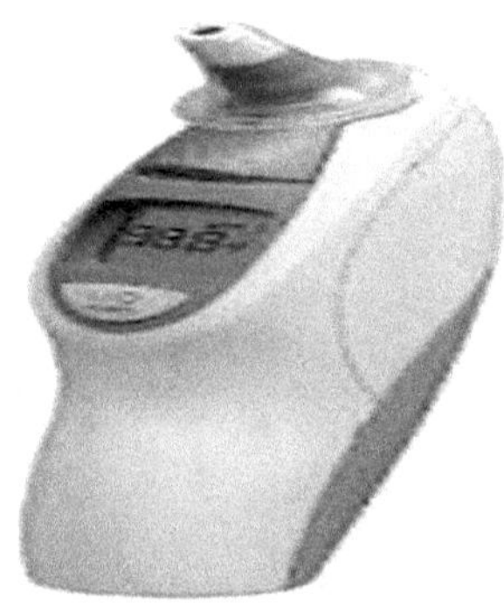

Contactless Flu Detection System

Contactless flu detection system is temperature measurement system using infrared thermometer or thermal imaging. An infrared thermometer is a thermometer which interprets temperature from the thermal radiation of object. The thermal radiation is also named the "Blackbody radiation" which emitted by the object being measured. By perceiving the volume of infrared energy emitted by the object and its emissivity, surface temperature of the object can be determined generally. The design of an infrared thermometer typically comprise of lens to aim attention at the infrared thermal radiation onto a detector. It then converts such radiant power into an electrical signal that can be displayed in temperature units after being offset for ambient temperature. This allows temperature measurement for an object can be done from a distance without physical contact with it.

The most common contactless infrared thermometers available in the market are the spot infrared thermometer and the infrared thermal imaging camera. Spot Infrared Thermometer or Infrared Pyrometer measures the temperature at a spot on a human body surface. The sample image of an infrared pyrometer is shown in Figure 5. The infrared pyrometer normally projects a visible red dot onto the center of the body part being measured. The red dot help identify the spot being measured, but it plays no part in the measurement. The measurement area is actually a relatively small area determined by the distance-to-spot ratio (D/S Ratio). The latest breakthrough in spot infrared thermometer technology is on Thermoflash no contact infrared thermometer (David, 2015), as shown in Figure 6. Thermoflash applies infrared mechanism to measure the heat access from the temporal artery. The tool is held between three to five centimeters away from the artery. The artery runs along the temple on the side of the head and the patient's skin surface must be dry. Thermoflash needs around

Figure 5. Infrared pyrometer

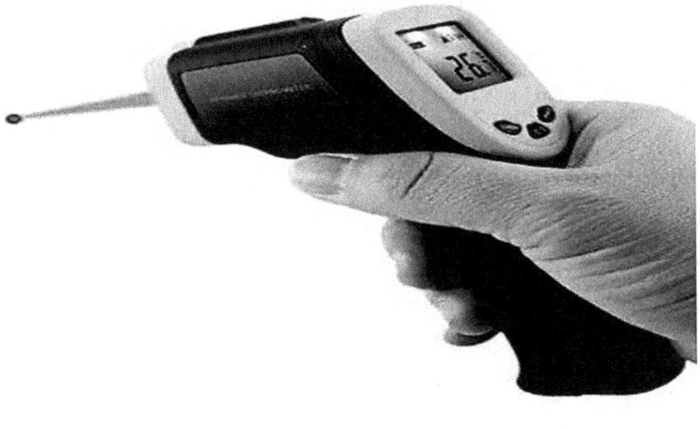

Figure 6. Thermoflash infrared thermometer

fifteen minutes to adjust when switching from one environment to another, and to be functioned properly, the tool measurement needs to be stay away from wind or powerful air conditioning units. Thermoflash have given doctors and nurses a way to quickly take a big amount of patients' temperatures in the Ebola outbreak zone in Africa, without the risk of transmitting Ebola though body contact. This kind of thermometer displays a reading in just a second. However, the Spot Infrared Thermometers' actual angular area being measured differs amongst instruments, and is not confined to the visible spot.

Infrared Thermal Imaging Cameras or Infrared Cameras are basically infrared radiation thermometer which measures the temperature on many points of a body part over a relatively large area to generate a 2-D image, so-called thermogram. Each pixel on the thermogram representing a temperature reading on a spot of the body part. The sample image of an infrared camera is shown in Figure 7. This technology is processor based and more software-intensive wise than infrared pyrometer. Infrared camera can detect the heat emitted by object through a remote sensor, transformed it into color images and displayed on a monitor. Based on the infrared ray intensity emitted from the object's surface, it will then converts with computer algorithm into different color palette. Infrared camera is normally applied in monitoring temperature for large areas. Typical applications include perimeter monitoring used by military or security personnel, inspection / process quality monitoring of manufacturing processes, and equipment or enclosed space hot or cold spot monitoring for safety and efficiency maintenance purposes.

Figure 7. Infrared camera

Image-Processing-Based Flu Detection System

The infrared thermal imaging is an imaging technique used for constructing thermal maps of emitting objects by detecting the infrared radiation emitted from the surface and by converting it into measurable signal (Surabhi, 2012). The advantage of using infrared thermal imaging for body temperature measurement is hygienic since it is contactless with human physical body. It is suitable for public screening and it is started for widespread used since the outbreak of H1N1 in 2006. This method is reliable, fast, more comfortable, no injury risk, and it may also help to limit the spread of viral disease. It is by the means of using image to trace out whether there is an existence of flu infected personal or not and further quarantine them if necessary. Image processing based flu detection system is widely in favor by many practitioners as compared to conventional physical contact flu measurement systems mainly due to that the conventional flu measurement approach depends on human operators to do temperature measurement to detecting the feverish individuals in crowd and required human operator to observe the measured data. However, conventional flu detection system is sometimes causes error due to negligence in human observer attention (James, 1998).

The weakness in conventional flu detection system has boosted the demand on smart flu detection system where it applies computer and pattern recognition technology to automatically analyze information from the images captured by infrared camera. One recent paper on image processing based flu detection system (Norulzahrah, 2015) explains a study on implementing a Physiology based Approach and proposes an algorithm for thermal screening system on flu detection. The study proposes the development of automatic body temperature detection system that incorporates face detection and object tracking techniques. However, the study focused on the whole face area for measuring the body temperature, it is not on the most reliable parts on the face.

The most reliable parts to measure core body temperature for thermal imaging is in the human facial which are at the forehead, ear region and inner canthus of eye. Based on (Ring, 2007; Mercer, 2009; Ng, 2004), the most accurate part to measure human body temperature is at inner canthus of eye, in which it is the most consistently warmest area on the head when ambient temperature change, and it is considered as the most stable area. This canthus region is fed by internal carotid artery with the blood circulation coming from brain whereas the rest of the face is supplied by external carotid artery, so the rest of the face is less reliable indicator of core body temperature. Previous thermal imaging temperature measurement method (Ng, 2004) is not automated fever screening since it needs human operator to manually operating the system. The result might not be accurate due to operator skills dependent or carelessness. Hence, in the following section, automated fever screening for flu detection using image processing techniques will be discussed. The algorithm and implementation of some image processing techniques on human object detection and extraction such as Background-foreground Segmentation (Radzi, 2011), Histograms of Oriented Gradients (Suard, 2006) and Head Curve Geometry technique (Wong, 2012) will be discussed.

HUMAN OBJECT EXTRACTION METHODS

An important step in automated fever screening for flu detection using image processing techniques is finding or extracting human object within a captured infrared image. There are three current works on human object extraction to be discussed below, which are Background-foreground Background-foreground Segmentation (Radzi, 2011), Histograms of Oriented Gradients (Suard, 2006) and Head Curve Geometry technique (Wong, 2012).

Background-Foreground Segmentation Method

The background-foreground segmentation method is purposed to extract the non-stationary objects in captured images by subtracting the new image frame from a stored background image. The background –foreground modeling algorithm can be classified into two categories, which are non-adaptive category and adaptive category. In non-adaptive category, the initialization background image is picking up reciprocally from one of the image in a captured sequence frames. There is no re-initialization process for the background image afterwards. Therefore, the error in the background will gather up over time. Hence, such method is only suitable for short time tracking process. In adaptive category, the training data images are gather up over time to produce the background prototype. Generally, the background obtained from adaptive category appears to be the current scene of the captured sequence.

Some adaptive background modeling algorithms are discussed below. It was first discover that (Stauffer, 1999) discussed modeling each pixel in captured image as a mixture of Gaussians and using an on-line approximation to update the model. Each pixel in the model is classified according to whether the Gaussian distribution that represents the pixel element most effectively is regarded as part of the background model or not. The proposed Gaussian distributions for such adaptive mixture model was developing a stable and real-time outdoor tracker that can deals with lighting changes, repetitive motions from clutter, and long-term scene changes. Since such method is flexible enough to manage the illumination variation, moving scene cluster as well as dynamic natural environment, many improvements based on this mixture of Gaussian model had been done, as further bring up in (Bouwmans, 2008). (Kiri, 2001) further developed constrained K-means Clustering (COP-KMEANS) algorithm for categorizing the information on each pixel from the Gaussians distribution into K-cluster. This method is developed to solve the real-world problem by automatically detecting road lanes using GPS data.

Kyungnam (2004, 2005) employed code-book model for segmenting background-foreground images. In the propose method, each sampled background pixel's values were quantized into code-book that symbolize a condensed form of background model in a long sequence frames. This method exhibits effective performance with the use of a long sequence training images. Nonetheless, this method cannot cast very slow illumination change in scene for long run. The work followed by (Mohamad, 2008) proposed a 2-layers code-book for improving the weakness of Kyungnam K. et al. single layer code-book. Mohamad H.S. and Mahmood F. 2-layer code-book method performed better than Kyungnam K. et al. single layer code-book in the condition where the illumination in the scene change very slowly in long time. However, such algorithm required much more training data.

(Marko, 2006) proposed the texture-based method for casting the background and detecting moving objects in a video sequence. In their work, each pixel is first modeled as a group of adaptive local binary pattern histograms, it then calculated over a circular region around the pixel. Such method rather has many parameters. Hence, it is hard to obtain a virtue set of parameter values for each modeled scene. (Chen, 2007) proposed an unsupervised K-Means background clustering technique to model each spatial position for the surveillance video frames temporal distribution in RGB domain by using Bayesian information criterion. The proposed background segmentation model incorporates noise reduction, cast shadow cancellation, and improved watershed transform to acquire object masks in satisfaction. The proposed method is utilizable on low-fame-rate, noisy video sequences in surveillance systems and it is with multi clusters classification capabilities. (Pradeep, 2008) formulated bimodal temperature distribution function and applying Bayesian framework to outline the facial tissues against the background images. It is established by known priori knowledge from the bimodal nature of scene use in face recognition field.

Hence, the recent work by (Sofiah, 2011) proposed a novel method for segmentating foreground from background by combining Gaussian Mixture model (Bouwmans, 2008), K-Means background clustering technique (Chen, 2007) and the bimodal nature of scene (Sofiah, 2011), so called the Bimodal Gaussian Mixture Model-Based Algorithm, used in thermal fever mass screening. The following subsections describe in details the Frame differencing method, and the Bimodal Gaussian Mixture Model-Based Algorithm.

Frame Differencing Method

The input to the image processing system is sequence images. Those images in RGB format are first converted to grayscale format. The intensity level of each pixel on coordinate (*x*,*y*): $I\left(x,y,t\right)$ from the sequence image at time *t* is accumulated:

$$\left\{G_1,\ldots,G_t\right\}=\left\{I\left(x,y,t\right):1<t<t_o\right\} \tag{1}$$

where G_t is the gray level of intensity of a pixel on coordinate $\left(x,y\right)$ at time *t*. The histogram of $I\left(x,y\right)$ is adopted:

$$h\left(I\left(x,y\right)\right)=n_d\left(x,y\right) \tag{2}$$

in which $n_d\left(x,y\right)$ is the number of intensity ranking in $\left[r_{d-1},r_d\right]$ and d is the rank number. Mixture of Gaussian method is used to evaluate the distribution of G_t. K-means clustering is used to classify the member of each cluster. The background model is modeled by averaging the intensity of the background cluster. By subtracting an image at time i with the background model, the foreground is extracted.

Bimodal Gaussian Mixture Model-Based Algorithm

Gaussian of mixture method is use to analyze the distribution of G_t. G_t is modeled by a mixture of K Gaussian distribution (Chen, 2007), with probability value:

$$P\left(G_t\right) = \sum_{i=1}^{K} w_{i,t} * \eta\left(\mu^2, \sigma^2\right) \tag{3}$$

where K is the number of distribution, and in infrared imaging as well as modeled using bimodal distribution (Sofiah, 2011), $K = 2$. $w_{i,t}$ is an estimate of weight for i-th cluster of the Gaussian distribution at time t, μ is the mean value and σ is the variance of i-th cluster using Expectation-Maximization algorithm technique:

$$z_{ij}^{(k)} = \frac{w_{s_i}^{(k)} \left(\sigma_{s_i}^{(k)}\right)^{-1} \exp\left\{-\frac{1}{2\left(\sigma_{s_i}^{(k)}\right)^2}\left(x_j - \mu_{s_i}^{(k)}\right)^2\right\}}{\sum_{t=1}^{2} w_{s_t}^{(k)} \left(\sigma_{s_t}^{(k)}\right)^{-1} \exp\left\{-\frac{1}{2\left(\sigma_{s_t}^{(k)}\right)^2}\left(x_j - \mu_{s_t}^{(k)}\right)^2\right\}} \tag{4}$$

$$w_{s_i}^{(k+1)} - w_{s_i}^{(k)} < \in,\ i = 1,2 \tag{5}$$

Expectation-Maximization algorithm supply iterative computation estimation based when the observed data are incomplete. This is well work for the case where the distribution of the temperature data weights, variances and means for the bi-modal clusters is unknown at the beginning. Initial crude estimates of the parameters are randomly presume and used for the following loop for $k = 0, 1, \ldots$ where $i = 1, 2$ (serve as the number of cluster–foreground and background) and $j = 1, \ldots, N_s$ (N is the total number of samples) until the condition for the terminating loop as shown in equation (5) reached.

The complete model for the Bimodal Gaussian Mixture Model-Based Algorithm (Sofiah, 2011) is shown in Figure 8. Thresholding is a common approach in background subtraction for infrared spectrum images. The initial Bimodal Gaussian Mixture Model (BGMM) has drawback of irremovable hot spot background. Hence a Thresholding Bimodal Gaussian Mixture Model (TBGMM) parameters method is implemented, where $x\%$ is introduced (from 10-50%) as threshold values drawn from the bimodal Gaussian distributions parameters (means μ, weights w and variances σ), tested on the samples and qualitatively observed the results. Both BGMM and TBGMM techniques are multiplied with results of Frame Differencing to produce the better image result A and B. At last, morphological operations are performed on result A and B to extract the foreground from the background and obtained the object of interest.

Figure 8. Bimodal Gaussian mixture model-based algorithm (Sofiah, 2011) for background-foreground segmentation

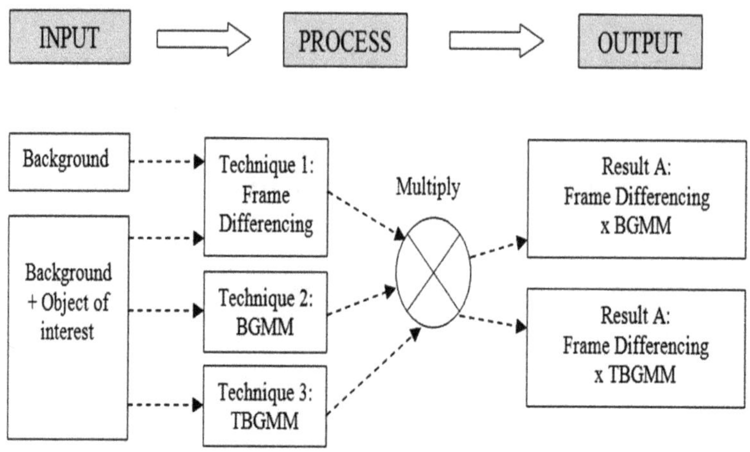

Histograms of Oriented Gradients Method

The histogram of oriented gradients (HOG) is an information data structure used in computer vision and image processing for object detection purposes. The method calculates appearances of gradient orientation on an image's localized portions. This method almost identical to that of edge orientation histograms (Levi, 2004), scale-invariant feature transform descriptors (Lowe, 2004), and shape contexts (Belongie, 2002), however, be different from that it is quantified on a dense grid of uniformly spaced cells and utilizes overlapping local contrast normalization to enhance accuracy.

HOG descriptor is first designate by (Navneet, 2005), first concentrated on pedestrian detection in static images. Later they broadened their tests to involve human detection in videos and on a variety of common animals and vehicles in static imagery. An object's appearance and shape inside an image can be interpreted by the distribution of gradients intensity and edge direction. Implementation of the HOG descriptor algorithm can be outlined as follows (Navneet, 2005):

1. **Gradient Computation:** Partition the image into cells (small connected regions), and calculate a histogram of gradient directions (or edge orientations) for the pixels within each cell.
2. **Orientation Binning:** Discretize each cell into angular bins corresponding to the gradient orientation. Each cell's pixel contributes weighted gradient to its corresponding angular bin.

3. **Descriptor Blocks Formation:** Groups of adjacent cells are regarded as spatial regions named blocks. The grouping of cells into a block is the basis for grouping and normalization of histograms.
4. **Block Normalization:** Normalized group of histograms serve as the block histogram. The set of these block histograms serve as the descriptor.

The details algorithm steps will be discussed below.

Gradient Computation

The first step of the HOG algorithm is the computation of gradients values. One dimensional centered point discrete derivative is applied in both the horizontal and vertical directions masking. Given an image I, the x and y derivatives is obtain using a convolution operation:

$$I_X = I * D_X \text{ and } I_Y = I * D_Y \tag{6}$$

where $D_X = \begin{bmatrix} -1 & 0 & 1 \end{bmatrix}$ and $D_Y = \begin{bmatrix} 1 \\ 0 \\ -1 \end{bmatrix}$ is the filter kernels.

The magnitude of the gradient is:

$$\left|G\right| = \sqrt{I_X^2 + I_Y^2} \tag{7}$$

and the orientation of the gradient is:

$$\theta = \arctan \frac{I_Y}{I_X} \tag{8}$$

Orientation Binning

The second step of the HOG algorithm requires generating the cell histograms. Each pixel inside the cell casted a "weighted vote" for an orientation based histogram channel depend on the values obtained in the gradient calculation. The cells are rectangular in shape and the histogram channels are uniformly cover 0 to 180 or 0 to 360, providing either the gradient is "signed" or "unsigned". (Navneet, 2005) discovered that unsigned gradients applied together with nine histogram channels achieved the best performance in their experiments. Regarding the vote weight, pixel contribution can be the gradient magnitude itself, or the square or square root of the gradient magnitude.

Descriptor Blocks Formation

The third step of the HOG algorithm is the formation of the descriptor blocks. By consider the changes in contrast and illumination, the gradient strengths ought to be locally normalized, and that needs to group the cells all at once, into a bigger and spatially connected blocks. Therefore the HOG descriptor is the joint-vector for the components of the normalized cell histograms from all of the block regions. These blocks normally overlay one another, means that each cell will commit more than once to the final descriptor. Generally there are two main block geometries exist: rectangular R-HOG blocks and circular C-HOG blocks.

R-HOG blocks are normally in square grids, characterized with three parameters: the number of pixels per cell, the number of cells per block and the number of channels per cell histogram. In (Navneet, 2005) human detection test, the excellent parameters were 16x16 pixel cells of 8x8 cell blocks with 9 histogram channels. Furthermore, using a Gaussian spatial window inside each block earlier than tabulating histogram votes can allow the user to minimize the weighting for the pixels all over the edge of blocks. (Navneet, 2005) also claimed that R-HOG blocks are comparable to the Scale-invariant feature transform (SIFT), with difference in R-HOG blocks are calculated in dense grid at some single scale without orientation alignment; while SIFT descriptor are calculated at sparse, scale-invariant key image points and with rotation alignment. Moreover, R-HOG blocks are applied mutually to encode spatial form information, whereas SIFT are used individually.

C-HOG blocks are normally in circular grids, comes with two variants: one with a single, central cell and another one with an angularly divided central cell. Moreover, it can be characterized with four parameters: the number of angular and radial bins, the radius of the center bin, and the expansion factor for the radius of additional radial bins. Navneet D. and Bill T. in their experiment (Navneet, 2005) discovered that the two mains variants contributed the same performance, and the two radial bins with four angular bins, a center radius of four pixels and an expansion factor of two gave the best performance. In addition, they found out that Gaussian weighting gave no benefit when applied mutually with the C-HOG blocks. In conclusion, C-HOG blocks turn out identical to the performance of Shape context descriptor (Belongie, 2002), however it varies much in that C-HOG blocks consist of cells with certain orientation channels, even though shape contexts only applies a single edge presence count in their establishment.

Block Normalization

The fourth step of the HOG algorithm is the block normalization. (Navneet, 2005) investigated four distinct methods for block normalization. Let v be the non-normalized vector consist of all histograms in a given block, $\|v\|_k$ be its $k - norm$ for $k = 1,2$ and e be any small arbitrary constant, therefore the normalization factor can be any of the following:

L2-*norm*:

$$f = \frac{v}{\sqrt{\|v\|_2^2 + e^2}}$$

L2-*hys*: L2-*norm* followed by clipping (limiting the maximum values of v to 0.2) and renormalizing.

L1-*norm*:

$$f = \frac{v}{\left(\|v\|_1 + e\right)}$$

L1-*sqrt*:

$$f = \sqrt{\frac{v}{\left(\|v\|_1 + e\right)}} \tag{9}$$

Navneet (2005) observed in their experiment that L2-*hys*, L2-*norm* and L1-*sqrt* schemes give similar performance. However, the L1-*norm* gives marginally less reliable performance. Overall, all the four methods exhibited better advancement over the non-normalized data.

After the HOG descriptor is build, object recognition can be done by putting up the descriptor into some supervised learning based recognition system. The support vector machine (SVM) (David, 2003) classifier is applied in (Navneet, 2005) work, whereby it is a binary classifier that seek for an optimal hyperlane in the role of a decision function. Once the SVM was trained on sequence images comprise with some specific object, it can make decision concerning the existence of an object, such as a human, in extra test images. Overall, the HOG model used in human detection can be summarized in the feature extraction and object detection chain as shown in Figure 9. The detector window is paved with a grid of overlapping block that extracted the HOG feature. The associated vectors are input to a linear SVM for the classification of object/non-object. The detection window is scanned over the images at all scales and positions. Conventional non-maximum suppression is continuing on the output pyramid to recognize object cases.

Head Curve Geometry Method

The head curve geometry method is a human object detection method in image processing by detecting the human head curve to identify whether an object in the image is a human or not. The algorithm is first applied in (Wong, 2012) for human face detection in infrared imaging. The object appear in the captured infrared image is extracted out by adopting background subtraction method, noise filtering, object detection and extraction (extract object boundary and save it as coordinate arrays). After the object had been extracted out from the image, it will further proceed with the following steps to identify whether it is a human object or not, by determine the head curve:

Step 1: Obtain the Mean x Coordinate in the object boundary. The starting point will be the highest point (smallest y coordinate) between intersection points of the boundary with the vertical x mean line. From the targeted starting point, compare $dVar$ proceeding pixels in clockwise and counter clockwise direction to obtain 2 peaks- Pc and Pcw. The highest point between these 2 peaks is refers as the head top point ($\mathrm{HeadPpoint}$). Refer to Figure 10(g) for better understanding.

Figure 9. HOG model used in human detection
(Navneet, 2005)

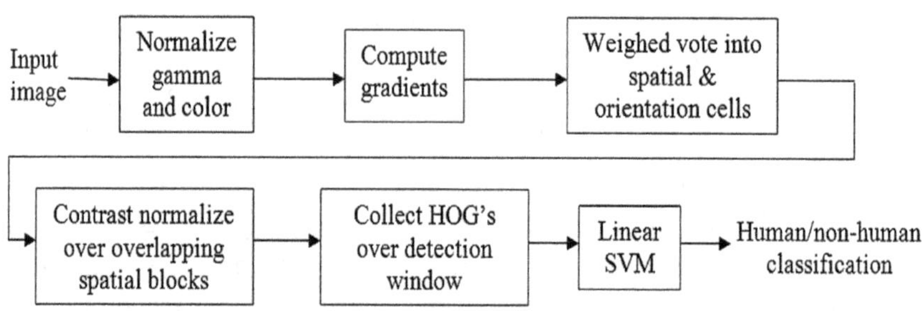

Step 2: Obtain the Mean y Coordinate in the object boundary. The intersection of the boundary with the horizontal y mean line commits 2 points- $\mathrm{LeftMean}$ and $\mathrm{RightMean}$.

Step 3: Find the Number of Next Pixel (L) using equation (10). Check for L pixel from the $\mathrm{LeftMean}$ to search for the $\mathrm{MostLeftPoint}$ in clockwise ($\mathrm{LeftMostOutCW}$ point) and counter clockwise ($\mathrm{LeftMostOutCCW}$ point) direction. The $\mathrm{MostLeftPoint}$ should be the point with the smaller x coordinate (most left) between those 2 points. Refer to Figure 10(e) for better understanding.

$$L\left(pixel\right) = \frac{x}{100} \times diffP \tag{10}$$

where $x = LeftRightOut\%$, $diffP$ is the difference between the index of $\mathrm{LeftMean}$ and $\mathrm{HeadPpoint}$

Step 4: Check for L Pixel From the $\mathrm{RightMean}$ to Search for the $\mathrm{MostRightPoint}$ in clockwise ($\mathrm{RightMostOutCW}$ point) and counter clockwise ($\mathrm{RightMostOutCCW}$ point) direction. The $\mathrm{MostRightPoint}$ should be the point with the higher x coordinate (most right) between those 2 points. Refer to Figure 10(b) for better understanding.

Step 5: Find the number of next pixel (K) using equation (11). Check for K pixel from the $\mathrm{MostLeftPoint}$ to search for the $\mathrm{LeftSouthPoint}$ in counter clockwise direction. The particular point should be situated at the most right (highest x coordinate) of the $\mathrm{MostLeftPoint}$. Refer to Figure 10(d) for better understanding.

$$K\left(pixel\right) = \frac{z}{100} \times pixelN \tag{11}$$

where $z = NeckPercent\%$, $pixelN$ is the number of pixel to build the object boundary.

Step 6: Check for K Pixel From the $\mathrm{MostRightPoint}$ to Search for the $\mathrm{RightSouthPoint}$ in clockwise direction. That point should be located at the most left (smallest x coordinate) of the $\mathrm{MostRightPoint}$. Refer to Figure 10(c) for better understanding.

Step 7: Determine the $\mathrm{LeftBreakTop}$ Point by divide the boundary for the top left curve test, and finding the number of next pixel (G) using equation (12). From the HeadPpoint, search for G pixel in counter clockwise direction to obtain LeftBreakTop coordinate. Refer to Figure 10(f) for better understanding.

$$G\left(pixel\right) = \frac{v}{100} \times pixelR \tag{12}$$

where v = HeadPercent%; pixelR = number of pixel to build the object boundary

Step 8: Determine the RightBreakTop Point by divide the boundary for the top right curve test, and finding the number of next pixel (G) using equation (12). From the HeadPpoint, search for G pixel in clockwise direction to have RightBreakTop coordinate.

Step 9: Perform Top Curve Test. It consists of Top Curve Left Test & Top Curve Right Test.

9.1 Top Curve Left Test: Extract the coordinate array (LeftHeadCCW) arranged from HeadPpoint to LeftBreakTop in counter clockwise direction. Use yFilter function to filter any repeated y value coordinate to produce LeftHeadNet. Refer to Figure 11 for an example.

Figure 10. (a) Right break top, (b) most right point, (c) right south point, (d) left south point, (e) most left point, (f) left break top (g) HeadPpoint

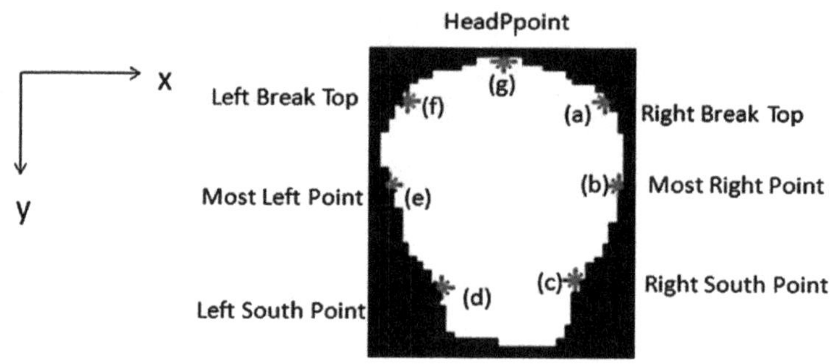

Figure 11. yFilter function filters repeated y coordinate

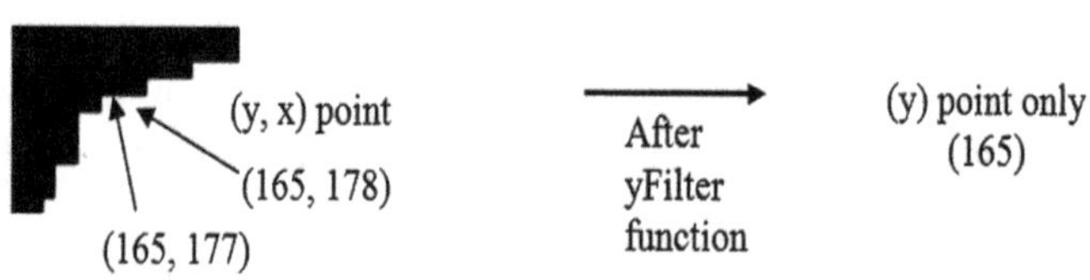

Consider the LeftHeadNet coordinate array has p element and the y value comparison is done by $(p-1)$ times.

$$countLeft = \begin{cases} 1, & y_{n+1} > y_n \| y_n < y_{n+1} < yTol \\ 0, & otherwise \end{cases} \tag{13}$$

where $n = n^{\text{th}}$ element in LeftHeadNet array; $yTol$ is the number of pixels that can be tolerated for error; || is OR condition.

If number of $countLeft$ with 1 is same with $p-1$, the line connecting the HeadPpoint and the LeftBreakTop is a curve.

9.2 Top Curve Right Test: The test is similar to the Top Curve Left Test by replacing the LeftHeadCCW with RightHeadCW coordinate array, which is arranged from HeadPpoint to RightBreakTop in clockwise direction. Later yFilter is used to generate RightHeadNet, followed by the y value comparison in each RightHeadNet element.

9.3 Combine Top Curve Left and Top Curve Right Test Result: Top Curve Test only confirms the line connecting from LeftBreakTop to HeadPpoint till RightBreakTop is a curve if the line passes Top Curve Left and Top Curve Right test.

Step 10: Perform Left Curve Test. It consists of Break To Out Test and Neck To Out Test

10.1 Break To Out Test: Extract the BreakLeftCCW coordinate array arranged from LeftBreakTop to Most Left Point in counter clockwise direction. Use xFilter function to filter any repeated x value coordinate to produce NorthLeftNet. Refer to Figure 12 for an example.

Figure 12. xFilter function filters repeated x coordinate: (a) left break top, (b) most left point, (c) left south point

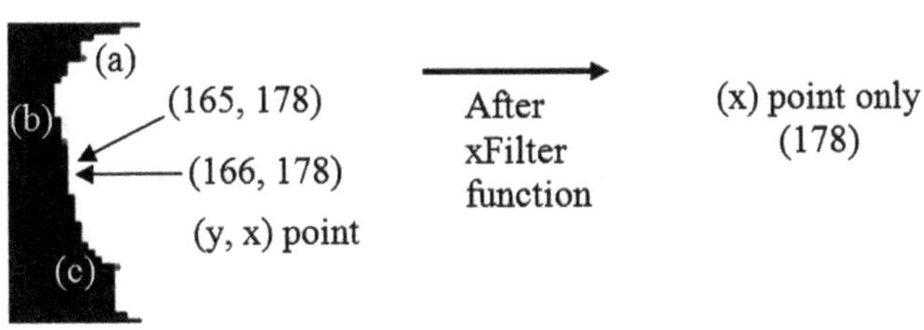

Consider the $\mathrm{NorthLeftNet}$ coordinate array has p element and the x value comparison is done by $p-1$ times.

$$coSideLeft = \begin{cases} 1, & x_{n+1} < x_n \| x_{n+1} - x_n < xTol \\ 0, & otherwise \end{cases} \tag{14}$$

where n = n^{th}element in $\mathrm{NorthLeftNet}$ array; $xTol$ is the number of pixels that can be tolerated for error; || is OR condition.

If number of *coSideLeft* with 1 is same with $p-1$, the line connecting the $\mathrm{LeftBreakTop}$ and the Most Left Point is a curve.

10.2 Neck To Out Test: The test is similar to the Break To Out Test by restoring the $\mathrm{BreakLeftCCW}$ with $\mathrm{SouthLeftCW}$ coordinate array, that is arranged from Left South Point to Most Left Point in clockwise direction. Later xFilter is applied to generate $\mathrm{SouthLeftNet}$, followed by the x value comparison in each $\mathrm{SouthLeftNet}$ element.

10.3 Combine Break To Out and Neck To Out Test Result: Left Curve Test only confirms if the line connecting from Left Break Top to Most Left Point till Left South Point is a curve, and there is a line passes Break To Out and Neck To Out Test.

Step 11: Perform Right Curve Test. It consists of Break To Out Test & Neck To Out Test

11.1 Break To Out Test: Extract the $\mathrm{BreakRightCW}$ coordinate array arranged from Right Break Top to Most Right Point in clockwise direction. Use xFilter function to filter any repeated x value coordinate to generate $\mathrm{NorthRightNet}$. Consider the $\mathrm{NorthRightNet}$ coordinate array has p element and the x value comparison is done by $p-1$ times.

$$coSideRight = \begin{cases} 1, x_{n+1} > x_n \| x_n - x_{n+1} < xTol \\ \quad 0, otherwise \end{cases} \tag{15}$$

where n = n^{th}element in $\mathrm{NorthRightNet}$ array; $xTol$ is the number of pixels that can be tolerated for error; || is OR condition.

If number of *coSideRight* with 1 is same with $p-1$, the line connecting the Right Break Top and the Most Right Point is a curve.

11.2 Neck To Out Test: The test is similar to the Break To Out Test by replacing the $\mathrm{BreakRightCW}$ with $\mathrm{SouthRightCCW}$ coordinate array, that is arranged from Right South Point to Most Right Point in counter clockwise direction. Later xFilter is applied to generate $\mathrm{SouthRightNet}$, followed by the x value comparison in each $\mathrm{SouthRightNet}$ element.

11.3 Combine Break To Out and Neck To Out Test Result: Right Curve Test only confirms if the line connecting from Right Break Top to Most Right Point till Right South Point is a curve, and there is a line passes Break To Out and Neck To Out Test.

Step 12: Perform Head Symmetric Test. Obtain wL and wR, in which $wL =$ width difference between Most Left Point and HeadPpoint and $wR =$ width difference between Most Right Point and HeadPpoint. Subsequently, determine the hL and hR, in which $hL =$ height difference between HeadPpoint and Left South Point and $hR =$ height difference between HeadPpoint and Right South Point. If the ${}^{wL}\!/_{wR} < 2$ or ${}^{wR}\!/_{wL} < 2$ or ${}^{hL}\!/_{hR} < 2$ or ${}^{hR}\!/_{hL} < 2$, the object passes the Head Symmetric Test.

Step 13: Perform Head Curve Test by combining the result from Top Curve Test, Left Curve Test and Right Curve Test. If the object passes any 2 of 3 tests, it passes the Head Curve Test.

Step 14: Decision Making. If the object passes either one of the Head Curve Test or the Head Symmetric Test or both, it is considered as human head and human object detected.

AUTOMATIC FLU DETECTION ADOPTING FUZZY-LOGIC-BASED VIOLA JONES METHOD

After human object was detected in a thermal image, the next task is on face area detection and followed by flu detection. For face area detection only (not recognition), Viola–Jones algorithm (Viola, 2001) can distinguish human faces from non-human faces. The algorithm has four stages: 1) Haar Feature Selection, 2) Creating an Integral Image, 3) AdaBoost Training, and 4) Cascading Classifiers.

Haar Feature Selection

The human facial characteristics needed by the detection structure generally comprise of the summation of image pixels inside some rectangular areas. Hence, they carry some similarity with Haar basis functions (Papageorgiou, 1998) that have been applied earlier in the image-based object detection domain. All human faces allocate some identical properties. These consistencies can be matched using Haar Features, based on a few properties typical to human faces, such as: human eye region is normally darker than the upper-cheeks and human nose bridge region is normally brighter than the eyes.

The composition of properties that forming match-able facial features are the location and size (between bridge of nose, mouth and eye) and their "Value" (oriented gradients of pixel intensities). It then forming the rectangle features:

- Value = Σ (pixels in black area) - Σ (pixels in white area).
- Three types: two-, three-, four-rectangles (Viola & Jones adopted the two-rectangle features).
- For example: the brightness differences between the white and black rect-angles over a particular region.
- Each feature is linked to a special location in the sub-window.

These four features matched by the algorithm are then sought in the image of a face.

Creating an Integral Image

An image description named the "summed area table" determines rectangular features in fixed time that provides them a large speed advantage over more complex different features. The integral image at position (x, y) is the sum of all the pixels above and to the left of (x, y), comprehensive:

$$I_{\sum}(x,y) = \sum_{\substack{x' \leq x \\ y' \leq y}} i(x', y') \tag{16}$$

Additionally, the summed area table can be calculated effectively in a single pass over the image, adopting the fact that the value in the summed area table on (x, y) is likely:

$$I(x,y) = i(x,y) + I(x-1,y) + I(x,y-1) - I(x-1,y-1) \tag{17}$$

After the summed area table calculated, the duty of determining the intensities above any rectangular area needs only four array references. This provides a constant evaluation time, which is independent of the rectangular area size. Specifically, adopting the written remarks in Figure 13, given, and, the sum of over the rectangle extended by and is:

$$\sum_{\substack{x_0<x<x_1 \\ y_0<y<y_1}} i(x,y) = I(D) + I(A) - I(B) - I(C) \tag{18}$$

Since each feature's rectangular area is ever neighboring to minimum one another rectangle, it leads to that any two-rectangle feature can be calculated in six array references, any three-rectangle feature in eight, and any four-rectangle feature in ten.

AdaBoost Training Algorithm

AdaBoost algorithm is employs to the object detect framework. The algorithm builds up a "strong" classifier as a linear combination of weighted simple "weak" classifier:

$$h(x) = sign\left(\sum_{j=1}^{M} \alpha_j h_j(x)\right) \tag{19}$$

Each weak classifier is a threshold function based on the feature:

$$h_j(x) = \begin{cases} -s_j & if\ f_j < \theta_j \\ s_j & otherwise \end{cases} \tag{20}$$

The threshold value and the polarity are discovered in the training, including the coefficients. A streamline of the training algorithm is summarized (Szeliski,

Figure 13. Computing a sum in the summed area table data structure

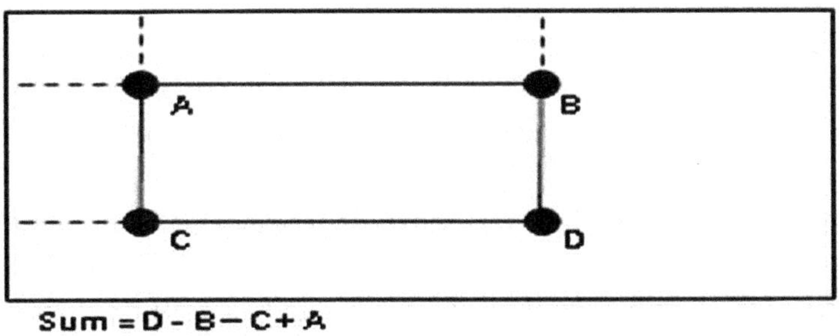

2010): Input are set of N positive and negative training images with their labels. There are two hypotheses:

Hypothesis 1: Image is a face.
Hypothesis 2: Image is not a face.

Step 1: Initialization by select a weight to each image
Step 2: For each feature with:

1. Renormalize the weights so that they total up to 1.
2. Implement the feature to each image in the training set, and seek for the optimal threshold and polarity, which reduces the weighted classification error. Specifically: where

$$\theta_j, s_j = \arg\min_{\theta,s} \sum_{i=1}^{N} w_j^i s_j^i \tag{21}$$

where $s_j^i = \begin{cases} 0 & if\ y^i = h_j\left(x^i, \theta_j, s_j\right) \\ 1 & otherwise \end{cases}$

3. Select a weight to which is inversely proportional to the error rate. In such way, the best classifiers are considered more.
4. The weights for the immediate iteration, that are, decrease for the images which were correctly classified.

Step 3: The final classifier is set to:

$$h\left(x\right) = sign\left(\sum_{j=1}^{M} \alpha_j h_j\left(x\right)\right) \tag{22}$$

Cascading Classifiers

In cascading part, each stage composes of a strong classifier. Hence, all the features are classified into a few stages, in which each stage has some amount of features. The task for each stage is to find out if a given sub-window is certainly a face or not a face. A sub-window is instantly removed as not a face if it misses up in any of the stages.

The cascade architecture has impressive significance performance for the particular classifiers. Since the provocation of each classifier relies completely upon the behavior of its predecessor, the false positive rate for an entire cascade is:

$$F = \prod_{i=1}^{K} f_i \tag{23}$$

Likewise, the detection rate is:

$$D = \prod_{i=1}^{K} d_i \tag{24}$$

Some important properties for cascading architecture are stated below:

- Only 0.01% of all sub-windows are positive (faces) on average.
- Equivalent calculation time is spend on all sub-windows
- Most time will spend on potentially positive sub-windows.
- A simple 2-feature classifier can reach approximately 100% detection rate with 50% false positive rate. Such classifier can work as a first later of a series to refine out most negative windows. The second later with ten features can handle negative windows well that withstand the first layer and so on.
- A cascade of increasingly more complex classifiers can work out better detection rates. The calculation of the strong classifiers produce by the learning process can be complete fast, but it is still not fast enough to run in real-time. Hence, the strong classifiers are organized in a cascade form according to their complexity, in which each consecutive classifier is trained only on those chosen samples that pass through the previous classifiers. If on any stage within the cascade, a classifier refuses the sub-window under inspection, no more processing is carry out and go on seeking the next sub-window. The cascade for that reason has the structure of a degenerate tree.

Fuzzy-Logic-Based Viola Jones Flu Detection Algorithm

The proposed flu detection algorithm (Nur Izzati, 2013) comprise of Viola Jones based eye detection, mean sample canthus region detection and fuzzy logic flu detection.

Viola-Jones-Based Eye Detection

The accuracy of current discovery (Surabhi, 2012) for automated flu detection that measure the overall human facial temperature may be questioned because (Ring, 2007; Mercer, 2009; Ng, 2004) declared that the most accurate part for measure human body temperature is on the inner canthus region of eye. Hence, further eye detection method on identified human object is necessary and (Nur Izzati, 2013) had proposed the Viola Jones based eye detection algorithm. The steps for Viola Jones based eye localization are:

Step 1: Construct Training Set. The face image is partitioned into patches and analyses for targeted facial component. The patches can be served as the coefficient vector since each coefficient is using a filter independently onto those patches. The constructed training set for those captured infrared spectrum image is unlike the digital vision spectrum captured image, due to high level clutter is exist in the infrared image (Martinez, 2010). For negative sample, rather than apply random image, regions which do not have human object facial image is occupied, in case if the clutter totally tarnish the face region, then less variables are applied and it can be modeled more easily. The positive samples categorized as right eye region for the targeted human object, whereas the negative samples are those non-eye region for human facial exist in the image.

Step 2: Initialize Weights. w for those positive samples and negative samples.

Step 3: Compute the Integral Image. The mean value and standard deviation value is get from this computation. Extend the range of mean value and standard deviation value; analyze the value of eye in the range. While the ranges attain the three conditions: 1) eye detection is 100%, 2) low non-eye detection and 3) total error is lower than 50%, the strong classifier is acquired.

Step 4: Set α of t . α values are in equation (19), which represent the coefficients of strong classifier. The greater the number of α , the better the classifier it is.

Step 5: Update Weights w . The weights w for each image are updated subsequently. Reduce the weights w for images that were recognized accurately.

Step 6: Conclude the Strong Classifier. The final strong classifier will be concluded after the training complete. The strong classifier will then apply to detect the right eye region in the input thermal image.

Canthus Region Detection

Canthus region detection is performed once the eye region in a human object for a thermal image has been detected. Suppose that the detected eye region will be cropped out as shown in Figure 15. The mean sample canthus region detection method is implemented on the detected box, and then searches the temperature range for fuzzy logic flu classification.

Figure 15. Cropped eye detected

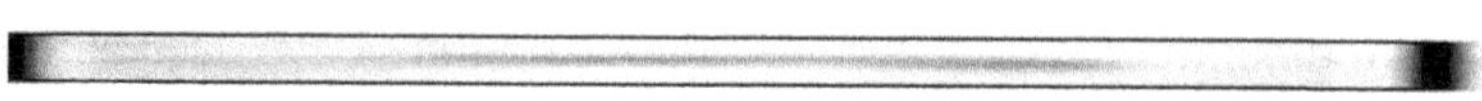

Figure 14. Eye detection

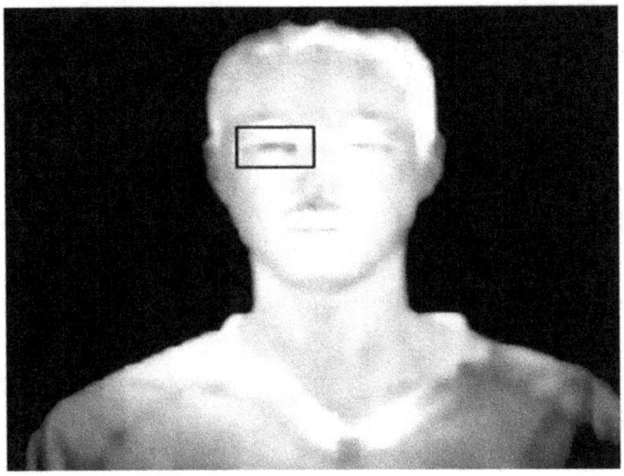

Step 1: Search the middle point of the cropped eye region's box by:

$$m = \left(y_2 - y_1\right) / 2 \tag{25}$$

Step 2: Search the quarter point of the cropped eye region's box by:

$$q = \left(x_2 - x_1\right) / 4 \tag{26}$$

Step 3: Search for the suitable temperature range in which the flu infected human subject will be decided according to pre-determined temperature range. Two methods proposed in searching the optimal temperature range: 1) based on grayscale pixel value and 2) based on color extraction pixel value. These two temperature range searching methods are manually determined using a database of thermal images captured on flu human subjects (in our experiment, it is 300 samples selected). The two searching methods are stated below:

1. **Grayscale Pixel Value:** The captured color based thermal image is first reconstructed into grayscale. Related to canthus region from in step 2, the searching of optimal temperature range is completed by manually computed the grayscale pixel value from the inputted thermal image according to the designated temperature. According to the temperature versus pixel value plot in Figure 16, the grayscale pixel value is linearly proportional to temperature.
2. **Color Extraction Pixel Value:** The color extraction pixel value method is by extracting the red and green elements from the input thermal im-

age and later sums them up. The summation value is proportional to the temperature value (refer Figure 16). The searching temperature range process is completed by manually computed the summation of red and green value in from step 2.

Grayscale pixel value method is preferable in this research due to its faster computational speed.

Step 4: Compute the mean value for that extracted region by:

$$a = \sum \left(q \middle/ _{Totalpixels} \right) \tag{27}$$

where is the canthus detection value (from step 2) and is the total pixels value of the detected box.

Fuzzy-Logic-Based Flu Detection

According to World Health Organization guidelines (Ring, 2008), in infrared thermal imaging, the temperature of 38°C for a human object is considered as fever. Hence, in fuzzy logic scheme, the fever range can be further categorized into ranges that

Figure 16. Temperature value vs. pixel value plot

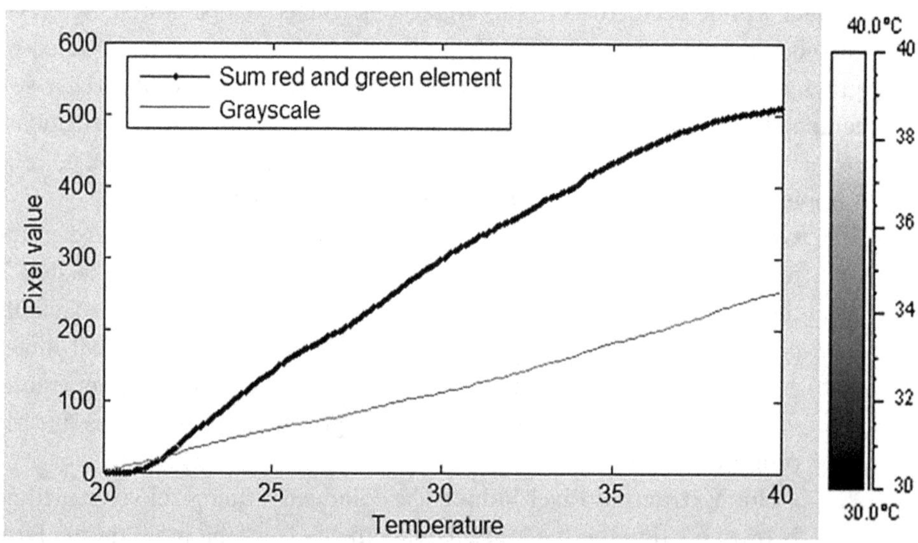

include non-fever, fever and high fever. As for fever and high fever, they can be further classified into flu symptoms.

For flu detection, the value of in equation (12) is adopted. If the value of is higher than the threshold value, the temperature reading for the human subject on thermal image is gets from the pixel value which signified temperature of. The fever classification is fall under the below 3 conditions:

$$\begin{aligned} &\textbf{Condition 1}: \textit{If } a < x_1, \textit{ the subject is classified as non} - \textit{fever.} \\ &\textbf{Condition 2}: \textit{If } x_1 < a < x_2, \textit{ the subject is classified as fever.} \\ &\textbf{Condition 3}: \textit{If } a > x_2, \textit{ the subject is classified as high fever.} \end{aligned} \tag{28}$$

in which x_1=pixel value corresponding to 37.5C and x_1=pixel value corresponding to 39.0C respectively. If Condition 2 and 3 fulfill, the particular human is identified as flu infected and will be pass to doctor/medical personnel for further detail inspection.

EXPERIMENTAL RESULTS

In this section, the application of the proposed thermal imaging system adopting fuzzy logic based Viola Jones method for flu detection is briefly illustrated, with some experimental results and analysis together. Thermal image is first captured by using thermal camera (our model is ThermoVision A-20M by FLIR, 320x240 pixels, captured temperature range -20C - 150C with refresh rate 50/60 Hz) on the site, then send for human object detection. After human object is identified inside the image, it will further proceed with human face region detection and narrow down to canthus region detection. After the canthus region successfully detected, flu detection can be take place. Therefore, the experimental testbed is divided into two main categories: 1) human object detection and 2) Flu detection.

Human Object Detection and Extraction

The results are reported according to 5,000 thermal images with arbitrary human object facing angle. Beside the 5,000 captured thermal images, another additional 100 frames were captured and used for the training data, both for testing the performance of Background-foreground Segmentation BGMM algorithm, Histogram of Gradient (HOG) algorithm and Head Curve Geometry algorithm.

For Background-foreground Segmentation BGMM algorithm, the parameters setting according to Lowe (2004) were by using threshold of mean $\mu = 2$, variance $\sigma = 2$, and $x\% = 50\%$. The number of cluster K=2 based on the clusters of background and foreground. For Histogram of Gradient (HOG) algorithm, the parameters were setting according to Navneet (2005). Out of the 100 training images, 50 of it was selected as positive training examples, another 50 of it was selected for initial negative set. For every detector and parameter combination, an initial detector was trained and the 50 negative training images were investigated thoroughly for 'False Positives'. The method was later re-trained adopting such improved set to generate the optimal detector. The set of 'False Positives' examples were subsampled to fit into the SVM training, until it reach 5% of the False Positive Per Window (FPPW) test. It was then passed for conventional non-maximum suppression continuing on the output pyramid to recognize human object. For Head Curve Geometry algorithm, there was not required training process since the algorithm was direct detection after some parameters set. The find out values for those 3 human body related parameters are: *HeadPercent%=14%, NeckPercent%=15%,* and *LeftRightOut%=3%.* They were used in the following experiments.

The three human object detection and extraction algorithms above were tested with the 5,000 captured thermal images with arbitrary human object facing angle. Out of the 5,000 tested images, Background-foreground Segmentation BGMM algorithm accurately completed 4,511 images with human object detected and fully extracted, with accuracy up to 90.22%. HOG algorithm accurately completed 4,503 images with human object detected and fully extracted, with accuracy upto 90.06%; whereas Head Curve Geometry algorithm accurately completed 4,511 images as well with human object detected and fully extracted, with accuracy up to 90.22%. Conclusively, the results for the three methods were comparable, however Background-foreground Segmentation BGMM algorithm required heavy multiplication process whereby either FD-BGMM or FD-TBGMM was chosen to extract the final human object, in which both these methods needed at least two classifiers. For greater training samples, the complexity is increased dramatically. HOG also required training and complex cascading classifiers. On the other hand, the proposed head curve geometry didn't required training and hence saving computational complexity and speed up the processing time. Hence, by this justification, Head Curve Geometry algorithm is nominated to be used in the proposed thermal imaging system with automated flu detection capability.

Flu Detection

For flu detection, 5000 thermal images for patients had been gathered at two clinics, namely: Polyclinic Gomez, Cyberjaya and Polyclinic Abdul Rashid, Melaka Malaysia. The disease types identified by doctor were varying from flu, headache, sore throat, cough, joint pain, diarrhea and others. Out of 5000 thermal images collected, 88 samples were identified as flu patients and 4,912 identified as non-flu patients.

After human object was identified inside the captured thermal image, it will further proceed with human face region detection and narrow down to canthus region detection. For human face region detection, only human object which facing front view is necessitate in the proposed automated flu detection system since only if human facing the camera, their frontal face image will be captured into the camera, and the canthus region is included there. By proceed with the continuum work on Head Curve Geometry algorithm, a face region extraction algorithm is formulated as below:

Step 1: Calculate the center point, *cenDraw* for the curve by utilizing the x and y coordinates from the Left Most Point and the Right Most Point. The mean value of the y coordinates (*cenDrawY*) from the two points gives the y coordinate of the *cenDraw*. Refer for Figure 17 for better understanding.
Step 2: Place point A and point B by expanding the horizontal y axis at the value of *cenDrawY* over the head boundary. Utilize the mean value for these two points to compute the x coordinate for *cenDraw*.
Step 3: Determine the interval from point A to *cenDraw* and from point B to *cenDraw*. The maximum value among these two interval is the *radius* of the neck curve.
Step 4: Utilize the *cenDraw* and the *radius* to mark the curve on the head boundary. Refer to Figure 18(b) for better understanding.
Step 5: After the curve is drawn, there would be two different objects. The object that has the HeadPpoint in the boundary will be chose to extract out the head shape from the human object. Refer to Figure 18(c) for better understanding.

Head curve geometry algorithm is now embedding with face detection feature. After human face detected in the image, the next step is to perform canthus region detection. This is done by using Viola Jones Based Eye Detection. To set up the Viola Jones training set, 1500 positive patches and 3000 negative patches each with size 25x25 was used. After the canthus region successfully detected, flu detection can be take place by applying Fuzzy Logic based Flu Detection. Parameter setting for this algorithm was on the temperature threshold value. This parameter was calculated with the measured of pixel value per temperature referred on the thermal camera's color bar. The chosen temperature threshold in equation (28) were x_1=37.5C and x_2=39.0C. The corresponding pixel values were 529 and 679 respectively.

All the gathered data was tested using Fuzzy Logic Based Viola Jones Flu Detection Algorithm. Throughout 88 gathered thermal images that were ascertained by doctor as flu patients, 71 samples were accurately detected by algorithm as flu case, with an accuracy of 80.68%. Throughout 4912 gathered thermal images ascertained by doctor as non-flu patients, 3965 samples were accurately detected by algorithm as non-flu case, with an accuracy of 80.72%.

Figure 17. (a) HeadPpoint, (b) left most point, (c) right most point, (d) left neck point, (e) right neck point, (f) LeftMean, (g) RightMean, (h) iDistance

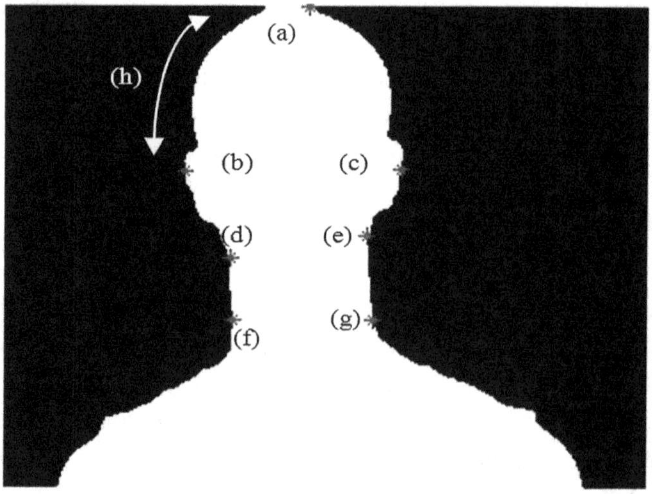

Figure 18. (a) Captured front face thermal image, (b) bwFil image with cenDraw point and a curve, (c) the extracted head shape in RGB image

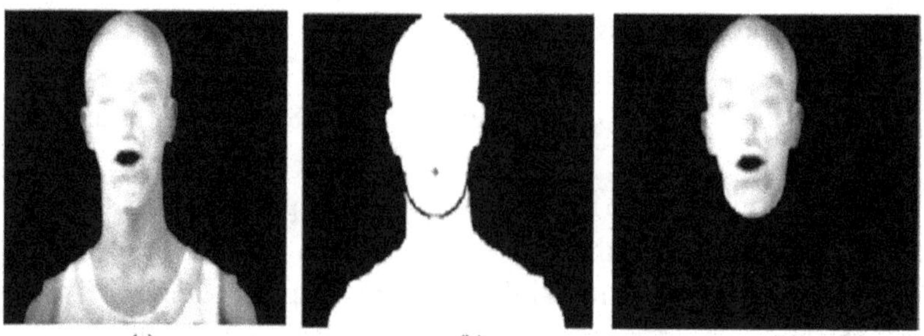

(a) (b) (c)

Among the 964 failed identified flu/non-flu cases by the algorithm, further investigation was done, with 384 false positive cases and 298 false negative cases tracked. The corresponding measured false positive rate was 39.83% and false negative rate was 30.91%. False positive case is defined as the condition when non-flu subject is falsely identified by the algorithm as flu. There are some possible factors which may cause false positive case, like the image captured during the patient was in stress mood or right after the patient performed some physical activities that might influence the patient's facial temperature. The aftereffect is time consuming because more examination required to reconfirm the patient is healthy. On the other hand, false negative case is defined as the condition when flu patient is falsely identified by the algorithm as non-flu. Such cases might occur due to the canthus region is falsely extracted and the temperature measurement is performed on other human facial parts that have lower temperature. False negative case might cause bad consequence because if the patient has infectious disease such as SARS, H1N1, etc., it might lead to the chance that the viral will infect more people. Summary of Flu detection results is shown in Table 1. Similarly summary of detected disease is shown in Table 2.

CONCLUSION AND FUTURE RESEARCH DIRECTIONS

This chapter demonstrated an intelligent thermal imaging system using fuzzy logic based Viola Jones method for flu detection. The thermal imaging system is claimed to be intelligent in the mean that it can automatically detect flu infected personal without required human operator to manually observe the monitor screen to trace out the target. In the proposed system, the thermal image is first captured on site with the thermal camera, and then sends for human object detection and extraction, human face region detection and extraction, followed by canthus region detection and eventually, the flu detection. All these developed algorithms are performed by image processing techniques. Three human object detection and extraction algorithms are

Table 1. Summary of flu detection results

Age/Types	Child	Youth	Adult	Older	Total
Flu identified by doctor	6	8	71	3	88
Flu identified by algorithm	6	6	56	3	71
Non-flu identified by doctor	184	321	4315	92	4912
Non-flu identified by algorithm	148	255	3472	90	3965

Table 2. Summary of disease detected

Disease/Types	Cough	Sore Throat	Headache	Diarrhea	Joint Pain	Other
Flu identify by doctor	21	26	25	3	12	1
Flu identify by algorithm	19	22	17	2	10	1
Non-flu identify by doctor	1505	330	725	105	100	2147
Non-flu identify by algorithm	1270	260	570	95	97	1673

studied and compared, namely the Background-foreground Segmentation BGMM algorithm, HOG algorithm and Head Curve Geometry algorithm.

Experiment results shown that Head Curve Geometry algorithm is of great interest since its human object detection and extraction accuracy is comparable with Background-foreground Segmentation BGMM algorithm and HOG algorithm, but it overwhelmed the two algorithms due to it does not required training and complex classifiers structure. Head Curve Geometry algorithm can also be embedded with a proposed face region extraction algorithm to help trace out the human face image. Viola Jones based eye detection method is proposed, follow by canthus region detection and end out with fuzzy logic based flu detection. The proposed thermal imaging system using fuzzy logic based Viola Jones method for flu detection can identify flu cases up to accuracy of 80.68% and non-flu cases up to accuracy of 80.72%.

Currently, the developed thermal imaging flu detection system is proposed for border pass gate to detect infectious disease human carrier, to prevent them entering a region/nation. The system at this moment is still a stand-alone system. In future, a similar and networked based system will be developed, in which it can trace flu suspects for more than one location automatically on the same time. For example in shopping complex or place where more people are crowded, to help trace out infectious disease personal to avoid outbreak. Furthermore, there is an intention to adopt microprocessor modules, for instance the ARM (Advanced RISC Machine) and FPGA (field programmable gate array) as image processing and analyzing tasks rather than a computer to efficiently lower the power consumption and hardware cost for the prospective system. These topics will be addressed in future works.

REFERENCES

Belongie, S., Malik, J., & Puzicha, J. (2002). Shape Matching and Object Recognition Using Shape Contexts. *IEEE Transactions on Pattern Analysis and Machine Intelligence*, *24*(24), 509–521. doi:10.1109/34.993558

Bouwmans, T., El Baf, F., & Vachon, B. (2008). Background Modeling using Mixture of Gaussians for Foreground Detection—A Survey. *Recent Patents Comput. Sci.*, *1*(3), 219–237. doi:10.2174/2213275910801030219

Buddharaju, P., & Pavlidis, I. (2008). Physiology-Based Face Recognition in the Thermal Infrared Spectrum. In Medical Infrared Imaging Handbook (pp. 13-1 - 13-14).

Chen, T. W., Hsu, S. C., & Chien, S. Y. (2007, July). Robust Video Object Segmentation Based on K-Means Background Clustering and Watershed in Ill- Conditioned Surveillance Systems. *Proceedings of theIEEE Int. Conf. on Multimedia and Expo* (pp. 787 – 7900). doi:10.1109/ICME.2007.4284768

David, M., Friedrich, L., & Kurt, H. (2003). The Support Vector Machine Under Test. *Neurocomputing*, *55*(1-2), 169–186. doi:10.1016/S0925-2312(03)00431-4

Egawa, S. (1997). *Clinical Radiation Thermometer*. WIPO patent: WO1997017887 A1.

Green, M. W. (1999). The appropriate and effective use of security technologies in U.S. schools, A guide for schools and law enforcement agencies (research report). Sandia National Laboratories, Albuquerque, NM, USA.

Heikkilä, M., & Pietikäinen, M. (2006). A Texture-Based. Method for Modelling the Background and Detecting Moving Objects. *IEEE TPAMI*, *28*(4), 657–662. doi:10.1109/TPAMI.2006.68

Heitz, D. (2015). *Infrared Thermometer to Fight Ebola Premiers at CES 2015*. Healthline Media. Retrieved from http://www.healthline.com/health-news/infrared-thermometer-to-fight-ebola-premiers-at-ces-010515#1

James, C. M., Matthew, L. S., & Michael, E. M. (1998). *Crew fatigue and performance on US coast guard cutters*. Springfield, VA: National Technical Information Service.

Kim K., Chalidabhongse T.H., Harwood D., Davis. L. (2004). Background Modeling and Subtraction by Codebook Construction. *Proceedings of the Int. Cont. on Image Processing (ICIP '04)* (Vol. 5, pp. 3061-3064).

Kim, K., Chalidabhongse, T. H., Harwood, D., & Davis, L. (2005). Real-time Foreground-background Segmentation Using Codebook Model. *Elsevier Real-time Imaging*, *11*(3), 172–185. doi:10.1016/j.rti.2004.12.004

Kiri, W., Claire, C., Seth, R., & Stefan, S. (2001). Constrained K-means Clustering with Background Knowledge.*Proc. of the 18th Int. Conf. on Machine Learning (ICML '01)* (pp. 577-584).

Levi, K., & Weiss, Y. (2004). Learning object detection from a small number of examples: the importance of good features.*Proc. of the IEEE Conference on CVPR*, 53–60. doi:10.1109/CVPR.2004.1315144

Lowe, D. G. (2004). Distinctive image feature from scale-invariant keypoints. *International Journal of Computer Vision*, *60*(2), 91–110. doi:10.1023/B:VISI.0000029664.99615.94

Martinez, B., Binefa, X., & Pantic, M. (2010). Facial Component Detection in Thermal Imagery. *Proceedings of the IEEE computer Society Conference on Computer Vision and Pattern Recognition Workshops (CVPRW)*, 48-54. doi:10.1109/CVPRW.2010.5543605

Mercer, J. B., & Ring, E. F. (2009). Fever Screening and Infrared Thermal Imaging: Concerns and Guidelines.Thermology International, 19(3), 67-69.

Mohamad, H. S., & Mahmood, F. (2008). Real-time Background Modeling/Subtraction using Two- Layer Codebook Model.Proc. of the Int. Multiconf. Of Engineers and Computer Scientists (IMECS '08).

Navleet, D., & Bill, T. (2005). Histograms of Oriented Gradients for Human Detection. *Proc. IEEE Conf. Computer Vision and Pattern Recognition CVPR '05*(pp. 886-893).

Ng, E. Y. K., Kaw, G. J. L., & Chang, W. M. (2004). Analysis of IR thermal imager for mass blind fever screening.Microvascular Research, 68(2), 104-109.

Norulzahrah, M. Z., Suzaimah, R., Kamarul, H. G., Talib, M. L., & Nor Asiakin, H. (2015). A Study on Implementing Physiology-Based Approach and Optical Flow Algorithm to Thermal Screening System for Flu Detection. *International Journal of Information and Electronics Engineering*, *5*(1), 31–34.

Nur Izzati, N. B. I., Wong, W. K., Jalil, B. M. D., Yante, B. M. N., Azlan, B. S., & Lim, H. S. (2013). Fuzzy Logic based Viola Jones Fever Detection Method in Thermal Imaging System. *Proceedings of the 2013 International Conference on Emerging Trends in Computer Science & Information Technology (ICETCSIT '13)*, 23-26.

Papageorgiou, C. P., Oren, M., & Poggio, T. (1998). A general framework for object detection. *Proceedings of theInternational Conference on Computer Vision* (pp. 555-562). doi:10.1109/ICCV.1998.710772

Radzi, S., Ghazali, K., Mohammad, A., & Naim, F. (2011). *Procedia Computer Science Using Bimodal Gaussian Mixture Model-Based Algorithm for Background Segmentation in Thermal Fever Mass Screening*. Academic Press.

Ring, E. F. J., Jung, A., Zuber, J., Rutowsk, P., Kalicki, B., & Bajwa, U. (2008). Detecting fever in Polish children by infrared thermography. *Proceedings of the 9th International Conference on Quantitative Infrared Thermography*,125-128. doi:10.21611/qirt.2008.03_07_17

Ring, F. (2007). Pandemic: Thermography for fever screening of airport passengers. *Thermology International*, *17*(2), 67.

Hampapur, A., Brown, L., Connell, J., Pankanti, S., Senior, A., & Tian, Y. (2003). Smart Surveillance: Applications, Technologies and Implications. *Information, Communications and Signal Processing*, *2*, 1133–1138.

Hampapur, A., Brown, L., Connell, J., Ekin, A., Haas, N., Lu, M.,... & Pankanti, S. (2005). Smart Video Surveillance. *IEEE Signal Processing Magazine*, 22(2), 38–51.

Sofiah, S., Radzi, M., Kamarul Hawari, G., Nazriyah, C. Z., & Faradila, N. (2011). Using Bimodal Gaussian Mixture Model-Based Algorithm for Background Segmentation in Thermal Fever Mass Screening. *Proceedings of the2nd World Conference on Information Technology*.

Stauffer, C., & Grimson, W. (1999). Adaptive Background Mixture Models for Real-time Tracking.*Proc. IEEE Conf. on Comp Vision and Patt. Recog. (CVPR '99)* (pp. 246-252). doi:10.1109/CVPR.1999.784637

Suard, F., Rakotomamonjy, A., & Bensrhair, A. (2006). *Pedestrian Detection using Infrared images and Histograms of Oriented Gradients*. Academic Press.

Surabhi, V., Spinello, D., & Neesulescu, D. (2012). Infrared Fever Body Identification using Shape and Temperature Filters. *Proceedings of the IEEE Instrumentation and Measurement Technology Conference (I2MTC)*, 1556-1560. doi:10.1109/I2MTC.2012.6229315

Szeliski, R. (2010). *Computer Vision, algorithms and applications*. Springer.

DG Topteches Electronic, Limited. (2015). *Walk Through Forehead Thermometer - Medical Thermometer Manufacturer*. curiousexpeditions.org Redirected from http://topteches1.sell.curiousexpeditions.org/pz5a00393-walk-through-forehead-thermometer-medical-thermometer-manufacturer.html

Viola, P., & Jones, M. (2001). Rapid object detection using a boosted cascade of simple features. *Proceedings of the 2001 IEEE Computer Society Conference on Computer Vision and Pattern Recognition (CVPR '01)* (Vol. 1, pp. 511-518). doi:10.1109/CVPR.2001.990517

Wong, W. K., Hui, J. H., Desa, J. B. M., Ishak, N. I. N. B., Sulaiman, A. B., & Nor, Y. B. M. (2012). Face Detection In Thermal Imaging Using Head Curve Geometry. *Proceedings of the 2012 5th International Congress on Image and Signal Processing (CISP '12)*, 1038-1041.

Chapter 2
The Brain Tumor Segmentation Using Fuzzy C-Means Technique:
A Study

Jyotsna Rani
NIT Silchar, India

Fazal A. Talukdar
NIT Silchar, India

Ram Kumar
NIT Silchar, India

Nilanjan Dey
Techno India College of Engineering, India

ABSTRACT

Image segmentation is a technique which divides an image into its constituent regions or objects. Segmentation continues till we reach our area of interest or the specified object of target. This field offers vast future scope and challenges for the researchers. This proposal uses the fuzzy c mean technique to segment the different MRI brain tumor images. This proposal also shows the comparative results of Thresholding, K-means clustering and Fuzzy c- means clustering. Dice coefficient and Jaccards measure is used for accuracy of the segmentation in this proposal. Experimental results demonstrate the performance of the designed method.

DOI: 10.4018/978-1-5225-2423-6.ch002

INTRODUCTION

Thermal Imaging is one of the most important part of infrared imaging science and technology. These techniques are usually applied in the fields of radiation detection, thermal mapping, chemical imaging, medical fields and various other fields. Imaging techniques involve various step by step processes such as restoration of images, enhancement of Image quality, color image processing as well as segmentation and clustering of images. Image segmentation is highly used in the fields of detection and segmentation of the diagnosed part of the affected organ thereby providing a great help in the field of medical imaging. Clustering techniques are widely used in detection and segmentation of brain tumor distinctly and with much better precision. Thresholding, K mean clustering and Fuzzy C mean are some of the important clustering techniques used in the Brain tumor segmentation process. Brain is the most vital organ contained in the skull of each and every type of vertebrates. It also acts as the most critical organ in most invertebrates. It is the coordinating center which coordinates the sensations, intellectual and all the nervous activities of an organ system. Brain is the centermost part of a nervous system. It is regarded as a complex and composite organ in a vertebrate's body which is located in the head. As the recreation of cells take place in an unchecked manner, an unusual growth of tissue starts taking place. It results in creation of tumor. If this unusual recreation of cells takes place inside a brain, then this unusual recreation of brain cells is regarded as brain tumor. Development of unusual recreation of brain cells is also called as intracranial neoplasm.

Tumors are of two types, malignant (cancerous tumors) and benign tumors. Cancerous tumors are again sub-divided into two further types. They are primary and secondary tumors. The root of primary tumors is inside the brain. Secondary tumors usually originate and expand from somewhere else. It is also known as brain metastasis tumors. Elemental indications of development of a brain tumor differ depending on the section of the brain convoluted. The basic signs and symptoms of a brain tumor affected patient include problems with eyesight, intense headache, vomiting and drastic cerebral changes. The headache is serious in the morning and gradually goes off after vomiting within a short span of time. Brain tumor may cause difficulty in walking, problems in speaking, intellectuals and sensations along with many more specific problems. As the diseases advances, mental incapability may occur with time. Apart from disclosure to vinyl chloride or ionizing radiation, there are no exactly known coincidental factors which are linked with the inducing causes of brain tumors. Variations and removal of tumor suppressor genes are also the causing factor for the creation of some of the types of brain tumors.

The human brain and other vertebrates are made up of a very delicate tissue which is a gelatin-like surface. A living brain tissue has a pink hue in color regarded as grey matter of the brain. On the other hand, the inside of a brain is almost complete white or called as white matter. Also there are some more faint variations in framework and structure of the brain. The brain's volume can be divided in three distinct brain areas. They are Telencephalon or commonly called as cerebral hemisphere or cerebrum, Mesencephalon or midbrain and Cerebellum. These areas are again classified in two broad classes of cells i.e. Neurons and Glia. Most of the part of brain is partitioned from the blood by a barrier called blood-brain-barrier generally named as BBB. This BBB restrains particles to penetrate through it. Many minute particles that reach tumors in the body can only penetrate there when there is some unusual breakage or irregularity of BBB. These irregularities taking place within the brain tumor are usually detected by a MRI and CT scan. This is the main analytic indicator for malignant gliomas and brain tumor. There are no specific clinical indications or signs for identifying brain tumors. Although, the existence of a particular combination of symptoms can be an indicator which pivots towards a diagnostic investigation and the occurrence of inter cranial neoplasm. Magnetic resonance imaging (MRI) is a medical imaging technique basically used to examine the anatomy and basic physiology of the body by the use of radiology. Although the strong magnetic field which is used do not generally put adverse effects, but the embedded medical devices containing metal may degrade and tend to various problems during an MRI diagnosis. Along with it, there is a probable risk of an allergic sensation after the material is injected causing a high chances of health disorder to the patient undergoing diagnosis. Computed Tomography (CT scan) is another medical imaging technique which uses the computer aided combinations of many X-ray images which are taken from various angles. It produces a cross-sectional or a topographic image along with various virtual slices of a targeted area which is being regulated to scanning. However, there is a bit of increased chances of cancer in future due to prolonged exposure to ionizing radiation. In addition to that, it also uses higher amount of radiation than normal X-ray. This tends to small risk which is generally considered to be greater than other medical imaging techniques. These unusual cells create specific issues or properties that follow closely to the characteristics of the organ they are in. This results in creation of a specific groups or clusters. These specific groups or clusters are partitioned using the digital image segmentation techniques. Medical image segmentation plays a very vital and instrumental role in clinical diagnosis in the field of medical sciences and specifically imaging technology. An ideal medical image segmentation technique should possess some specified properties which include minimum user interaction. They create specific characteristics that follow closely to the properties of the organ they are in. They specifically form specific groups or clusters. These specific groups or clusters can be segmented using the image segmentation techniques.

Medical image segmentation plays an instrumental role in clinical diagnosis. An ideal medical image segmentation scheme should possess some specific properties such as minimum user interaction, fast computation, precise and robust segmentation results. Thereby providing a new insight into the field if medical imaging.

BACKGROUND

Image is regarded as an artifact that depicts or records visual perception. Let us consider an example of a two dimensional picture, we find that it has a similar appearance to some subject, or a physical object or a person, these all provides a depiction of the specified image. Images contain plenty of objects and patterns. These patterns are actually attached with valuable information which are required for medicine, biology, photography areas or domains. In analyzing of an image, inputs taken are images whereas outputs taken are either an image or a set of characteristics or parameters related to that of the image. The final motive of an image analysis is to acquire information from the gathered data through recognized or classified objects or classes or attributes and then take action accordingly to the acquired information. For a digital image representation, an image may be represented as a two-dimensional function where x co-ordinate and y co-ordinate are spatial or planar co-ordinates, and the amplitude of the particular function at any pair of co-ordinates are regarded as the intensity of the image at that particular point.

A new descriptive name is introduced which is regarded as gray level which refers to the intensity of monochromatic images. In general, Color images are formed by a specified combinational mixture of individual images. For example, in the RGB color system, a color image is composed of three individual monochromatic images. They are denoted as the red (R), green (G) and blue (B) for primary images. An image may be continuous with respect to the spatial or planar co-ordinates. Its intensity may be continuous in terms of amplitude of the particular image. To convert such an image into a digitalized from we need to digitalize the co-ordinates, as well as the amplitude too. The process of getting the co-ordinate values digitalized are termed as sampling and the amplitude values is termed as quantization. Thus, when the spatial and planar co-ordinates along with the amplitude values of the particular two dimensional functions are all finite and discrete quantities, we regard the image as a digital image. In general a digital image can be categorized into four groups or types. They are:

- Gray scale image: A gray scale image is a two dimensional matrix barcode which encodes the raw received information. Its values are represented as the shades of gray having integer values ranging from [0, 255] or [0, 65535] respectively.

- Binary images: A binary image is a digital image. It has only two possible values allotted to each and every specified pixel. The two colours generally taken into consideration for a binary image are black and white. But it is not compulsory to choose the two fixed colours. Any two colours can be taken into consideration in order to represent a binary image. The colour used for the targeted object in the image is the foreground colour and the rest of the image is the background colour.A binary image can be referred as a logical set of arrays consisting of 0s and 1s.
- RGB Images: RGB Images are the combination of Red (R), Green (G), and Blue (B) color in a specified proportion. The RGB colour mode is regarded as an add-on colour model in which the light of all the three colours are added together in every possible ways as per requirement to generate a broad array of colours.
- Indexed images (Gonzalez, et.al 2002).

Generally, segmentation subdivides an image into its constituent areas or targeted regions.

Image segmentation is an image division process that aims at partitioning an image into several areas. Sub-division takes place on the basis of the criteria of homogeneity. Image segmentation is a very difficult process, which is accomplished through the benefits from computer assistance. No particular algorithm has been specified that would help us to reach perfect segmentation either independent from the input data type. It is now been altered by the active improvement of a wide range of very specialized and sophisticated techniques.

Image segmentation is regarded as partitioning of a digital image into multiple segments. It aims in order to separate a digital image into certain regions by finding specific boundaries. These boundaries correspond to objects related in the image. These boundary areas are partitioned either based on some common characteristics or on the differences between the edges at objects' contours. The accuracy of objects recognition or classification is based upon the quality of segmentation implemented. The aim of a segmentation technique is to simplify or modify the nature of an image to one that is more meaningful and much easier to examine.

Thresholding is one of the most important technique in image segmentation. It is the simplest method of image segmentation. From a grayscale image, thresholding can be used to create binary images. The simplest thresholding method refers to replacing each pixel in an image with a black pixel if the image intensity, l_{ij} is less than some fixed constant T suppose (that is, $l_{ij} < T$), or a white pixel if the image intensity exceeds T. Moreover, some binarization methods for segmentation, via thresholds are also taken into consideration which is based on the distribution of pixel properties such as intensity values. Clustering is one another way of determining the optimal threshold or a boundary value. Optimal threshold in case of classification allows classifying different classes of objects amongst them. Two threshold methods are examined as a starting point for the classification process. The first one is based on adaptive thresholding and global binarization and second one is based on dendrograms. The basic flow diagram of brain image segmentation process is shown in Figure 1.

THRESHOLDING

Often, we have to examine the foreground and /or background of images because they divide an image into regions that correspond to the objects. This is straightforward matter and is called image segmentation based on thresholding.

Simple Thresholding

For a grayscale input image $f(x,y)$ the threshold image $g(x,y)$ threshold T is

$$g(x,y) = \{1, iff(x,y) \geq T\} \quad or \quad g(x,y) = \{0, iff(x,y) \leq T\} \tag{1}$$

where, T is the boundary limit or the specified threshold constant taken into consideration.

Figure 1. Basic flow diagram of brain image segmentation process

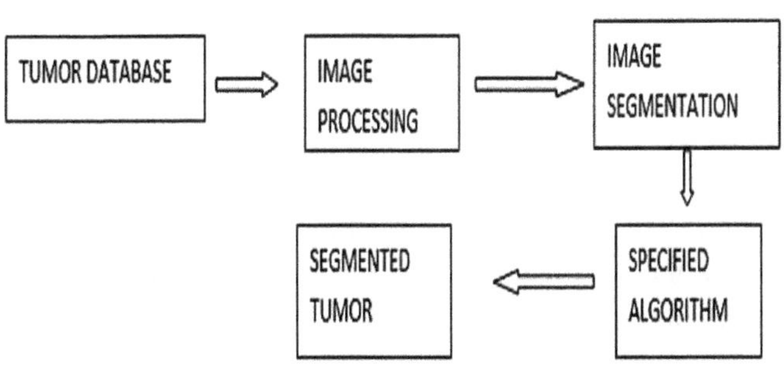

Adaptive Thresholding

Non-uniform illumination due to image acquisition technologies may dramatically change the histogram, so it becomes impossible to segment an image based on a single threshold value. Adaptive thresholding segmentation separates the foreground and background based on non-uniform illumination.

Apart for the global threshold method, the adaptive thresholding computes the individual threshold values for small regions of the image (Chandhok et al., 2012). It examines the intensity values of local neighborhood of each pixel based on the mean of the local intensity distribution and median statistic functions (Bradley et al., 2007)

Histogram Thresholding

Image histograms are tools to visualize the distribution of grayscale intensity values in the entire image. The histogram of an image f with m lines and n columns and the gray levels in the range $\left[0, L-1\right]$ is given by a discrete function h having the following properties:

$$h(i) = \frac{1}{MN}\sum_{m \to 0}^{M-1}\sum_{n \to 0}^{N-1}\delta(i, f(m,n)) \tag{2}$$

where $i = 0, \ldots, L-1$, and

$$\sum_{i \to 0}^{L-1} h(i) = 1 \tag{3}$$

where,

$$\begin{aligned} &\delta(x,y) = \left\{1, ifx = y\right\} \\ &or \\ &\delta(x,y) = \left\{0, ifx \neq y\right\} \end{aligned} \tag{4}$$

Image histogram includes peaks and valleys and the valley point can be used as the threshold so that the image is separated into a number of objects. The peaks and valleys are examined and the thresholds were chosen in two ways. First, binary image corresponds to the thresholds chosen as maximum (the higher number of pixels at this value) or minimum (the lower number of pixels at this value) gray-level values. It has to be noted that these values do not necessary correspond to the background or foreground. The second approach uses a threshold value that was manually determined as the minimum value (or valley) that separates the two peaks.

However, for real applications, the threshold value should be determined automatically, via computer programs, by using algorithms based on discriminant analysis or unsupervised classification methods.

CLUSTER ANALYSIS

Cluster analysis is used in situations where little prior knowledge on the raw data exists. Often, clustering is used for the segmentation of images and might be implemented using various features extracted from textured images. These selected features are grouped into clusters corresponding to different classes. The distance between classes is the main criterion of separation and the similarity of the data into the cluster is the building criterion of one cluster. Various algorithms have offered from cluster analysis. Here, two algorithms namely *K*-means clustering and fuzzy c-means clustering are presented.

K-Means Clustering

The image segmentation based on *K*-means clustering algorithm is one of the most widely used methods. It is an unsupervised clustering algorithm that performs the partition of an image into *k* clusters using the distance between multiple classes existing in image. Jumb, et al. (2014), Tatiraju, S et al. (2011), Chandhok, et al. (2012)

The distance $V = \sum_{i-1x_i}^{k} \sum_{eS_i} (x_j - \mu_j)^2$ minimizes between analyzed points (observations or features) that clusters around centroids μ_i, $\forall i = \overline{1, k}$ allows clustering. Here k is the number of cluster and μ_i is the centroid of all the point where, $x_j \in S_i$. The iterative algorithm works as follows:

1. Compute the histogram of a digital image as a distribution of its discrete intensity levels.
2. Initialize *k* class centers randomly.
3. Cluster the points/features using distance minimization relation:

$$c^{(i)} = \arg\min \left\| x^{(i)} - \mu_j \right\|^2$$

4. Recalculate the new position of centroids:

$$\mu_i = \frac{\sum_{i=1}^{m}\{c_{(i)} = j\}x^{(i)}}{\sum_{i=1}^{m}\{c_{(i)} = j\}x^{(i)}}$$

5. Repeat steps (3) and (4) until the cluster labels of the image does not change.

Fuzzy C-Means Clustering

Fuzzy c-means clustering algorithm is mainly used in medical image processing for pattern texture exploration, anomalies detection or salient features identification (Xin-bo, et al., 2009; Yang et al., 2007). This clustering approach allows that some data belong to two or more clusters with different values of the membership coefficient (Dunn et al., 1973; Bezdek et al., 1981; Kwon et al., 2004). It is an iterative method based on minimization of the functional called c-means functional.

$$J_{FCM} = \sum_{k=1}^{n}\sum_{i=1}^{c}(\mu_{ik})^p d^2(x_k, v_i) \tag{5}$$

$X = \{x_1, x_2, ..., x_n\} \subseteq \Re^p$ is data set in the p-dimensional vector space, c is the number of the clusters and n is number of data items $(2 \leq c < n)$, μ_{ik} is the fuzzy partition of x_k in the ith cluster, v_i is a cluster prototype (or center), $d^2(x_k, v_i)$ is distance measure between object x_k and cluster center v_i. The cluster center v_i has to be determined.

IMAGE SEGMENTATION OUTPUTS

The experimental outputs have been taken using various image segmentation techniques which are explained in detail.

Image Segmentation by Thresholding

Thresholding is the simplest technique used for image segmentation. It is accomplished by separating pixels in different categories based on their pixels grey levels. A thresholding technique denotes an intensity value generally referred as the threshold, which separate the pixels into desired categories. The segmentation is accomplished by taking threshold value. By taking into consideration of the threshold value, pixels are categorized accordingly. The one with intensity greater than the threshold value is categorized into one class and the rest pixels grouped into separate class. The main drawback of this categorization technique is that, in the simplest form only two categories are created which cannot be applied to multichannel images.

In this technique, image having only two values either black or white is taken into consideration. MR image contains 0 to 255 grey values. Therefore, MR images ignore the tumour cells after thresholding technique is applied to it. Jay Pateland et al. (2014)

Image Segmentation by K-Means Clustering

K-means clustering is a self-organizing or unsupervised algorithm that deals with clustering problem. The steps followed for k mean clustering algorithm is a very simple and easy. It segments the image using basic knowledge of cluster value. In k-means, initially we randomly assign k centroids. The assigning of this k centroid is done in a specified way because different location generates different cluster groups. Therefore, it is better to place centroid value which is as much as far away from each other. Now we calculate the separation between each pixel from selected centroid of a particular cluster. Each pixel of the cluster then compares with k clusters centroids and calculates distance using distance formula. If the pixel calculates shortest distance amongst all other pixels, then it moves to the particular cluster. This process is repeated till all pixels compare to the allotted cluster centroids. This process terminates till a particular convergence criteria are fulfilled.

The *K*-means algorithm for image segmentation is highly affected by the initially selected random centroids and also to the number of clusters. This drawback can be overcome if the algorithm is ran multiple times. The initial centroids are chosen at some particular points randomly for different number of clusters.

Image Segmentation by Fuzzy C-Means Clustering

Fuzzy C-means is a coinciding clustering technique. A particular pixel value may depend on more than one cluster centres. Fuzzy C-mean clustering is also called as soft clustering method. One of the most popularly utilized fuzzy clustering algorithms is the Fuzzy C-means (FCM) algorithm. Bezdek, et al. (1981)

In case of fuzzy clustering, each point has a degree of association to more than one cluster, as referred in fuzzy logic. It does not just associate with one cluster completely. Therefore, points which are lying on the edge of a cluster is associated in the cluster to a lesser degree of association than points lying in the center of cluster. For each point x, we have a coefficient giving the degree of association in the k_{th} cluster $\mu_k(x)$. Hakeem Aejaz Aslam et al. (2013)

Usually, the sum total of all these coefficients for any mentioned x is specified to be one.

$$\sum_{k=1}^{c} \mu_k(x) = 1 \tag{6}$$

In case of fuzzy c-means, the calculated centroid of a particular cluster is the mean of all points, which is weighted by their degree of association to the cluster.

$$centre_k = \frac{\sum_x \mu_k(x)^m x}{\sum_x \mu_k(x)^m} \tag{7}$$

The degree of association is related to the inverse of the distance between the point and the cluster center.

$$\mu_k(x) = \frac{1}{d(centre_{k_i} x)} \tag{8}$$

The coefficients are assigned and fuzzyfied with a real parameter $m > 1$ so that their sum is one. So,

$$\mu_k(x) = \frac{1}{\sum_j \left(\frac{d(centre_{k_i} x}{d(centre_{k_i} x} \right)^{\frac{2}{m-1}}} \tag{9}$$

For m equal to 2, this is equivalent to assigning the coefficient linearly to make their sum one. When m is close to 1, then cluster center closest to the point is given much more weight than the others, and the algorithm is similar to k-means. The fuzzy c-means algorithm is quite identical to the k-means algorithm. N. Dey et al. (2012)

EXPERIMANTAL RESULTS

The experimental results for various sample images using three image segmentation techniques which are:

- Thresholding,
- K-means clustering,
- Fuzzy C-means clustering.

The results are obtained are obtained accordingly after implementing the codes in MATLAB2013a software. In this study we have used 2D MRI images and for comparison purposes, the same test image is used across this work. MATLAB programming language is used to demonstrate some practical image processing techniques. The operations have been implemented using Image Processing tools for MATLAB R2013a software. A comparison table has been presented in order to show a comparative study of various outputs obtained during the implementations of the image processing operations.

Following are the comparative outputs of a particular original image and the corresponding outputs using various image segmentation techniques.

In the above Figure 2, the segmentation results of an MRI brain tumor image of size 132 × 132 is shown. Figure 6(a) is the original image of brain tumor which is of low intensity. As the segmentation method is applied to the original image, the contour starts searching for the specified object which is the brain tumor. Figure 6(b) is the segmentation result after implementation of thresholding technique. Figure 6(c) is the output image obtained after implementation of K-means clustering technique Figure 6(d) is the output image obtained after implementation Fuzzy C-mean clustering technique. The segmented tumor in Figure 6(d) obtained is clear and segmented distinctly.

The tumor segmented out using Fuzzy C-mean algorithm is the most distinct and having a comparatively higher intensity of the segmented part than the original MRI image.

Figure 3 shows the segmentation results of an MRI brain tumor image with blur edges of 132 × 132 size.

Figure 3(a) shows an original image having a tumor which has blur edges along with faint intensity and is difficult to segment out from the rest of the brain. When the segmentation method is applied to the original image, the search for the targeted object begins which is the brain tumor as the targeted part. Figure 3(b) shows the segmentation result after implementing thresholding technique. Figure 3(c) is the output image obtained after implementing K-means technique Figure 3(d) is the output image obtained after implementing Fuzzy C-mean technique.

Figure 2. (a) Original image having low intensity, size 132x132; (b) image obtained after implementation of thresholding; (c) image obtained after implementation of K-means clustering technique; (d) image obtained after fuzzy C means clustering technique

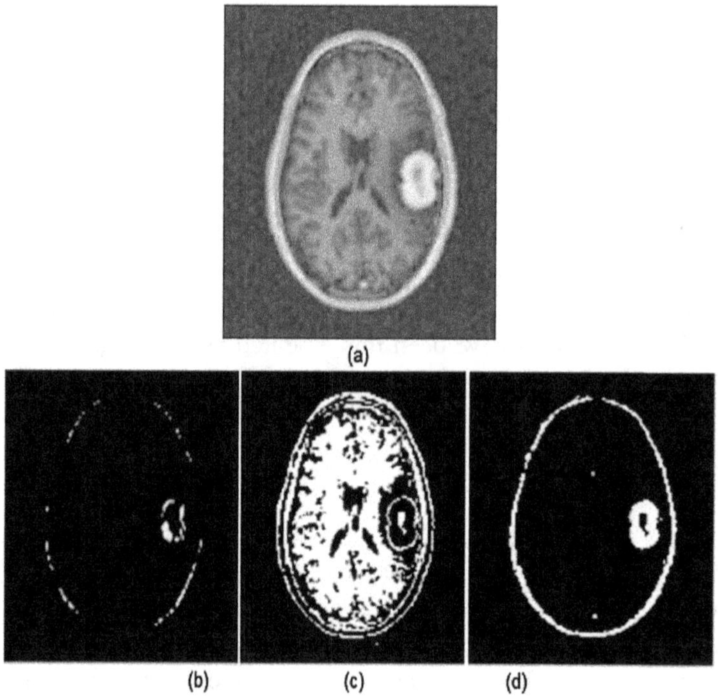

The segmented tumor in Figure 3(d) which is obtained after implementing Fuzzy C means technique again the most distinct, evident and recognizable one.

Figure 4 shows the segmentation results of an MRI brain tumor image sample of size 132 × 132.

Figure 4(a) is the original image having very low intensity. As we apply the segmentation method to the Figureplace.

Figure 4(b) shows the segmentation result after implementing thresholding technique. Figure 4(c) is the output image obtained after implementing K-means technique. Figure 3(d) is the output image obtained after implementing Fuzzy C-mean technique.

The segmented tumor obtained in Figure 4(d) is having comparatively higher intensity than the previous all image sample obtained and it is distinct.

Figure 5 shows the segmentation results of a brain tumor of 132 × 132 size.

Figure 5(a) is the original MRI image sample. Now the segmentation method is applied to the original image. The contour again begins in order to search for the

Figure 3. (a) Original image with blur edges, size 132x132; (b) image obtained after implementation of thresholding; (c) image obtained after implementation of K-means clustering technique; (d) image obtained after fuzzy C means clustering technique

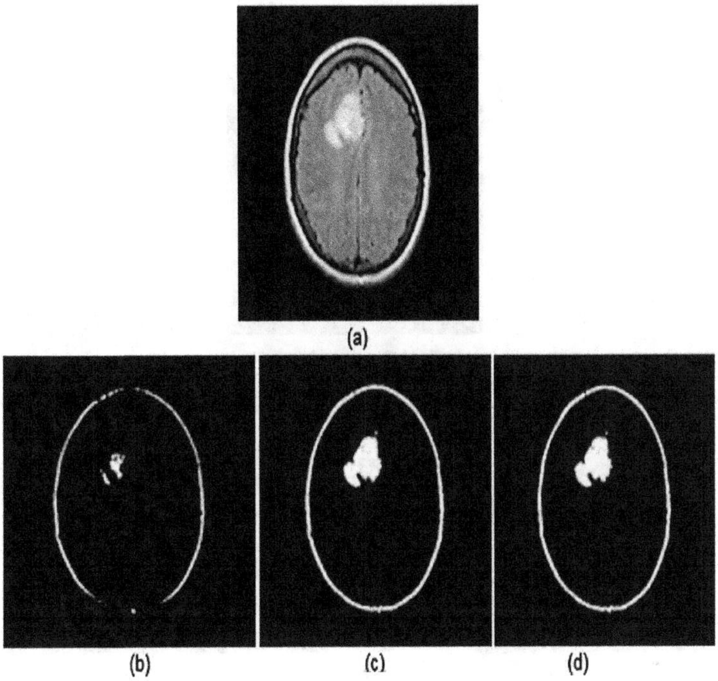

targeted object which is the brain tumor. Figure 5(b) shows the segmentation result after implementing thresholding technique. Figure 5(c) is the output image obtained after implementing K-means technique. Figure 5(d) is the output image obtained after implementing Fuzzy C-mean technique.

The segmented tumor obtained in Figure 5(d) is segmented distinctly and is more easier to analyze compared to other segmentation technique.

Figure 6 shows the segmentation results of an MRI brain tumor image with bur edges as well as image having tumor and rest of the boundary with same intensity. Size of the sample image is 132 × 132. Figure 6(a) is the original sample image taken.When the segmentation method is applied to the original image, the contour again begins and searches for the targeted object which is the brain tumor. Figure 6(b) shows the segmentation result after implementing thresholding technique. Figure 6(c) is the output image obtained after implementing K-means technique. Figure 6(d) is the output image obtained after implementing Fuzzy C-means technique. The segmented tumor in Figure 6(d) is clear and segmented distinctly along with higher intensity of the tumor compared to the original image.

Figure 4. (a) Original image with tumor of highly blur edges and very low intensity, size 132x132; (b) image obtained after implementation of thresholding; (c) image obtained after implementation of K-means clustering technique; (d) image obtained after fuzzy C means clustering technique

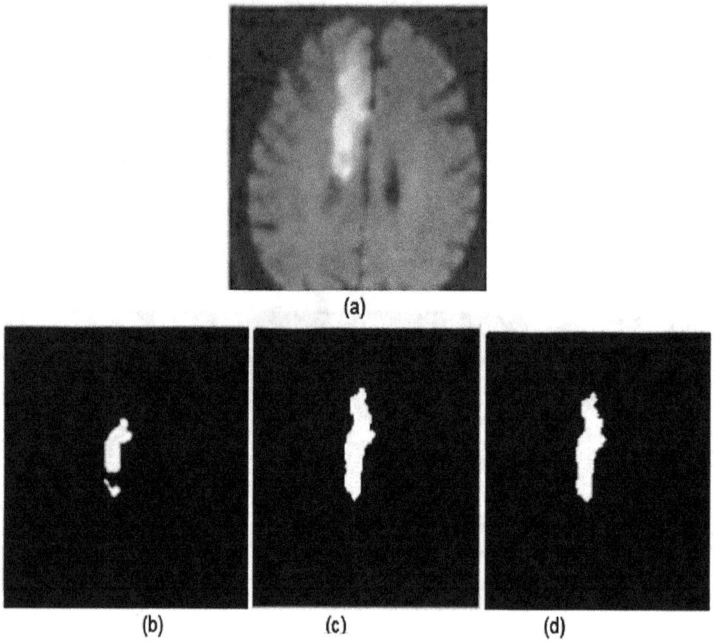

Figure 7 shows the segmentation results of a brain tumor of 132 × 132 size. Figure 7(a) is the original image. When the segmentation technique is applied to the original image, the contour begins and searches for the targeted object which is the brain tumor. Figure 7(b) is the segmentation result obtained after implementing thresholding technique. Figure 7(c) is the output image obtained after implementing K-means technique and Figure 7(d) is the output image obtained after implementation of Fuzzy C-mean technique.

The segmented tumor in obtained after applying Fuzzy C means clustering technique is found out to be more clear and segmented distinctly.

Jaccards' Measure

The special overlap between the pair of segmentations defined by Q_M and Q_A can be measured by using Jaccards coefficient. It can also expressed as a ratio in percentage as specified below where Q_M and Q_A are arrays of tumor which are seg-

Figure 5. (a) Original image with tumor of very less intensity, size 132x132; (b) image obtained after implementation of thresholding; (c) image obtained after implementation of K-means clustering technique; (d) image obtained after fuzzy C means clustering technique

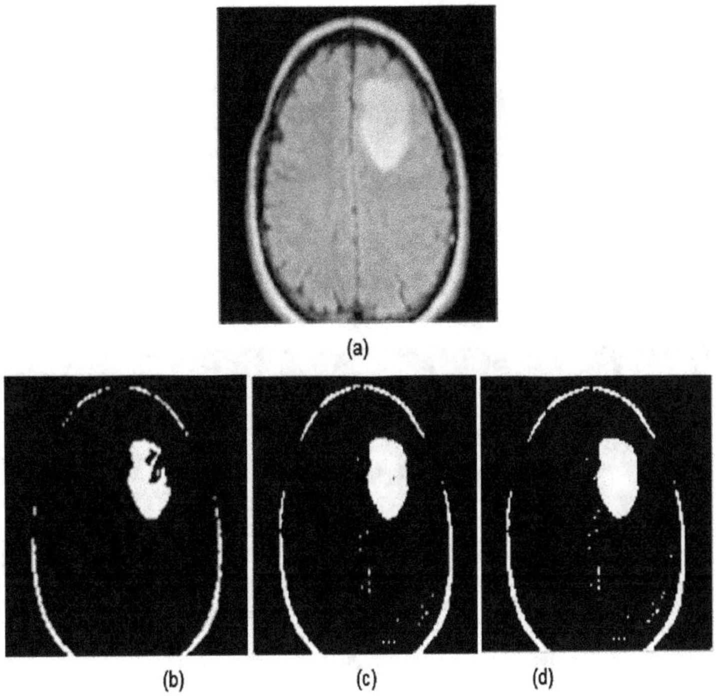

mented by different methods and by human calculations respectively. Jaccards' measure and is calculated by

$$JM = \frac{|Q_M \cap Q_A|}{|Q_M \cup Q_A|} \tag{10}$$

Dice Coefficient

Dice Coefficient measures the correspondence or resemblance between one set and another. This is one of a number of measures of the degrees of structural intersection amongst the two binary images. This coefficient is widely used in approving the performance of image segmentation and its value may range from zero to one, where zero indicates no overlap and one denotes excellent resemblance. Dice coefficient is calculated by

Figure 6. (a) Original image with blur edges and intensity similar with both tumor and rest of the boundary, size 132x132; (b) image obtained after implementation of thresholding; (c) image obtained after implementation of K-means clustering technique; (d) image obtained after fuzzy C means clustering technique

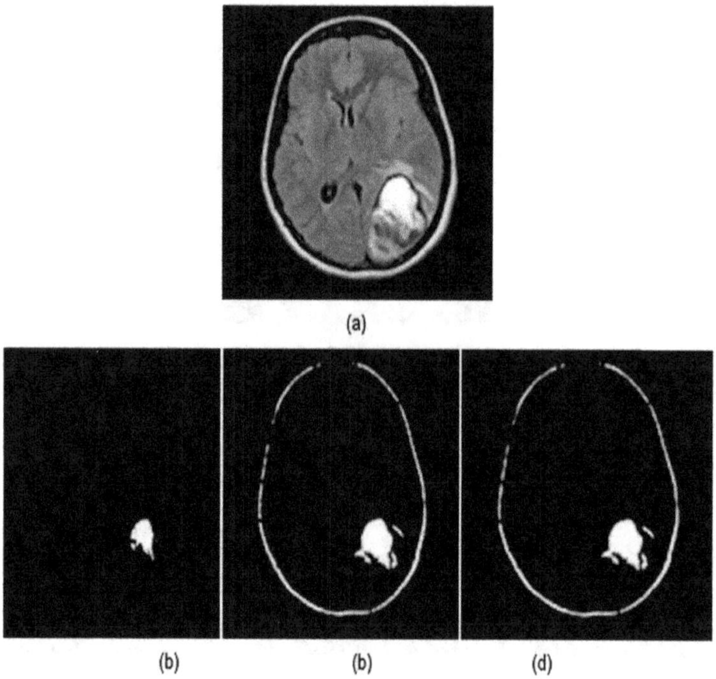

$$DC = \frac{2\left|Q_M \cap Q_A\right|}{\left|Q_M\right| + \left|Q_A\right|} \tag{11}$$

where, Q_M and Q_A are the tumor voxels segmented manually and the different methods respectively. In this section, we carried out several experiments using various kinds of images in order to demonstrate the performance of the proposed method. The proposed method is compared subjectively and objectively, in terms of effectiveness.

COMPARATIVE ANALYSIS

Taking into consideration, both fuzzy c means and k means algorithm, we wind up with the following results and analysis. As the number of records increases, the

Figure 7. (a) Original image, size 132x132; (b) image obtained after implementation of thresholding; (c) image obtained after implementation of K-means clustering technique; (d) image obtained after fuzzy C means clustering technique

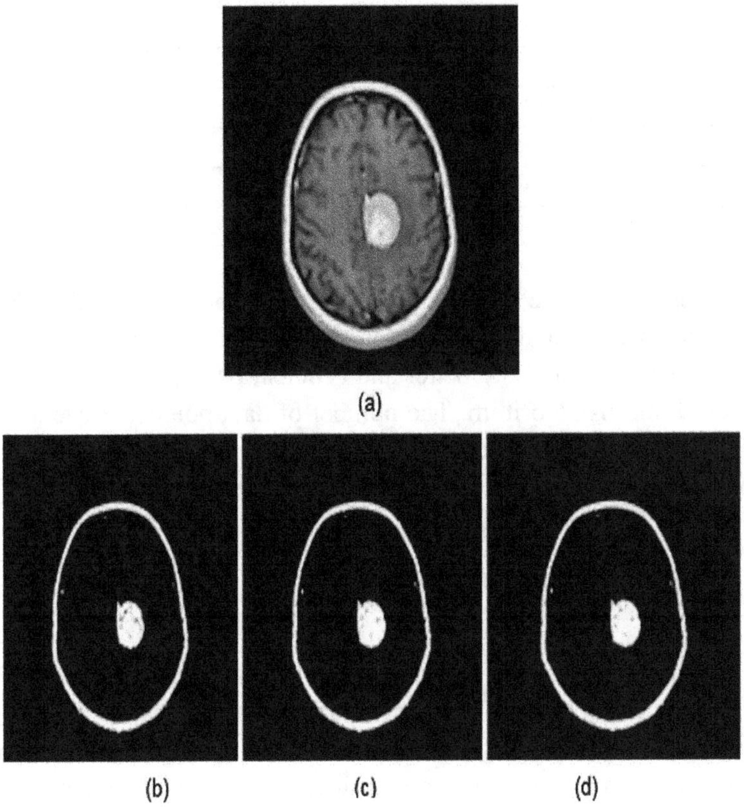

Table 1. Jaccards' measure

Image Index	Thresholding	K Mean	Fuzzy C Mean
Image 2	0.34	0.36	0.51
Image 3	0.28	0.34	0.44
Image 4	0.36	0.41	0.59
Image 5	0.31	0.37	0.42
Image 6	0.42	0.46	0.63
Image 7	0.31	0.35	0.52

Table 2. Dice coefficient

Image Index	Thresholding	K Mean	Fuzzy C Mean
Image 2	0.35	0.47	0.63
Image 3	0.38	0.45	0.58
Image 4	0.42	0.51	0.69
Image 5	0.38	0.42	0.52
Image 6	0.42	0.53	0.71
Image 7	0.41	0.49	0.65

time execution of both the technique also increases but we find that the fuzzy c means performance is found to be better than the k means algorithm. The precision, recall and f measure values are better and efficient on applying fuzzy c means as compared to k means algorithm. The number of data points are evenly delivered in fuzzy c means algorithm. Fuzzy c means algorithm is adept for larger data set than compared to small data set and is highly applicable for requirement clustering.

CONCLUSION AND FUTURE ASPECTS

In this contemporary period, countless category of diseases are discovered by mankind whose diagnosis require a precise technique. And in the medical science, Image segmentation method is one of such method. Image-segmentation method defined region of an image which are partition into different segment. Due to increase in death rate because of mistaken diagnosis done by manual method, image segmentation method is preferred over manual diagnosis. Therefore, doctors favor image segmentation method for efficiently drawing out the information of the infected region of an affected patient so that speedy and successful treatment can be achieved. Because of this features, this proposed method is recommended as one of the best approach for pulling out the statistics of brain tumor from the MR images.

Image segmentation is a very important tool in image analysis. It has an assuring future as the broad segmentation algorithm and has become the center of attraction of present-day research. Fuzzy c-means has been a very vital technique for image segmenting in grouping similar characteristic objects in a particular object of an image. Fuzzy C means algorithm is a unique clustering algorithm and is being used in varied range of engineering as well as scientific arena such as thermal imaging, medicine imaging, pattern detection, data mining and bio-informatics domains. The main aim of image segmentation and analysis is to provide steady, precise and fast algorithms as a way to clarify and upgrade the automatic analysis methods. This will direct us to calculable and productive applications in medical image segmentation arena.

This segmentation technique can prove to be very beneficial and productive in the various fields of the field of imaging technology.

ACKNOWLEDGMENT

We acknowledge NIT Silchar for helping us in providing a good platform to work in this area. We are also thankful to Dr Sweta Rani (Nobel medical college) who has shared her medical knowledge to us and motivated us to work in the field of Biomedical engineering.

REFERENCES

Aslam, Ramashri, & Ahsan. (2013, March). A New Approach to Image Segmentation for Brain Tumor detection using Pillar K-means Algorithm. *International Journal of Advanced Research in Computer and Communication Engineering*.

Bezdek, J. C. (1981). *Pattern Recognition with Fuzzy Objective Function Algorithms*. New York: Plenum Press. doi:10.1007/978-1-4757-0450-1

Bradley, D., & Roth, G. (2007). Adaptive thresholding using the integral image. *Journal of Graphics, GPU, and Game Tools*, *12*(2), 13–21. doi:10.1080/2151237X.2007.10129236

Chandhok, C., Chaturvedi, S., & Khurshid, A. A. (2012). An approach to image segmentation using k-means clustering algorithm. *International Journal of Information Technology*, *1*(1), 11–17.

Dey, N., Ashour, A. S., Beagum, S., Pistola, D. S., Gospodinov, M., Gospodinova, E. P., & Tavares, J. M. R. S. (2015). Parameter Optimization for Local Polynomial Approximation based Intersection Confidence Interval Filter Using Genetic Algorithm: An Application for Brain MRI Image De-Noising. *Journal of Imaging*, *1*(1), 60–84. doi:10.3390/jimaging1010060

Dey, N., Roy, A. B., Pal, M., & Das, A. (2012). *FCM based blood vessel segmentation method for retinal images*. arXiv preprint arXiv: 1209.1181

Dunn, J. C. (1973). A fuzzy relative of the ISODATA process and its use in detecting compact well-separated clusters. *Journal of Cybernetics*, *3*(3), 32–57. doi:10.1080/01969727308546046

Gonzalez, R. C., & Richard, E. (2002). *Woods Digital Image Processing* (2nd ed.). Upper Saddle River, NJ: Prentice hall.

Gordillo, N., Montseny, E., & Sobrevilla, P. (2013, October). State of the art survey on MRI brain tumor segmentation. *Magnetic Resonance Imaging*, *31*(8), 1426–1438. doi:10.1016/j.mri.2013.05.002 PMID:23790354

Jumb, V., Sohani, M., & Shrivas, A. (2014). Color image segmentation using k-means clustering and Otsu's adaptive thresholding. *International Journal of Innovative Technology and Exploring Engineering*, *3*(9), 72–76.

Kwon, S. H. (2004). Threshold selection based on cluster analysis. *Pattern Recognition Letters*, *25*(9), 1045–1050. doi:10.1016/j.patrec.2004.03.001

Tatiraju, S., & Mehta, A. (2011). Image Segmentation using k-means clustering, EM and Normalized Cuts. In *ICIAP'11 Proceedings of the 16th International Conference on Image Analysis and Processing* (vol. 2, pp. 229-240). Genova, Italy: Springer-Verlag.

Xin-bo, Z., & Jiang, L. (2009). An image segmentation algorithm based on fuzzy c-means clustering. *Digital Image Processing*, *1*, 22–26.

Yang, Y. (2007). Image Segmentation by fuzzy c-means clustering algorithm with a novel penalty term. *Computing and Informatics*, *26*, 17–31.

KEY TERMS AND DEFINITIONS

Clustering: The task of categorizing a set of objects in a particular manner such that objects in the same group or cluster have similar characteristic to each other than to those in different groups or clusters.

Fuzzy C Mean Clustering: A soft clustering technique in which data elements can belong to more than one clusters and they are associated with each element according to a set of membership levels. It is a process of assigning these membership levels thereby using them to assign data elements to one or more clusters.

Image Processing: The processing of images using some specific mathematical operations or some form of signal processing for which the input is an image and the obtained output of image processing may be either an image or a set of characteristics or parameters which refers to the image.

K-mean Clustering: A vector quantization method which aims to partition n observations into k clusters in which each observation has a higher degree of belonging to the cluster with the nearest mean.

Segmentation: The decomposition of a particular targeted object into different components in order to facilitate the task at higher levels such as pattern detection and recognition.

Thresholding: A process in which a grayscale input image is converted into a binary output image by using an optimal level of threshold.

Chapter 3
Computational Intelligence Foundations and Principles

Murugan Sethuraman Sethuraman
Wolkite University Ethiopia, India

ABSTRACT

AI has been defined in different ways, including the abilities for abstract thought, understanding, communication, reasoning, learning, retaining, planning, and solving. Intelligence is most widely studied in humans, but has also been observed in animals and plants. AI is the intelligence of machines or the simulation of intelligence in machines. AI is both the intelligence of machines and the branch of Computer Science which aims to create it, through the study and design of intelligent agents or rational agents, where an intelligent agent is a system that perceives its environment and takes actions which maximize its chances of success. Achievements include constrained and well-defined problems such as games, crossword-solving and optical character recognition. Among the traits that researchers hope machines will exhibit are reasoning, knowledge, planning, learning, communication, perception, and the ability to move and manipulate objects. In the field of AI there is no consensus on how closely the brain should be simulated.

DOI: 10.4018/978-1-5225-2423-6.ch003

INTRODUCTION

Computational Intelligence is the study of the design of intelligent agents. An intelligent agent is a system that acts intelligently, it is flexible to changing environments and changing goals, it learns from experience, and it makes appropriate choices given perceptual limitations and finite computation.

Bezdek defined Computational Intelligence (CI) is when it: deals with only numerical data, has pattern recognition components does not use knowledge in the artificial intelligence sense. It shows:

- Computational adaptively.
- Computational fault tolerance.
- Speed approaching human-like turnaround.
- Error rates that approximate human performance.

In the intelligence intrusion detection field, supervised learning usually produces classifiers for misuse detection from class-labeled training datasets. Classifiers are basically viewed as a function mapping data samples to corresponding class labels. Unsupervised learning distinguishes itself from supervised learning by the fact that no class-labeled data is available in the training phase. It groups data points based upon their similarities. Unsupervised learning satisfies the requirement of anomaly detection hence it is usually employed in anomaly detection.

Artificial Intelligence(AI) techniques like as data mining, artificial neural networks, fuzzy logic and expert systems can be integrated with traditional procedural and statistical methods to analyze the collected data by sensors, recognize reconnaissance patterns, filter and correlate events to support security event management and prevention of intrusions. These techniques improve the ability of security management systems to correlate events generated by a diversified suite of modern tools used for network management and security monitoring (Hentea, 2005a). Statistical methods have been used for building intrusion and fault detection models (Manikopoulos & Papavassiliou, 2002), but these models lack the capability to learn and adapt in time.

Expert systems are the most common form of AI applied today in manufacturing, telecommunications, business and other areas. The systems which are based on expert system and inference techniques are not well-organized and accessible because they mainly rely on human expertise, known facts and statistics implemented in rules for a specific host or network and their capability is limited. However the expert systems evolved to a new trend of integration with the traditional information processing such that in the early nineties, the expert systems merged to a new infrastructure based on knowledge based technology.

Knowledge-based systems, artificial neural networks and fuzzy logic are the most promising approaches of AI for applications such as faults and events monitoring, detection, isolation, diagnosis, supervisory and adaptive control, direct control (Rodd, 1992). Adaptive control refers to the capability of the system to adapt itself to meet a desired output despite shifting control objectives and process conditions or unordered uncertainties in process dynamics. The techniques associated with intelligent control include fuzzy, expert, and neural control (Hentea, 1997; Passino & Ozguner, 1996).

AI techniques enhance agent comptabilities. Intelligent agents and multi-agent systems are among the most rapidly growing areas of research and development. Vulnerability assessment and intrusion detection based on agent approach are discussed in (Cardoso & Freire, 2005). Multi-agent framework design and implementation issues for the autonomous database administration system and security are described in (Ramanujan & Capretez, 2005). Strategies for security information analysis and data mining techniques to discover hidden information about possible cyber threats are discussed in (Yao, Wang, Zeng & Wang 2005).

A multi-agent system is designed and implemented as several interacting agents. Multi-agent systems are ideally suited to representing problems that have multiple problem solving methods and multiple perspectives. Intelligent agents take initiative where appropriate and socially interact with other artificial agents and humans in order to complete their own problem solving and to help others with their activities.

Although techniques based on AI are evolving to support information security management, these are still focused on a limited scope. Recently AI methods to create robust intrusion detection and prevention systems have been explored. Several techniques and examples of applications for intrusion detection and prevention systems are discussed in (Hentea 2005b). Intelligent systems for network management support functions such as monitoring, diagnosing or managing specific network resources are discussed in (Berenji 1994; HenteaTurban, Aronson & Liang, 2005). Artificial neural networks techniques are suggested for biometric identification applications (Kung, Mak & Lin 2005). A system based on neural networks agent for mail server management is discussed in Willow (2005).

Computational Intelligence is different from the well-known field of AI. AI handles symbolic knowledge representation, while CI handles numeric representation of information; AI concerns itself with high-level cognitive functions, while CI is concerned with low-level cognitive functions. Furthermore AI analyzes the structure of a given problem and attempts to construct an intelligent system based upon this structure, thus operating in a top-down manner, while the structure is expected to emerge from an unordered beginning in CI, thus operating in a bottom-up manner.

Although there is not yet full agreement on what computational intelligence exactly, there is a widely accepted view on which areas belong to CI: artificial neural networks, fuzzy sets, evolutionary computation, artificial immune systems, swarm intelligence, and soft computing. These approaches, except for fuzzy sets are capable of autonomously acquiring and integrating knowledge, and can be used in either supervised or unsupervised learning mode.

DIFFERENT AI TYPES

There are several different IIDPS techniques and algorithms that can be successfully used to detect Unknown malware.

These techniques include:

- Fuzzy logic,
- Probabilistic reasoning,
- Neural networks,
- Genetic algorithms,
- Data mining,
- Association rules,
- Fuzzy association rules.

Combinations of these can also be used. Neural networks are basically sets of individual cells that have weighted connections to other connected cells.

FUZZY LOGIC

Fuzzy concepts derive from fuzzy phenomena that commonly occur in the natural world. For instance “walking” is a fuzzy statement of”Today walking heavily” since there is no clear boundary between “walking” and “heavy walking”. In intrusion detection suppose want to write a rule as given below need a reason about a quantity such as the number of different destination IP addresses in the last 2 seconds. IF the number of different destination addresses during the last n seconds was high THEN an unusual situation exists.

Fuzzy logic, which is utilized, is a superset of conventional logic that has been extended to handle the concept of partial truth, which lies between completely true and completely false.

DATA MINING METHODS

Data mining is one of the hot topics in the field of knowledge extraction from database. Data mining is used to automatically learn patterns from large quantities of data.

Mining can efficiently discover useful and interesting rules from large collection of data. Fuzzy logic provides a powerful way to categorize a concept in an abstract way. The advantage of fuzzy logic is that it allows representation of overlapping categories. Combining techniques from fuzzy logic and data mining for anomaly detection and it helps us to create more abstract patterns.

Data mining methods are used to automatically discover new patterns from a large amount of data. Association rules have been successfully used to mine audit data to find normal patterns for anomaly intrusion detection.

ASSOCIATION RULES

The association rule mechanism proposed by Agrawal is a most popular tool. Agrawal and Srikant established the Apriori Algorithm for mining association rules. The Apriori Algorithm needs minimum confidence and represent minimum support. These two values conclude the degree of association that must hold.

Fuzzy Association Rules

To professionally use the Apriori algorithm of Agrawal and Srikant for mining association rules, quantitative variables should be partitioned into discrete categories. This leads to "sharp boundary problem" in which a very small change in value causes an abrupt change in category. To address this problem fuzzy association rules was developed by Kuok, Fu and Wong. Modify the algorithm by introducing a normalization factor.

GENETIC ALGORITHMS

The training process of a neural network consists of setting weights for each connection and comparing the output with the desired output. This is iterated until the desired accuracy with a test set of data has been achieved. Genetic algorithms can be used to keep the number of iterations as small as possible. Genetic algorithms first randomize values for each set. Then they select a few of the best sets and mix and differentiate values from each of these. This is continued until the desired result is reached or the maximum number of generations has passed.

PROBABILISTIC REASONING

Gowadia, Farkas and Valtorta have studied a probabilistic reasoning based AI in IDS use in their article. They used Bayesian networks to evaluate the probability of an intrusion. They had multiagent system in which agents in separate computers could communicate and tell each other threat estimates from each one's point of view. The agents concentrated on detecting different intrusions on separate domains. The authors concluded that this kind of approach is feasible. The system offers a possibility to select the percentage that an event must reach in order to be shown.

Knowledge Discovery in Databases (KDD) is an automatic, exploratory analysis and modeling of large data repositories. KDD is the organized process of identifying valid, novel, useful, and understandable patterns from large and complex data sets.

Data Mining (DM) is the core of the KDD process, involving the inferring of algorithms that explore the data, develop the model and discover previously unknown patterns. The model is used for understanding phenomena from the data, analysis and prediction.

The accessibility and abundance of data today makes KDD a matter of considerable importance and necessity. Given the recent growth of the field, it is not surprising that a wide variety of methods is now available to the researchers and practitioners. No one method is superior to others for all cases.

KDD PROCESS

The knowledge discovery process is iterative and interactive, consisting of nine steps. Note that the process is iterative at each step, meaning that moving back to adjust previous steps may be required. The process has many "artistic" aspects in the sense that one cannot present one formula or make a complete taxonomy for the right choices for each step and application type. Thus it is required to deeply understand the process and the different needs and possibilities in each step. Taxonomy for the DM methods is helping in this process.

The process starts with determining the KDD goals and "ends" with the implementation of the discovered knowledge. As a result, changes would have to be made in the application domain. This closes the loop, and the effects are then measured on the new data repositories, and the KDD process is launched again. Following is a brief description of the nine-step KDD process starting with a managerial step:

1. Understanding of the application domain - understands and defines the goals of the end-user and the environment in which the knowledge discovery process will take place.

2. Selecting and creating a data set on which discovery will be performed. It starts with the best available data set and later expands and observes the effect in terms of knowledge discovery and modeling.
3. Preprocessing and cleansing. In this stage, data reliability is enhanced. It includes data clearing, such as handling missing values and removal of noise or outliers.
4. Developing prediction model through which missing data can be predicted.
5. Choosing the appropriate DM task for classification, prediction or clustering.
6. Choosing the DM algorithm- selecting the specific method to be used for searching patterns.
7. Each algorithm has parameters and tactics of learning.
8. Employing the DM algorithm - employ the algorithm several times until a satisfied result is obtained.
9. Evaluation and interpretation of mined patterns with respect to the goals defined in the first step. In this step the discovered knowledge is also documented for further usage.
10. Using the discovered knowledge. The knowledge becomes active in the sense that changes to the system and measure the effects.

DATA MINING (DM) METHODS

There are many methods of DM used for different purposes and goals. Taxonomy is help in understanding the variety of methods, their interrelation and grouping. It is useful to distinguish between two main types of DM: verification-oriented and discovery oriented. Discovery methods are those that automatically identify patterns in the data. The discovery method branch consists of prediction methods versus description methods. Descriptive methods are oriented to data interpretation, which focuses on understanding the way the underlying data relates to its parts. Prediction-oriented methods aim to automatically build a behavioral model, which obtains new and unseen samples and is able to predict values of one or more variables related to the sample. It also develops patterns, which form the discovered knowledge in a way which is understandable and easy to operate upon. Some prediction-oriented methods can also help provide understanding of the data.

Most of the discovery-oriented DM techniques are based on inductive learning, where a model is constructed, explicitly or implicitly, by generalizing from a sufficient number of training examples. Verification methods, on the other hand, deal with the evaluation of a hypothesis proposed by an external source. These methods include the most common methods of traditional statistics, like goodness of fit test, tests of hypotheses and analysis of variance (ANOVA). These methods are less associated with Data Mining than their discovery-oriented counterparts, because most Data Mining problems are concerned with discovering a hypothesis rather than testing a known one. Much of the focus of traditional statistical methods is on model estimation as opposed to one of the main objectives of Data Mining: model identification and construction which is evidence based.

Another common terminology used by the machine-learning community refers to the prediction methods as supervised learning as opposed to unsupervised learning. Unsupervised learning refers to demonstrating the distribution of instances in a typical high-dimensional input space. Unsupervised learning refers mostly to techniques that group instances without a pre specified dependent attribute. Thus the term "unsupervised learning" covers only a portion of the description methods presented it covers Clustering methods but not visualization methods. Supervised methods are methods that attempt to discover the relationship between input attributes and a target attribute sometimes referred to as a dependent variable. The relationship discovered is represented in a structure referred to as a model. Usually models describe and explain phenomena, which are hidden in the data set and can be used for predicting the value of the target attribute knowing the values of the input attributes. The supervised methods can be implemented on a variety of domains, such as marketing, finance and manufacturing. It is useful to distinguish between two main supervised models: classification models and regression models. The latter map the input space into a real-valued domain. For instance, a regressor can predict the demand for a certain product given its characteristics. On the other hand, classifiers map the input space into predefined classes.

The art expands but so does the understanding and the automation of the nine steps and their interrelation. For this to happen researcher need better characterization of the KDD problem spectrum and definition. The terms KDD and DM are not well-defined in terms of what methods they contain, what types of problem are best solved by these methods, and what results to expect. KDD / DM compared to statistics, machine learning, operations research, etc. If subset or superset of the above fields Or an extension of them Or a separate field by itself, In addition to the methods – which are the most promising fields of application and what is the vision KDD / DM brings to these fields, Certainly already see the great results and achievements of KDD / DM, but cannot estimate their results with respect to the potential of this field. All these basic analyses have to be studied and see several trends for future research and implementation, including:

Active DM – closing the loop, as in control theory, where changes to the system are made according to the KDD results and the full cycle starts again. Stability and controllability which will be significantly different in these types of systems need to be well-defined.

- **Full Taxonomy:** For all the nine steps of the KDD process. DM taxonomy will contain methods appropriate for each step (even the first one) and for the whole process as well.
- **Meta-Algorithms:** Algorithms that examine the characteristics of the data in order to determine the best methods, and parameters (including decompositions).
- **Benefit Analysis:** To understand the effect of the potential KDD / DM results on the enterprise.
- **Problem Characteristics:** Analysis of the problem itself for its suitability to the KDD process.
- **Mining Complex Objects of Arbitrary Type:** Expanding Data Mining inference to include also data from pictures, voice, video, audio, etc. This will require adapting and developing new methods.
- **Temporal Aspects:** Many data mining methods assume that discovered patterns are static. However, in practice patterns in the database evolve over time. This poses two important challenges. The first challenge is to detect when concept drift occurs. The second challenge is to keep the patterns up-to-date without inducing the patterns from scratch.
- **Distributed Data Mining:** The ability to seamlessly and effectively employ Data Mining methods on databases that are located in various sites. This problem is especially challenging when the data structures are heterogeneous rather than homogeneous.

Expanding the knowledge base for the KDD process including not only data but also extraction from known facts to principles.

Expanding Data Mining reasoning to include creative solutions not just the ones that appear in the data but being able to combine solutions and generate another approach.

EXPERT SYSTEMS

Expert Systems are used primarily by knowledge-based intrusion-detection techniques. The expert system contains a set of rules that describe attacks. Audit events are then translated into facts carrying their semantic signification in the expert system, and the inference engine draws conclusions using these rules and facts. This method increases the abstraction level of the audit data by attaching semantic to it. It also encodes knowledge about past intrusions, known system vulnerabilities and the security policy. As information is gathered, the expert system determines whether any rules have been satisfied.

Rule-based languages are a natural tool for modeling the knowledge that experts have collected about attacks. This approach allows a systematic browsing of the audit trail in search of evidence of attempts to exploit known vulnerabilities. They are also used to verify the proper application of the security policy of an organization.

Model-based reasoning, which also uses expert systems but has additional properties, has been introduced by Garvey and Lunt. The knowledge about the behavior of an attacker is described by the attacker's goals, the actions he takes to reach these goals, and his usage of the system which sometimes reveals a certain level of paranoia. The tool then scans the audits for evidence of these actions and transitions.

KNOWLEDGE ENGINEERING

It is difficult to extract knowledge about attacks. It is even more difficult to translate this knowledge into production rules using audits as inputs. Sometimes the information required is not available in the audits. Also there may be many ways to exploit a given vulnerability, thus leading to as many rules.

BACKGROUND FOR INTELLIGENCE

Neural network (NN) is a system loosely modeled on the human brain. This area of study has several terminologies such as connectionism, parallel distributed processing, neuro computing, machine learning algorithms and Artificial Neural Networks (ANN).

The power and speed of modern digital computers are truly astounding. No human can ever hope to compute a million operations at second. However, there are some tasks for which even the most powerful computers cannot compete with the human brain, perhaps not even with the intelligence of an earthworm. NN approaches such problem by trying to mimic the structure and function of our nervous system. Many researchers believe that AI and NN are completely opposite in their approach. Conventional AI is based on the symbol system hypothesis. A symbol system consists of indivisible entities called symbols, which can form more complex entities, by simple rules. The hypothesis then states that such a system is capable of and is necessary for intelligence.

Most of current approaches are process of detecting intrusions utilize some form of rule-based analysis. Rule-Based analysis relies on sets of predefined rules that are provided by an administrator, automatically created by the system, or both. Expert systems are the most common form of rule-based intrusion detection approaches. The early intrusion detection research efforts realized the inefficiency of any approach that required a manual review of a system audit trail. While the information necessary to identify attacks was believed to be present within the voluminous audit data, an effective review of the material required the use of an automated system. The use of expert system techniques in intrusion detection mechanisms was a significant milestone in the development of effective and practical detection-based information security systems.

An expert system consists of a set of rules that encode the knowledge of a human expert. These rules are used by the system to make conclusions about the security related data from the intrusion detection system. Expert systems permit the incorporation of an extensive amount of human experience into a computer application that then utilizes that knowledge to identify activities that match the defined characteristics of misuse and attack. Unfortunately, expert systems require frequent updates to remain current. While expert systems offer an enhanced ability to review audit data, the required updates may be ignored or performed infrequently by the administrator. At a minimum, this leads to an expert system with reduced capabilities. At worst, this lack of maintenance will degrade the security of the entire system by causing the system's users to be misled into believing that the system is secure, even as one of the key components becomes increasingly ineffective over time.

Rule-based systems suffer from an inability to detect attacks scenarios that may occur over an extended period of time. While the individual instances of suspicious activity may be detected by the system, they may not be reported if they appear to occur in isolation. Intrusion scenarios in which multiple attackers operate in concert are also difficult for these methods to detect because they do not focus on the state transitions in an attack, but instead concentrate on the occurrence of individual elements. Any division of an attack either over time or among several seemingly unrelated attackers is difficult for these methods to detect.

Rule-based systems also lack flexibility in the rule-to-audit record representation. Slight variations in an attack sequence can affect the activity-rule comparison to a degree that the intrusion is not detected by the intrusion detection mechanism. While increasing the level of abstraction of the rule-base does provide a partial solution to this weakness, it also reduces the granularity of the intrusion detection device. A number of non expert system based approaches to intrusion detection have been developed in the past several years. While many of these have shown substantial promise, expert systems remain the most commonly accepted approach to the detection of attacks.

NN simulations appear to be a recent development. However, this field was established before the advent of computers, and has survived at least one major setback and several eras.

Many important advances have been boosted by the use of inexpensive computer emulations. Following an initial period of enthusiasm, the field survived a period of frustration and disrepute. During this period when funding and professional support was minimal, important advances were made by relatively few researchers. Currently, the NN field enjoys a resurgence of interest and a corresponding increase in funding.

NN Model

NN can learn various tasks from training examples; classify phenomena, and model nonlinear relationships. However, the primary features that are of concern in the design of the network are problem specific. Despite the availability of some guidelines, it would be helpful to have a computational procedure in this aspect, especially for the optimum design of an NN. The gradient descent algorithms have reported difficulties in learning the topology of the networks whose weights they optimize.

ANN are networks of interconnected simple units that are based on a greatly simplified model of the brain. ANN are useful learning tools by being able to compute results quickly interpolating data well. There are two main types of ANN, feed forward networks and recurrent networks. Perceptron is a special case of feed forward networks with only input and output nodes. Three main Perceptron learning algorithms are covered: mistake bound Perceptron algorithm, Perceptron training rule and the Delta rule. The delta rule uses gradient descent, which makes it easy to compute what changes are needed to optimize the network. The back propagation-learning algorithm is widely used for multi-layer feed forward network, uses gradient descent as well.

Architecture of NN

Feed-forward ANN allow signals to travel one way only; from input to output. There is no feedback (loops) i.e. the output of any layer does not affect that same layer. Feed-forward ANN tend to be straight forward networks that associate inputs with outputs. They are extensively used in pattern recognition. This type of organization is also referred to as bottom-up or top-down.

Feedback networks can have signals traveling in both directions by introducing loops in the network. Feedback networks are very powerful and can get extremely complicated. Feedback networks are dynamic; their ‘state’ is changing continuously until they reach an equilibrium point. They remain at the equilibrium point until the input changes and a new equilibrium needs to be found. Feedback architectures are also referred to as interactive or recurrent, although the latter term is often used to denote feedback connections in single-layer organizations.

The commonest type of artificial neural network consists of three groups, or layers, of units a layer of input units is connected to a layer of hidden units, which is connected to a layer of output units. The activity of the input units represents the raw information that is fed into the network.

The activity of each hidden unit is determined by the activities of the input units and the weights on the connections between the input and the hidden units. The behavior of the output units depends on the activity of the hidden units and the weights between the hidden and output units.

This simple type of network is interesting because the hidden units are free to construct their own representations of the input. The weights between the input and hidden units determine when each hidden unit is active, and so by modifying these weights, a hidden unit can choose what it represents.

Distinguish single-layer and multi-layer architectures. The single-layer organization, in which all units are connected to one another, constitutes the most general case and is of more potential computational power than hierarchically structured multi-layer organizations. In multi-layer networks, units are often numbered by layer, instead of following a global numbering.

PERCEPTRON

The most influential work on neural nets in the 60's went under the heading of perceptrons a term coined by Frank Rosenblatt. The perceptron turns out to be an MCP model neuron with weighted inputs and some additional, fixed, pre-processing. Units labeled A1, A2, Aj, Ap are called association units and their task is to extract specific, localized featured from the input images. Perceptrons mimic the basic idea behind the mammalian visual system. They were mainly used in pattern recognition even though their capabilities extended a lot more.

Single layer perceptron could not do some basic pattern recognition operations like determining the parity of a shape or determining whether a shape is connected or not. Researcher did not realize, until the 80's, is that given the appropriate training, multilevel perceptrons can do these operations.

Structure of the Nervous System

For our purpose, it will be sufficient to know that the nervous system consists of neurons which are connected to each other in a rather complex way. Each neuron can be thought of as a node and the interconnections between them are edges.

Such a structure is called as a directed graph. Further, each edge has a weight associated with it which represents how much the two neurons which are connected by it can interact. If the weight is more, then the two neurons can interact much more - a stronger signal can pass through the edge.

Functioning of the Nervous System

The nature of interconnections between two neurons can be such that - one neuron can either stimulate or inhibit the other. An interaction can take place only if there is an edge between two neurons. If neuron A is connected to neuron B as below with a weight w,

then if A is stimulated sufficiently, it sends a signal to B. The signal depends on the weight w, and the nature of the signal, whether it is stimulating or inhibiting. This depends on whether w is positive or negative. If sufficiently strong signals are sent, B may become stimulated.

Note that A will send a signal only if it is stimulated sufficiently, that is, if its stimulation is more than its threshold. Also if it sends a signal, it will send it to all nodes to which it is connected. The threshold for different neurons may be different. If many neurons send signals to A, the combined stimulus may be more than the threshold.

Next if B is stimulated sufficiently, it may trigger a signal to all neurons to which it is connected. Depending on the complexity of the structure, the overall functioning may be very complex but the functioning of individual neurons is as simple as this. Because of this try to simulate using software or even special purpose hardware knowledge.

ANN have been applied to an increasing number of real-world problems of considerable complexity. Their most important advantage is in solving problems that are too complex for conventional technologies.Problems that do not have an algorithmic solution or for which an algorithmic solution is too complex to be found. In general, because of their abstraction from the biological brain, ANNs are well suited to problems that people are good at solving, but for which computers are not. These problems include pattern recognition and forecasting.

STRUCTURE OF A NEURON

Our "artificial" neuron will have inputs (all N of them) and one output: The neuron has set of nodes that connect it to input, output or other neurons also called synapses. A Linear Combiner, which is a function that takes all inputs and produces a single value. A simple way of doing it is by adding together the input (double Input (dInput) represents the floating point number) multiplied by the Synaptic Weight dWeight:

```
for(int i = 0; i < nNumOfInputs; i++)
 dSum += dInput[i] * dWeight[i];
```

For simplicity, (as well as the rest of the world) will replace the threshold with an extra input, with weight that can change during the learning process and the input fixed and always equal (-1). The effect, in terms of mathematical equations, is exactly the same, but the programmer has a little more breathing room.

A NEURAL NET

A single neuron by itself is not a very useful pattern recognition tool. The real power of NN comes when combine neurons into the multilayer structures called neural networks.

There are three layers in our network can make it more but if make it less will have a few capable net. Making four layers is sometimes useful when looking for a non-evident things. There are N neurons in the first layer, where N equals number of inputs. There are M neurons in the output layer, where M equals number of outputs.

Any number of hidden layers, the quality of a prediction will drop and the net doesn't have enough "brains". And if make it too many it will have a tendency to "remember" the right answers, rather than predicting them. Then researcher neural net will work very well on the familiar data but will fail on the data that was never presented before. Finding the compromise is more of an art than science.

One of the most popular Artificial Neural Network (ANN) architecture is the Multilayer Feed forward (MLFF) network with back propagation (BP) learning. This type of network is sometimes called multilayer perceptron because of its similarity to perceptron networks with more than one layer. Back propagation is a systematic method of training multilayer artificial neural networks.

PERCEPTRON MODEL

For the perceptron in the first layer, the input comes from the actual inputs of the problem, while for the perceptron in the second layer the inputs are outputs of the first layer. The perceptron of the second layer do not know which of the real inputs from the first layer were on or off.

It is impossible to strengthen the connections between active inputs and strengthen the correct parts of the network. The actual inputs are effectively masked off from the output units by the intermediate layer. Hence, the network is unable to determine which of the input weights should be increased and which one should not and so, it is unable to work to produce a better solution next time. The way to go around the difficulty using the step function as the thresholding process is to adjust it slightly and to use a slightly different nonlinearity.

The Solution: If the threshold function is smoothen so that it more or less turns on or off as before but has a sloping region in the middle that will give us some information on the inputs, will be able to determine when need to strengthen or weaken the relevant weights. Now, the network will be able to learn as required. An artificial neuron is developed to mimic the characteristics and functions of a biological neuron. Biological neuron, an artificial neuron receives much input representing the output of the other neurons. Each input is multiplied by the corresponding weights analogous to synaptic strengths. All of these weighed inputs are then summed up and passed though an activation function lo determine the neuron output

$$U(t) = W_1I_1 + W_2I_2 + \ldots + W_nI_n = \langle W \rangle \{ I \}$$

Considering threshold θ the relative input to the neuron is given by

$$U(t) = W_1I_1 + W_2I_2 + \ldots + W_nI_n - \theta = \sum_{i=1}^{n} W_i I_i ;$$

where $W_0 = -\theta$ and $I_0 = 1$.

Model for Multilayer Perceptron

The adapted perceptions are arranged in layers and so the model is termed as multi-layer perception. This model has three layers; an input layer, and output layer, and a layer in between not connected directly to the input or the output and hence, called the hidden layer- For the perceptron in the input layer, use linear transfer function, and for the perceptron in the hidden layer and the output layer, use sigmoidal or squashed-S functions. The input layer serves to distribute the values they receive to the next layer and so, does not perform a weighted sum or threshold. Because modified the single layer perceptron by changing the nonlinearity from a step function to a sigmoidal function and added a hidden layer, forced to alter the learning rules as well. So now, a network that should be able to learn to recognize more complex things.

The activity of neurons in the input layers represents the raw information that is fed into the network. The activity of neurons in the hidden layer is determined by the activities of the neurons in the input layer and the connecting weights between input and hidden units. Similarly, the activity of the output units depends on the activity of neurons in the hidden layer and the weight between the hidden and output layers. This structure is interesting because neurons in the hidden layers arc free to construct their own representation of their input.

BACK PROPAGATION LEARNING

Input Layer Computation

Consider linear activation function the output of the input layer is the input of input layer (considering g = tan ϕ = 1). Taking one set of data. $O_I = H_{I.}$ The hidden neurons are connected by synapses to the input neurons and V_{ij} is the weight of the arc between i[th] input neuron to j[th] hidden neuron. The input to the hidden neuron is the weighted sum of the outputs of the input neurons to get I_{Hp} (input to the pth hidden neuron) as

$$I_{Hp} = V_{1p} O_{I1} + V_{2p} O_{I2} + \ldots + V_{1p} O_{Im}$$

Denoting weight matrix or connectivity matrix between input and hidden neurons as [V], get an input to hidden neuron as: $I_H = V^T O_I$

Hidden Layer Computation

Considering sigmoidal function or squashed-S function, the output of the pth hidden neuron is given by:

$$O_{Hp} = \frac{1}{(1 + e^{-\lambda(I_{Hp} - \theta_{Hp})})}$$

where O_{Hp} is the output of the pth hidden neuron, I_{Hp} is the input of the pth hidden neuron, and θ_{Hp} is the threshold of the pth neuron. A non-zero threshold neuron is computationally equivalent to an input that is always held at -1 and the non-zero threshold becomes the connecting weight values. The input to the output neurons is the weighted sum of the output of the hidden neurons. To get I_{Oq} (the input to the qth output neuron).

$$I_{Oq} = W_{1q} O_{H1} + W_{2q} O_{H2} + \ldots + W_{mq} O_{Hm}$$

Denoting weight matrix or connectivity matrix between hidden and output neurons as [W], get an input to output neuron as: $I_O = W^T O_H$

Output Layer Computation

Considering sigmoidal function, the output of the qth output neuron is given by

$$O_{Oa} = \frac{1}{(1 + e^{-\lambda(I_{Oq} - \theta_{Oq})})}$$

where, O_{Oq} is the output of the qth output neuron, I_{Oq} is the input to the qth output neuron, and θ_{Oq} is the threshold of the qth neuron.

CALCULATION OF ERROR

Considering any r^th^ output neuron

$E_r = ½ (T-O)^2$

where E_r = 1/2 second norm of the error in the rth neuron for the given training pattern. The square of the error is considered since irrespective of whether error is positive or negative, consider only absolute values. The Euclidean and for the training example calculated the output 'O' for which the target output 'T' is calculated. Hence, the error norm in output for the rth output neuron is given by: norm of error E for the first training pattern is given by

$$E = \frac{1}{2}\sum_{r=1}^{n}(T_{Or} - O_{Or})^2$$

TRAINING OF NN

The synaptic weighting and aggregation operations performed by the synapses and soma respectively provide a 'similarity measure' between the input vector I and the synaptic weights [V] and [W] (accumulation knowledge base). When a new input pattern that is significantly different from the previously learned pattern is presented to the neural network, the similarity between this input and the existing knowledge base is small. As the neural network learns this new pattern, by changing the strengths of synaptic weights, the distance between the new information and accumulated knowledge decreases. In other words, the purpose of learning is to make W and V very similar to given pattern I. Most of the neural network structures undergo 'learning procedures' during which synaptic weights W and V are adjusted. Algorithms for determining the connection strengths lo ensure learning are called 'learning rules'. The objective of learning rules depends upon applications. In classification and functional approximation problems, each cycle of presentation of all cases is usually referred as 'learning epoch' However, there has been no generalization as to how a neural network can be trained. Based on error signal, neural network modifies its synaptic connections to improve the system performance. In this scheme, it is always assumed that the desired answer is known "a priori". In this back propagation neural network, researcher use the procedure of supervised learning.

NN Description

The first prototype neural network was designed to determine if a neural network capable of Identifying specific events that are indications of misuse. Neural networks had been shown to be capable of identifying TCP/IP network events in, but our prototype was designed to test the ability of a neural network to identify indications of misuse. The prototype utilized a Multi Layer Perceptron(MLP) architecture that consisted of four fully connected layers with nine input nodes and two output nodes. While there are a number of architectures that could be used to address this problem a feed-forward neural network architecture was selected based on the flexibility and applicability of the approach in a variety of problems.

ARTIFICIAL IMMUNE SYSTEMS

The AIS is biological-inspired method which appears owing to the biological immune system. Before the start consideration of AIS's mechanisms let's describe shortly basic principles of the BIS.

The biological immune systems revive The biological immune system is based on capability of antibodies to distinguishes between self (cells of own body) and nonself (antigens, foreign substance). For complete and successful detection of wide variety of antigens the BIS must generate a large variety of detectors (B-lymphocytes and T-lymphocytes). Lymphocytes are formed from the bone marrow stem cells and initially incapable of antigens detect. In order to acquire immunological ability they have to go through maturation process. T-lymphocytes are mature in thymus, a glandular organ of vertebrates, consisting in man of two lobes situated below the thyroid, and B- lymphocytes are mature in lymph nodes, bean-shaped masses of tissue, situated along the course of lymphatic vessels.. Mature lymphocytes have on the own surface detectors which able to react on specific antigens. They circulate in the body and perform function on antigens detection When some lymphocyte detects an antigen the process called clonal selection is occurred. The clonal selection process consists in proliferation those lymphocytes who detected a virus; thereby a large population of identical detectors for quick virus eliminating is formed.

Another important process in the BIS is immune memory. After elimination of all antigens owing to clonal selection most of cloned lymphocytes died, however some of them move to so-called memory cells. Population of such cells forms the immune memory. By repeated infection antigens can be detected quickly sine the BIS already has lymphocytes which react on this infection. Described basic principles of BIS are underlined to AIS activity.

The AIS is founded on the same processes as BIS: detectors generation, detectors maturation, detection process, detectors cloning and mutation, immune memory creation. Process of detectors creation in computer system represents a random generation of detectors population. Each of them can be binary string of fixed size, or neural network.After generation detectors undergo a selection process. Since detector generations is randomly process, defend the computer system from undesirable detectors. During the selection process unsuitable detectors are eliminated and survive only those which able to distinguish between self and nonself.

According to negative selection algorithm detectors are compared with set of self files. If detector is similar to self files, it is reputed as negative and destroyed. Only those detectors survive which are structurally different from self files. For matching between detectors and files can be applied different rules: bit-by-bit comparison, r-contiguous matching and r-chunk matching. Mature detectors structurally different from self files therefore react only against malicious code.

Mature detectors circulate in computer systems. For maintenance of wide variety of structurally different detectors, each detector has a lifecycle. Lifecycle is a time during detectors can be found in the computer system. When the life time ends the detector is destroyed but if the detector detected malicious code then lifecycle is prolong. Lifecycle mechanism allows the AIS to unload from weak detectors and permanently provide a space for new various detectors.

When malicious code enters the computer system it often infects a large quantity of files. For quick reacting and eliminating virus manifestation need a great number of similar detectors. A detector which found malicious code undergoes a cloning mechanism. Cloning means a large quantity of similar detectors creation. This mechanism allows the AIS infection elimination in a short space of time. Along with cloning a mutation mechanism is used. Mutation process means small random changes in detectors structure thereby as much as possible similar structure to finding virus acquires. When the malicious virus is eliminated then most of cloning detectors die. However the fittest of them are kept as memory detectors. A set of such detectors are formed an immune memory. The immune memory keeps information about all malicious code which a computer system infects. The same as BIS the immune memory allows the AIS to quickly react on repeated infection and to fight against it.

Neuronet Immune Detectors

In the artificial immune system the detectors act as main element for malware detection. The immune detectors circulate through file system and RAM of computer and detect malware. Successful malware detection depends to a large extent on choice of detectors structure.

In this case each immune detector represents one neural network. A set of such neuronet immune detectors forms a population of detectors in the artificial immune system. Malicious software performs a function as full or partial data destruction, user information steal, computer system infection for it further appliance in net attacks. Since they contain destructive or spy function therefore malicious viruses structurally (by instruction set) differ from non infected files (operating systems files, different software, documents) which haven't destructive function. Using difference between malware and software can detect computer viruses. Owing to generalizing ability of neural network the immune detectors can find differences between self files (different software) and malware and detect it. The results described in the next section show the unique ability of one neuronet immune detector detect several malware. Trained by ad hoc algorithm detectors of the AIS allow to detect unknown, previously unseen malware.

AIS approach for malware detection and unique ability of the neuronet immune detectors to detect a several malware. The AIS is able to discern between non infected file of operation system and malware. The future of the AIS consists in capability for unknown malicious code detection. The detectors trained on malware owned to one class able to detect similar viruses as well as malware owned to another classes. As experiments show, neuronet immune detectors trained on well-known malware able to unknown malware detect. If a computer system with outdated antivirus bases can be unprotected in the face of new malicious code attack then the AIS gives a high probability detect it. Applying of the AIS for malware detection will expand the potentialities of existing antivirus software and will increase level of computer systems security.

Computer Immune Systems

The security in the field of computing may be considered as analogous to the immunity in natural systems. In computing, threats and dangers may arise because of malfunction of components or intrusive activities (both internal and external).

The idea of using immunological principles in computer security started since 1994. Stephanie Forrest and her group at the University of New Mexico have been working on a research project with a long-term goal to build an artificial immune system for computers. This immunity-based system has much more sophisticated notions of identity and protection than those afforded by current operating systems, and it is suppose to provide a general-purpose protection system to augment current computer security systems. The security of computer systems depends on such activities as detecting unauthorized use of computer facilities, maintaining the integrity of data files, and preventing the spread of computer viruses. The problem of protecting computer systems from harmful viruses is viewed as an instance of the more general problem of distinguishing self (legitimate users, uncorrupted data, etc.) from dangerous viruses unauthorized users and other malicious agents. This method called the negative-selection algorithm is intended to be complementary to the more traditional cryptographic and deterministic approaches to computer security. As an initial step, the negative selection algorithm has been used as a file-authentication method on the problem of computer virus detection

CONCLUSION

The computing world to gain lot from neural networks. Furthermore there is no need to devise an algorithm in order to perform a specific task. They are also very well suited for real time systems because of their wild response and computational times which are due to their parallel architecture.

Neural networks also contribute to other areas of research such as neurology and psychology. They are regularly used to model parts of living organisms and to investigate the internal mechanisms of the brain.

Perhaps the most exciting aspect of neural networks is the possibility that some day 'conscious' networks might be fashioned. There is a number of scientists arguing that consciousness is a research and development of intrusion detection systems has been ongoing since the early 1980's and the challenges faced by designers increase as the targeted systems because more diverse and complex. Misuse detection is a particularly difficult problem because of the extensive number of vulnerabilities in computer systems and the creativity of the attackers. Neural networks provide a number of advantages in the detection of these attacks. The early results of our tests of these technologies show significant promise, and our future work will involve the refinement of this approach and the development of a full-scale demonstration system.

The new methods can be implemented using the swarm intelligent algorithms like Particle Swarm Optimization(PSO), Ant colony Optimization(ACO), Article Bee Colony Algorithms(ABC).

Other popular algorithm single-handed not provide solution, use of blend of that provide lot of better solution than alone. Recent techniques and algorithm to produce still better results. A new operating security system cater with platform independency techniques fulfill the issues of unknown malware or new player of abnormal activities

REFERENCES

Chebrolu, S., Abraham, A., & Thomas, J. P. (2005). Feature detection and ensemble design of intrusion detection systems. *Computers & Security*, *24*(4), 295–307. doi:10.1016/j.cose.2004.09.008

Chen, Y., Abraham, A., & Yang, J. (2005). Feature deduction and intrusion detection using flexible neural trees. *Second IEEE International Symposium on Neural Networks.*

Chen, Z., Gao, L., & Kwiat, K. (2003). Modeling the Spread of Active Worms. *INFOCOM. Twenty-Second Annual Joint Conference of the IEEE Computer and Communications Societies*. IEEE. doi:10.1109/INFCOM.2003.1209211

Chou, T.-S., & Yen, K. K. (2007). Fuzzy belief k-nearest neighbors anomaly detection of user to root and remote to local attacks, The2007IEEE workshop on information assurance, 207–213. doi:10.1109/IAW.2007.381934

Cohen, F. (1987). Computer viruses: Theory and experiments. *Computers & Security*, *6*(1), 22–35. doi:10.1016/0167-4048(87)90122-2

Ghosh, A.K., & Schwartzbard, A. (n.d.). *A study in using neural networks for anomaly and misuse detection*. In Usenix security symposium, Washington, DC.

Ghosh, A. K., Schwartzbard, A., & Schatz, M. (1999). *Learning program behavior profiles for intrusion detection*. Workshop on intrusion detection and network monitoring, Santa Clara, CA.

Livadas, C., Walsh, B., Lapsley, D., & Strayer, T. (2006). *Using machine learning techniques to identify botnet traffic*. Second IEEE LCN workshop on network security (WNS), Tampa, FL.

Meystel, A. M., & Albus, J. M. (2002). *Intelligent systems architecture, design, and control*. New York: John Wiley & Sons, Inc.

Mukkamala, S., Janoski, G., & Sung, A. (2001). Monitoring systsem security using neural networks and support vector machines. *International workshop on hybrid intelligent systems*, 121–138.

Mukkamala, S., Sung, A., & Abraham, A. (2007). Hybrid multi-agent framework for detection of stealthy probe. *Appl Soft Computing J*, *7*(3), 631–641. doi:10.1016/j.asoc.2005.12.002

Mukkamala, S., Sung, A., & Abraham, A. (2004a). Designing intrusion detection systems: architectures and perspectives. The international engineering consortium (IEC) annual review of communications, 57, 1229–1241.

Mukkamala, S., Sung, A. H., & Abraham, A. (2003). Intrusion detection using ensemble of soft computing paradigms. *Third international conference on intelligent systems design and applications, advances in soft computing*. Springer. doi:10.1007/978-3-540-44999-7_23

Mukkamala, S., Sung, A. H., & Abraham, A. (2004b). Modeling intrusion detection systems using linear genetic programming approach. *17th international conference on industrial and engineering applications of artificial intelligence and expert systems*, 633–642. doi:10.1007/978-3-540-24677-0_65

Ourston, D., Matzner, S., Stump, W., & Hopkins, B. (2004). Coordinated internet attacks: Responding to attack complexity. *Computers & Security*, *12*(2), 165–190. doi:10.3233/JCS-2004-12201

Shah, K., Dave, N., Chavan, S., Mukherjee, S., Abraham, A., & Sanyal, S. (2004). Adaptive neuro-fuzzy intrusion detection system. IEEE international conference on ITCC'04, 1, 70–74.

Wang, J., Liu, Y., Tian, D., & Wei, D. (2007). Internet Worm Early Detection and Response Mechanism. *Journal of China Universities of Posts and Telecommunications*, *14*(3), 79–84. doi:10.1016/S1005-8885(07)60153-8

Wang, W., Gombault, S., & Guyet, T. (2008). Towards fast detecting intrusions: using key attributes of network traffic. *The third international conference on internet monitoring and protection*. IEEE Press.

Section 2
Thermal Imaging in Human Health Care and Diagnosis System

Images in different modality play a major role for diagnosis in medical field. During imaging process, instruments come in contact to human body that may lead to severe pain, example MRI images. Similarly human body may also be exposed to radiation. Thus to avoid human contact and radiation, thermal imaging can be considered as complementary to other modality of images. These images are used to measure thermal energy radiated from human body. The measurement of heat helps us to diagnosis disease to greater extent and its cost is also very low compare to other images of modalities. Keeping this as objective, in this section the role thermal images in medical science is discussed elaborately. In addition, a novel approach using iris feature for human health diagnosis is also presented.

Chapter 4
Thermal Imaging in Medical Science

Nilanjan Dey
Techno India College of Technology, India

Amira S. Ashour
Tanta University, Egypt

Afnan S. Althoupety
Taif University, Saudi Arabia

ABSTRACT

Thermal imaging is a non-destructive, non-contact and rapid system. It reports temperature through measuring infrared radiation emanated by an object/ material surface. Automated thermal imaging system involves thermal camera equipped with infrared detectors, signal processing unit and image acquisition system supported by computer. It is elaborated in wide domains applications. Extensive focus is directed to the thermal imaging in the medical domain especially breast cancer detection. This chapter provided the main concept and the different applications of thermal imaging. It explores and analyses several works in the light of studding the thermograph. It is an effective screening tool for breast cancer prediction. Studies justify that thermography can be considered a complementary tool to detect breast diseases. The current chapter reviews many usages and limitations of thermography in biomedical field. Extensive recommendations for future directions are summarized to provide a structured vision of breast thermography.

DOI: 10.4018/978-1-5225-2423-6.ch004

INTRODUCTION

Temperature measurement is a significant characteristic in any infrared thermography, industrial process and various applications that revolutionized temperature measurement concept. Typically, traditional temperature measurements instruments such as thermometers, thermistors, resistance temperature detectors and thermocouples can only determine the temperature at explicit points. Most of these instruments required direct contact with the object/material. Conversely, infrared thermal imaging is a non-contact, rapid, non-invasive and nondestructive monitoring of processes, which provides temperature mapping of the object/material (Puri *et al.*, 2005). It allows the creation of new ways to examine operations that were not even accessible before.

Thermal imaging is a technique that converts the invisible radiation pattern of an object into visible images (Ishimwe *et al.*, 2014). Afterward, feature extraction and analysis are applied to these thermal images for further processing. Consequently, the use of infrared thermal imaging become extensively increasing in numerous fields. It can be applied in all applications where temperature differences are required to support diagnosis, evaluation or analysis of a product or process. Thus, thermal imaging has established to be a precious tool for solving a wide range of scientific real time problems, such as in medicine (Ring, & Ammer, 2012; Izhar, & Petrou, 2012), veterinary (McDannold *et al.*, 2000; Infernuso *et al.*, 2010), industry (Zhang *et al.*, 2003), civil engineering (Clark *et al.*, 2003) and aerospace (Winfree, & Heath, 1998).

Recently, with the improvements in the camera technology, temperature is considered useful marker of various diseases (Ring, 2010) using infrared (IR) imaging. The camera is used to capture the natural thermal radiation emitted by any organ/ object above absolute zero (Jones, & Plassmann, 2002). Therefore, the body surface thermal distribution can be depicted using the IR imaging. This thermal distribution depends mainly on the complex associations describing the heat exchange processes between the skin tissue, inner tissue, local vasculature, and metabolic activity (Kakuta *et al.*, 2002).

Medical thermal imaging is also known as Medical Infrared Imaging or Digital Infrared Thermal Imaging (DITI). For high-resolution diagnostic body images, the DITI is equipped with ultra-sensitive thermal imaging cameras along with sophisticated computers to detect, to analyze and to produce accurate diagnosis. The thermal cameras are the ideal tools in countless applications. For early pre-clinical control/ diagnosis during treatment, the medical Thermography is used. It is free from any limitations or contra- indications.

Generally, several modalities can be used to test the body anatomy in order to measure its structure, such as X- rays, computed tomography (CT) scans, ultrasound and the magnetic resonance imaging (MRI). Such modalities employ several medical image processing techniques in different applications (Tapia, & Corchado, 2009; Janssen *et al.*, 2013; Hore *et al.*, 2015; Fouad *et al.*, 2016; Beagum *et al*, 2016; Hemalatha, & Anouncia, 2016; Dey *et al.*, 2016; Samanta *et al.*, 2016; Goswami *et al.*, 2016; Fadlallah *et al.*, 2016; Kotyk *et al.*, 2016; Saba *et al.*, 2016; Ahmed *et al.*, 2016; Trabelsi *et al.*, 2017; Dey *et al.*, 2017; Ashour *et al.*, 2017). On the contrary, the thermography is the unique imaging technique that has the ability to measure the physiological changes of the body metabolic processes (Bronzino, 2006). It provided safe imaging service using high-technology, revolutionary and digital thermal imaging cameras. In the past decade, due to various innovations, thermal imaging in medicine becomes the forefront of diagnosis. Some of the widespread applications that the thermography can be used are: breast pathologies (Arora *et al.*, 2008), neuro musculoskeletal (Tkáčová *et al.*, 2010), extracranial vessel disease (Okada *et al.*, 2007), lower extremity vessel disease (Huang *et al.*, 2011) and the vertebrae (Kim *et al.*, 2003).

Breast cancer is the most recurrently diagnosed cancer of women. It remains the second leading cause of death, despite the progress in treatment which reduced the mortality. Generally, numerous well established techniques are used for breast cancer screening including ultrasound, mammograms and clinical breast exams. Recently, thermography introduced its potential as a screening tool that provides an efficient, non-invasive and low risk tool for the early detection of breast cancer (Smith, 2005).

Consequently, the motivation of the current chapter is to provide an extensive focus on the thermal imaging in detecting the breast cancer. The main concept of the thermography is introduced.

BACKGROUND

Thermal Imaging: Theory and Measurement

All materials/ objects that are at temperature above -273 degrees C (0 degrees Kelvin) emanate infrared energy. The emitted infrared (IR) energy from the measured object is converted into an electrical signal by imaging sensor (microbolometer) (Davis, & Lettington, 1988). This imaging sensor is in the camera which connected to computer for displaying on a monitor the converted signal as a color or monochrome thermal image. The IR is a form of electromagnetic radiation that has wavelength 'band' of 0.78 to 1000µm (micrometers), which are longer than the wavelength of visible light and shorter than the radio waves. Several characteristics of the infrared radiation are similar to visible light, such as: the IR radiation can be refracted, focused, reflected and transmitted. The absorptivity, emissivity, transmissivity, and reflectivity of infrared radiation vary for different objects/ materials. Good absorber objects for the infrared radiation are also good emitters (Colthup, 2012).

Kirchhoff's law stated the relationship between the absorptivity (η), reflectivity (ς) and transmissivity (ξ) of an object as follows:

$$\eta + \varsigma + \xi = 1 \quad (1)$$

At the thermal equilibrium of an object, the absorption is equal to emission. This law can be simplified in several thermographic applications, for opaque objects equation (1) can be expressed as:

$$\eta + \varsigma = 1 \text{ or } \psi + \varsigma = 1 \quad (2)$$

where, ψ is the emissivity of the object. In a thermal camera, the infrared sensors receive the total IR radiation emitted from the objects' surface. The total amount of radiation emitted by an object per unit area is directly related to the object's emissivity and its temperature. It is given according to Stefan-Boltzmann law as follows:

$$E = \sigma \psi T^4 \quad (3)$$

where, E is the total amount of radiation emitted by an object per square meter, σ is the Stefan-Boltzman constant = 5.67 x 10^{-8} (W m^{-2} K^{-4}) and T is the temperature of the object in Kelvin (K).

Consequently, the object's temperature can be calculated if the emissivity of the object and total radiation emitted are known. Since, the human skin emissivity is extremely high (within 1% of that related to a black body). Thus, the measurements of infrared radiation emitted by the skin can be converted directly into accurate temperature values. In the thermal camera, infrared detectors sense the radiation emitted from the surface of the object in the spectral range of 3-5 μm (short wave) or 8-12 μm (long wave) (Anonymous, 2002).

Thermal Imaging Cameras

Thermal imaging refers to improve objects visibility even in a dark environment by recognizing the objects' infrared radiation as well as creating an image based on that information. Since, all objects emit heat (infrared energy) as a function of their temperature. This emitted infrared energy is known as the object's heat signature (Gaussorgues,& Chomet, 2012). Generally, the emitted radiation is directly proportional to the object temperature (i.e. the hotter an object is, the more radiation it emits). Therefore, a heat sensor is required to detect tiny differences in temperature. This thermal sensor is essentially a thermal infrared (IR) camera (thermal camera) that collects the infrared radiation on the surface of the objects under investigation and produces an electronic image based on the temperature differences information (Vollmer,& Möllmann, 2010). Additionally, thermal camera can detect different objects in the same view based on the objects' different temperature, which make them appear as distinct objects in a thermal image.

Typically, captured thermal images are gray scale in nature, where white objects represent hot objects, while black objects refer to cold ones, while the variation between the two states is indicated by the depth of gray scale (Heriansyah,& Abu-Bakar, 2009). However, some thermal cameras identify objects at different temperatures by adding color to the thermal images. Generally, thermal cameras can be divided into imaging cameras and measurement cameras. Thermal imaging cameras demonstrated only the color map of approximate temperature field (passive manner), while the thermal measurement cameras are calibrated by the manufacturer and can be used for temperature measurements (Minkina,& Dudzik, 2009). Since, the contrast in the thermal image will be poor if the differences between analyzed objects' temperatures and ambient temperature are small, when using passive thermal imaging. Thus, using active thermal imaging is preferable, where passive thermal imaging analysis only the existing temperature distribution of the object without imposing an additional heat flow to the object.

Thermal imaging applications are based on capturing the thermal image then analyzing the recorded temperature distribution using static thermal imaging techniques. However, some processes illustrate temperature change rates which cannot be estimated as static. Thus, dynamic thermal imaging (DTI) is applied to monitor and measures the temporal changes in temperature over the areas of interest (object) (Diakides,& Bronzino, 2008). The thermal changes can be monotonic (areas of interest are hot or cold), or periodic (visible as temperature modulation). Therefore, thermal image processing techniques are required.

Thermal Image Processing

Significance in digital image processing (DIP) methods and algorithms stems from two main application areas: pictorial information improvement for further analysis and processing of image data for storage, transmission, and representation for automated machine perception. Recently, numerous algorithms and techniques were developed for more accurate and representative information from the images using DIP (Kumar *et al.*, 2011; Santhi *et al*, 2013; Saxena *et al.*, 2014; Ashour *et al.*,2015; Dey *et al.*, 2015; Ghosh *et al.*, 2015; Hore *et al.*, 2015; Nandi *et al.*, 2015; Cheriguene *et al.*, 2016). Thermal images are one type of the digital images that requires image processing. Meanwhile, thermal image processing follows several steps of capture, preprocessing, and segmentation as illustrated in Figure 1.

Initially, thermal camera is used to capture the thermal images. Since intrinsic thermal white noise (low-frequency noise) arises because of the heat exchange fluctuations. Thus, filtering is used as a pre-processing step to reduce the noise. Several techniques including histogram of the gray levels for the whole image was proposed by Wiecek *et al.* (1999) to improve further image processing steps.

Infrared breast thermograms are usually captured in a larger area. Therefore, it is necessary to remove the unnecessary regions and background from the thermogram images before further processing. Then, convert the thermal images into grayscale images. Grayscale conversion of the thermograms is followed by segmentation of the region of interest (ROI). In the thermal images analysis, the ROI segmentation plays a vital role. Still, the ROI extraction from the thermal infrared image is a very challenging task. It requires tremendous effort in the case of vague nature of the images (Bezerra *et al.*, 2013). Thus, in order to locate objects and boundaries in the thermal images, segmentation is performed to partition the thermal image into multiple segments as a set of pixels or super-pixels.

After segmentation, features extraction specifies quantifiable property and important characteristics of the thermal image. The extracted features from the image can assist the discrimination of various regions in an image. These features can be either low-level features or high level features. The low-level features are directly extracted from the image intensity values (Saber,& Tekalp, 1998). For computational time reduction, the selected features vectors have to be reduced to contain the most significant features using features selection algorithms. Finally, the shape

Figure 1. Infrared thermal images processing

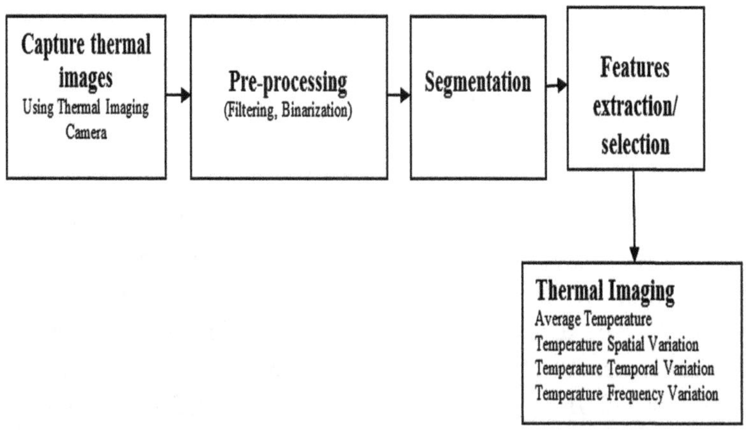

may require to be recognized in a series of thermal images (registration) to measure temporal variations in temperature. The temperature average, spatial variation, temporal variation and frequency variation are used and can be compared for the different regions of the same object (Jones, & Plassmann, 2002).

Thermal Imaging Applications

Thermography is a simple, non-invasive, noncontact and inexpensive imaging technique that widely relevant for different thermal imaging domains such as in industry, medical and research. Passive thermography is used for various applications such as tumor progress evaluation and monitoring (Poljak-Blazi *et al.*, 2009) and thermal monitoring of computer systems (Kolarić *et al.*, 2011). In the absence of significant natural temperature differences, active thermography methods are used (Vollmer, & Möllmann, 2010). These active methods are based on heating/ cooling of the monitoring the temperature change through time in the object's surface. In the medical domain, thermography is considered completely safe as it does not use radiation. Typically, thermography technique is computer based which equipped usually with IR camera and a standard computer. Consequently, it is able to store huge number of images for further image analysis.

Thermal imaging has numerous applications such as: i) in industry, various new products have been developed with the assist of the thermal imaging camera, ii) in designing printed circuit boards, which are challenged with managing the heat dissipation, iii) in characterizing material properties and supply rapid non-contact temperature measurement, iv) in thermal imaging microscope (thermal imaging camera combined with a microscope), v) in thermal and dynamic analysis of jet engine turbine blades, supersonic projectiles, and explosions, due to the high-speed thermal imaging, vi) in inspecting the state of the insulation material surrounding the pipework, vii) in night vision monitoring and security, viii) in thermal signatures IR signatures, which measure a target's apparent infrared brightness as a function of wavelength. IR signatures are important tools in the design of sensor, vehicle, and camouflage systems, ix) in tracking systems, where the tracking thermal imaging camera systems match the video tracking systems to increase visibility in low light, x) Short Wave Infrared (SWIR) Short Wave IR (SWIR) imaging provides nondestructive quantitative analysis of pharmaceuticals, crops, agricultural products, and lasers and xii) in medical thermography for accurate, non-contact and quantifiable diagnosis used to visualize and quantify changes in surface temperatures using high performance thermal imaging cameras. Several medical applications of thermal imaging include vascular evaluation, tumorous tissue identification, muscle strain assessment, and bleed point detection.

Thermal Imaging in Medical Diagnosis

Various medical modalities such as X-Ray, computed tomography (CT), Ultrasound, and Magnetic resonance imaging (MRI) are used to test the human body anatomy. Hildebrandt *et al.* (2010) indicated that based on the electromagnetic spectrum, the diagnostic imaging modalities utilize portions of the spectrum as illustrated in Figure 2.

Meanwhile, Stefan-Boltzmann law of emissivity can be applied to the body surface temperature, where the human body is an excellent radiator. Basically, there is an association between the human body temperature and diseases. After detecting the heating rays beyond the visible red, John Herschel in 1840, made the first thermal image from the sunlight using the evaporograph procedure. He used the term thermogram to depict the thermal image (Ring, 2000). Thus, medical thermal imaging is unique in its potential to demonstrate physiological change and metabolic processes. It is an easy and quick test to scan the affected area or the entire body.

For detecting and monitoring diseases and injuries, Digital Infrared Thermal Imaging (DITI) is used as a non-invasive clinical imaging procedure. The DITI illustrated the thermal body abnormalities. It is used to support diagnosis and prognosis, in addition to monitoring therapy progress (President, 2011), including for example: Vascular Disease, Breast Disease, Skin Cancer, Carpal Tunnel Syndrome, Nerve Damage, Back Injuries, Disc Disease, Digestive Disorders, Inflammatory Pain, Artery Inflammation, Rheumatology and Stroke Screening.

Medical Digital Infrared Thermal Imaging is a reliable tool of graphical mapping that display the skin surface temperature. It assists doctors to visualize in color the temperature changes in the skin surface. The color spectrum indicates both cold and hot responses. Generally, infrared thermal imaging is considered a non-invasive diagnostic technique, which does not involve radiation exposure. Figure 3 demonstrated several uses in the medical domain such as: i) in early detection

Figure 2. Association between the electromagnetic spectrum and the medical modalities

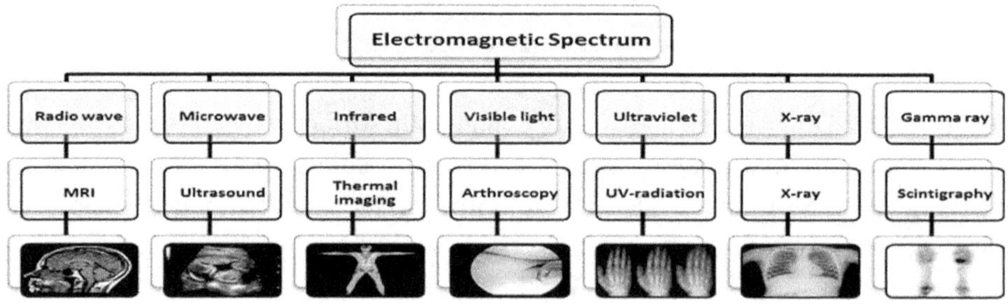

Figure 3. Thermal imaging applications in the medical domain

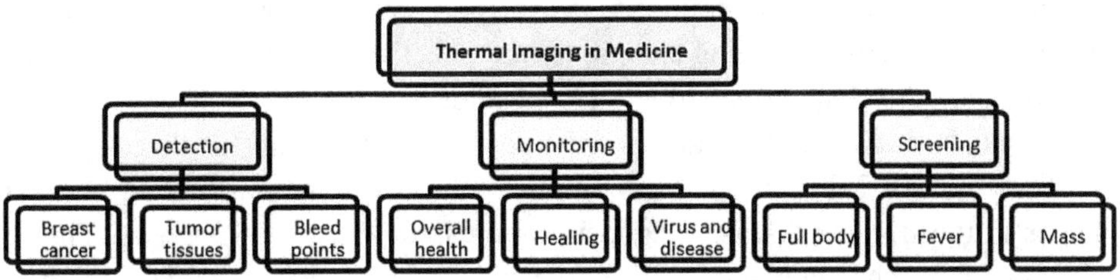

of the breast cancer, ii) monitoring healing processes, iii) monitoring changes in the overall health, monitoring Virus and disease, iv) full body screening, v) fever screening and vi) mass screening.

Bagavathiappan *et al.* (2009) studied peripheral vascular diseases using a non-invasive Infrared Thermography diagnosis. Temperature gradients were monitored in the affected regions of patients with vascular disorders. The thermal images indicated abnormal blood flow in the affected region. Increased temperature profiles were observed in certain areas on the affected limbs due to inflammation and underlying venous flow changes. The experimental results depicted that the temperature contrast in the affected regions is about 0.7 to 1° C above the normal regions due to slow blood circulation. Figure 4 (a) illustrated the thermal image of the photograph of the affected patient's leg shown in Figure 4 (b). Figure 4 (a) clearly indicated a lower temperature was noted at the distal portion along the toe tips due to sluggish blood circulation in the toes. A comparison between the abnormal area temperatures to the temperature of the surrounding normal area of the same patient's

Figure 4. (a) Isothermal image and (b) affected patient's leg photo
Bagavathiappan et al., 2009.

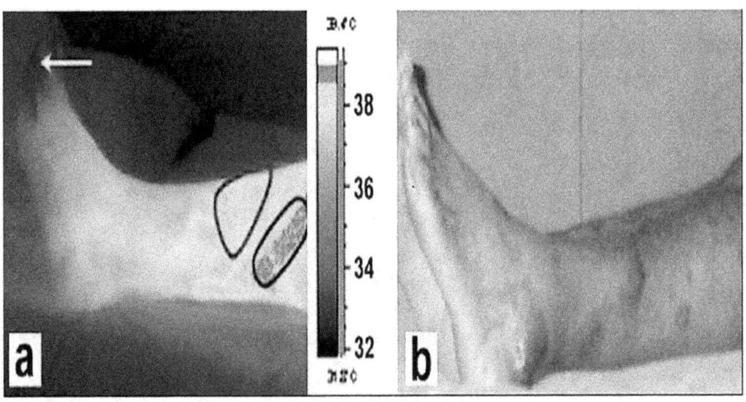

leg indicated the differences in temperature. The temperature in the abnormal marked regions had an average temperature of 0.7 to 1°C above the normal regions. It was suggested that thermal imaging techniques are effective for detecting small human body temperature changes due to vascular disorders.

Sivanandam *et al.* (2012) examined the potential of Infrared (IR) thermography in predicting type 2 diabetes and its complications as well as diagnosing. Thermal images of the skin for disease diagnosis were captured using IR camera. The receiver operating characteristic (ROC) curve was plotted to set threshold temperature for statistically significant body regions under t test. For both the control and diabetic subject, Figure 5 (a, b) illustrated the thermogram of the palm region; respectively. In a control subject, due to the normal metabolic function of the body and good perfusion of the blood, the skin temperature of the palm indicates 35.2 ° C. While in a diabetic subject, the palm temperature was lower the control subject due to poor blood perfusion and decreased metabolism. The skin temperature was 33.0° C in the diabetic subject. Similarly, the thermogram of the foot region was investigated by measuring the mean skin temperature of the metatarsal, mid foot and the heel regions. Consequently, the thermography can be used for diabetes diagnosis as well as prognosis.

MAIN FOCUS OF THE CHAPTER

The DITI is used in oncology, where tumors have an increase in the blood supply as well as an increased metabolic rate. This sequentially translates into increased

Figure 5. The thermogram of the palm region in: (a) control subject and (b) diabetic subject

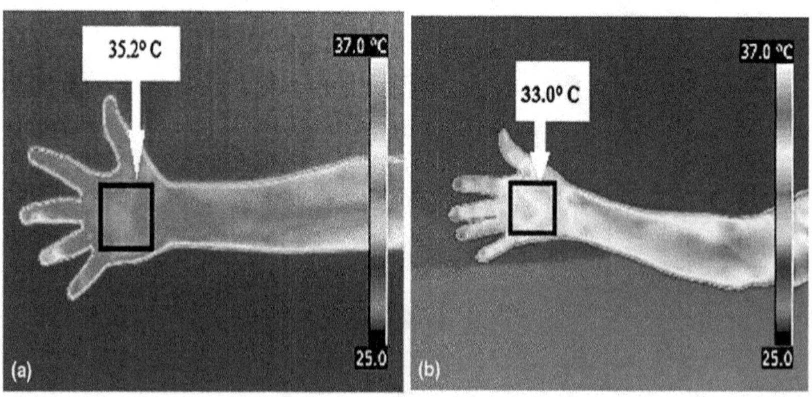

temperature gradients compared to adjacent normal tissues (Carmeliet,& Jain, 2000). Detecting these infrared hotspots and gradients can thereby assist to diagnose and identify malignancy. Additionally, previous thermography studies were limited by resolution, equipment and sensitivity capabilities. However, the supplementary sophisticated imaging and analytical techniques available today allow the use of DITI and artificial neural networks to detect malignancy with up to 100% sensitive (Ng EY,& Kee, 2008). Recently, digital infrared thermography cameras are able to sensing changes in temperature at 0.08°C without any patient contact. Thus, DITI achieved significant impact in medicine.

Thermal Imaging: Breast Cancer

Breast cancer is a primary cause of death among women, so researches are interested to reach its early diagnosis. Furthermore, cysts and benign neoplasms pathologies for example deserve investigation. Recently, the infrared thermography becomes a promising technique for early diagnosis of breast pathologies. Studies justify that thermography can be considered a complementary exam to detect breast diseases. Several papers on the use of infrared imaging for breast screening can be found in the current medical literature. The following survey explores and highlights previous studies conducted on breast cancer detection using thermal imaging. Consequently, the comments are organized according to the main steps of pattern recognition systems.

Temperature distribution is inherently connected to physiological and biological processes related to some disease. Breast cancer cells generate nitric oxide (NO), which impedes with the normal neuronal control of the blood vessel flow. This leads to a local vasodilatation in the cancerous growth early stages which causes the angiogenesis in later stages (Anbar *et al.*, 2001). Consequent, the increased blood flow in the abnormal area results in a temperature increase compared to the normal breast temperature. Even deep breast abnormal regions have the ability to induce changes in the skin temperature. In addition, breast cancer metabolic processes can contribute to increase the detected heat.

Generally, the temperature is symmetrical across the midline of the body in the healthy individuals (Ng *et al.*, 2001). Subjective analysis of several diagnostic imaging modalities including infrared thermographic images relies on the underlying philosophy, where normal images are comparatively symmetrical, while any abnormalities are asymmetries (Usuki *et al.*, 2002). Asymmetric temperature distributions are often related to physiologic changes as well as neoangiogenesis progress. Therefore, by identifying asymmetric temperature distribution areas on the breast surface, the IR thermography can be able to detect a breast disease (Qi,& Diakides, 2009). One of the most imperative signs for the breast cancer is the existence of asymmetry between the breasts, which can be measured and quantified the differences comparing both breasts. In addition to, the determination of areas within the breasts that have a high level of blood perfusion or vessels is a strong indicator.

SOLUTIONS AND RECOMMENDATIONS

Extensive studies were conducted to improve and analysis the thermal imaging technique for breast cancer detection. Recent literatures reported various image processing techniques for breast cancer detection from conventional breast thermograms. Yahara *et al.* (2003) discovered a correlation between microvessel density of breast malignancies and thermographic hot spots. It was concluded that the advantage of thermography lies in its noninvasive feature without requiring radiation. Results achieved that for detecting breast tumors thermography has 10% false-negative rate, which is similar to that obtained by mammography or ultrasound. Figure 6 demonstrated a comparison between the area of the diagnosed pathologically tumor and the hot area measured by thermography. Thus, the authors suggested a mechanistic explanation for the DITI use in cancer diagnosis.

Numerous segmentation algorithms have been proposed to extract regions of interest (ROI) with moderate success rates. NG *et al.* (2000) depicted the concurrent use of thermography along with artificial neural networks (ANN) for breast cancer diagnosis. It has been reported that in the younger patients, breast thermography could detect breast cancer up to 10 years prior to the conventional golden methods such as mammography. Though, the thermography accuracy depends on many factors such as the temperature stability and symmetry of the breasts' temperature. The thermography accuracy in women whose thermal images are captured in a proper period (5th-12th and 21st day of menstruation) is higher (80%) than the total population of patients (73%). This is due to the fact that woman's body temperature is

Figure 6. Comparison between the pathological area (on the right figure) and the thermographic area (on the left figure) of the tumor

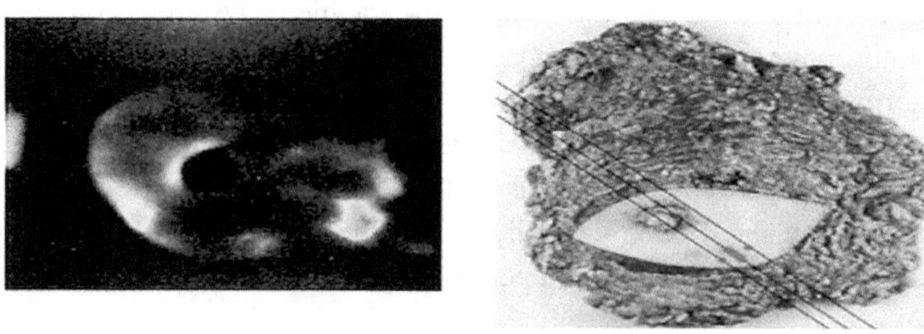

stable in certain periods after menstruation, where the body temperature stability depends on physiological state. The authors investigate the ANN use to complement the infrared heat radiating from the body surface with other physiological data. Four Back-propagation neural networks were proposed and trained using patients' physiological data and thermographs. Diagnose with only 61.54% accuracy was obtained for breast cancer cases owing to the inaccuracies in thermography and the low population size gathered.

Eddie *et al.* (2006) developed an intelligent diagnostic system based on thermography for the detection of tumors in breast. A segmentation process is performed for the thermograms breast cancer images for further extraction of the useful regions from the background. One-thousand thermograms images were obtained for contact and 800 thermograms for non-contact approaches. Standard ambient conditions were used for all examinations. The thermograms obtained were analyzed, where the first step in processing these thermograms was image segmentation to discern the ROI from the background. Two different techniques were conducted to extract the objects from the background. After analyzing these thermograms, a classification of the thermograms images was done to identify the normal or abnormal thermograms. Additionally, the temperature distribution pattern and the menstrual cyclic variation of temperature with time were evaluated. These obtained results can be used to detect breast cancer.

Kapoor, and Prasad (2010) proposed an automatic thermograms's breast image segmentation approach for analysis of infrared thermal images. Edge detection and Hough transform were outlined for asymmetry analysis of heat patterns in contralateral breasts. Subsequently, based on pattern classification and pixel distribution, asymmetries were identified for the breasts, where the pathological changes of the breasts can be used for diagnosis. Thus, this approach is effectual and feasible for diagnosing the asymmetric abnormalities for breast using infrared images.

Kapoor *et al.* (2012) investigated first order statistical parameters effectiveness in the breast asymmetry detection from infrared breast thermograms. A manual segmentation technique was used to avoid erroneous analysis of breast thermograms. Canny edge detector is used for the ROI segmentation with a threshold value to get the outstanding lower parabolic boundary of each breast in a breast thermogram. An individual mask is designed using these edges for each breast thermogram. The breast masks proposed one for healthy breast thermogram and another for an abnormal breast thermogram as demonstrated in Figure 7. The corresponding ROI of each breast thermogram segmented after the application of breast mask is illustrated in Figure 7 (c) and (f) related to the normal and abnormal images; respectively. The steps depicted in Figure 7 are the same shown in Figure 1 for the thermal image processing procedure.

Figure 7. (a) A normal breast thermogram; (b) its corresponding breast mask; (c) segmentation of breast region from background; (d) an abnormal breast thermogram; (e) its corresponding breast mask and (f) segmentation of breast region from breast thermogram
Kapoor et al., 2012.

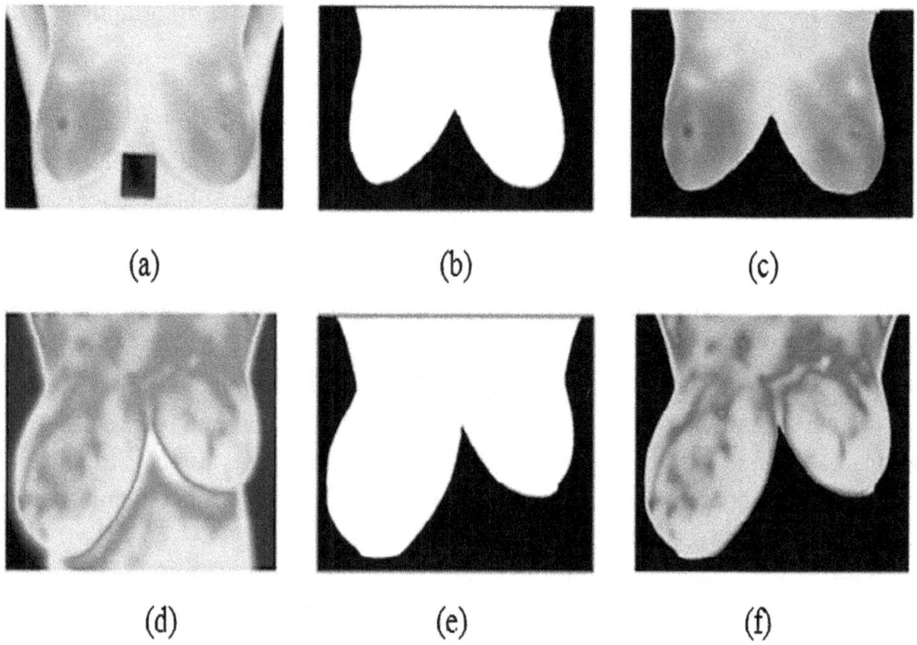

The results depicted that the feature values of left and right breast of a normal breast thermogram were quite similar while, in an abnormal breast thermogram cases, there is a significant difference in the feature values of both breasts. Statistical features were such as the mean difference, skewness, kurtosis, and entropy that shown noticeable difference between the left and right breast of an abnormal breast thermogram compared to other features.

Golestani *et al.* (2014) compared three various breast thermograms images segmentation methods, namely k-means, fuzzy c-means and level set. Since, shape, size and borders of the images hottest regions can assist to extract features which are used to detect abnormalities. Three segmentation methods were tested for fibrocystic and inflammatory cancer cases. From the thermal breast images, the hottest regions in all cases were extracted and compared to the original images. Experimental results established that level set method is a more accurate approach and has potential to extract almost exact shape of tumors.

Mahmoudzadeh *et al.* (2015) proposed a novel extended hidden Markov model (EHMM) to optimize breast thermogram images segmentation for more effective image interpretation and easier analysis of Infrared (IR) thermal patterns. EHMM method had a competitive advantage due to handling random sampling of the breast IR images with re-estimation of the extended hidden Markov model parameters. The experimental results illustrated the algorithm performance by applying EHMM segmentation method on the images of IUT_OPTIC database and compared to the K-means, Fuzzy C-means, Lloyd–Max, self-organizing map (SOM) and standard HMM algorithms. Figure 8 illustrated the segmentation results of two examples breast thermography images using the EHMM compared to the mentioned methods. Results indicated the proposed approach remarkable capabilities. It is worth noting that the suggested algorithm is able to map semi hot regions into distinct areas. Additionally, it extracted the regions of breast thermal images significantly, while the execution time is reduced.

Figure 8 depicted the segmented image, where the regions' colored with white, red, orange, yellow, green, purple, blue and black based on the regions temperature from high to low; respectively. Figure 8(g) included the hottest regions as they included white and red colors. Figure 8(b) and (e) depicted that both Lloyd–Max and SOM approaches didn't determine the hot regions properly. As well Figure 8(c) and (d) demonstrated that the hottest regions are not determined by both the K-means and fuzzy c-means. While, Figure 4(f) illustrated that the standard HMM algorithm

Figure 8. (a) Breast thermography images of breast; (b) segmented image using Lloyd–Max; (c) segmented image using K-means; (d) segmented image using Fuzzy C-means; (e) segmented image using SOM; (f) segmented image using standard HMM; (g) segmented image using EHMM approach (proposed) by Mahmoudzadeh et al. (2015)

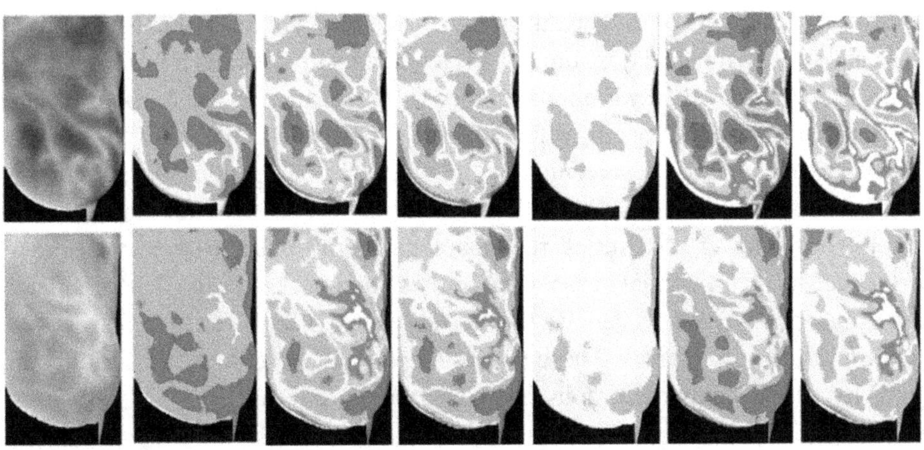

cannot determine the exact borders shape. Conversely, the EHMM Figure 8(g) can successfully determine the hottest regions of the thermal image. In addition, it was able to properly detect the exact borders of the hottest regions which can be used to determine some useful features. Thus, Mahmoudzadeh *et al.* concluded that none of the other segmentation methods can separate the first hottest regions from the second hottest regions, except their proposed approach. Furthermore, it can be noticed that in Figure 8(c), (d) and (f), an extracted vessel is determined as multi regions (one region for the innermost pixels and others for the border), instead of one region in the segmented images using K-Means, Fuzzy C-means and standard HMM (only top image). Accordingly, this multi regions segmentation is ineffective for interpretation and only causes more complexity. Conversely, the EHMM approach in Figure 8(g) extracted the vessel as one region.

Since, breast cancer is the most frequently diagnosed form of cancer in women. Thus, the preceding survey discussed several studies that were concluded for medical thermal imaging to detect breast cancer. The thermal imaging detection process is mainly based on the identification of the local temperature increase caused by the high metabolic activity of the cancer cells. Additionally, thermography has been established to provide a proficient screening modality for detecting breast cancer as it is able to detect small tumors and hence can lead to earlier diagnosis. Thus, it is more suitable to detect tumors in their early stage using the thermography along with other modalities such as the mammography.

FUTURE RESEARCH DIRECTIONS

Recently, thermal sensors sensitivity and spatial resolution have improved significantly, holding out promise for infrared (IR) imaging techniques based image processing. Thus, thermal imaging applications in industry, civil engineering, security monitoring, night vision and healthcare fields realized a high level of technical performance. In the medical domain, there is a relationship between body temperature and diseases based on the variations in blood flow, which affect the skin temperature. Consequently, thermal imaging offers a useful, simple, non-contact and non-invasive approach to the diagnosis, screening and treatment (as therapeutic aids) of numerous disorders/ diseases, in particular in the areas of breast cancer, dermatology, vascular disease of the lower extremities, rheumatology, neuro-musculo-skeletal disorders, orthopaedics, extra-cranial vessel disease (head and neck vessels), and circulatory abnormalities. Physicians can use the thermal image map to verify the presence of abnormal hot or cold areas. These hot/ cold areas can relate to a number of conditions for which the thermography procedure was approved.

Generally, in breast cancer detection, hot and cold spots are of an underlying surface are strong indicators for abnormities. Thus, thermogram interpretation is based on exaggerate hot spots, vascularisation and on asymmetries between breasts. Thermal images can be demonstrated in grey scale or in false-colour by several types of palettes depending on the used camera or software. The number of colors and the RGB (red-green-blue) values of each color vary in consistent with the used palette. Temperature related to each point of breast surface can be subject to analysis and processing to detect different kinds of diseases.

High-quality scientific research along with modern technologies coupled with well understanding of the skin blood flow regulation causes a revolution that improved the capability of thermal imaging in different medical applications. Along with the wide array of infrared sensor types and optics provides the facility for thermal imaging indispensable in various research environments. Thermal imaging engineers are able to simply visualize and quantify heat patterns in the devices that they create. However, still several recommendations and challenges that face the researchers/ engineering designers for more improvements in the thermal imaging in medicine including:

- Thermal imaging camera has several limitations such as:
    - Cannot deal with underwater objects or gather information through walls. In addition, it cannot capture images through glass due to the thermal energy can be reflected off shiny surfaces.
    - High quality thermal imaging cameras are required for more accurate thermal imaging. The existing quality cameras are expensive, while cheaper cameras have only resolution of 40X40 till 120X120 pixels.
- In some cases, the existence of different surfaces, accurate temperature measurements are obstructed by reflection and emissivity from these surfaces.
- Thermal imaging techniques are able only to directly detect surface temperatures. In addition, it is affected by the obesity and the depth of the region to be scanned and monitored.
- The biological nature of thermal signals and consistent thermal alterations of different specific injuries should be further addressed.
- During breast cancer detection using thermal imaging: high density of vessels and blood flow causes the tumor and its surrounding tissue to produce signals. In contrast, several factors may lead to noises, including the age, obesity, distance from the skin to the tumor, menopausal status, menstrual cycle, and diabetes mellitus or liver cirrhosis can cause abnormal peripheral vascular hemodynamic behavior.

The intensity effect of these noises may be strong, and the procedure for reading images of a thermogram is complicated and subjective, it is still difficult to apply thermography as a diagnostic procedure clinically.

- A main challenge is to combine the anatomical and physiological information demonstrated by the thermal pattern of the skin. Thus, it should not be used as the only deciding factor, but it should be combined with other instruments, such as the mammography or ultrasound tests should be used in adjacent to the DITI.

Another limitation to the DITI is the fact that inflammation or infection of breast parenchyma, can also alter temperature recordings and lead to false positive findings. In addition, morbidly obese women and breast size greater than DD preclude accurate recording of temperature from the inferior aspect (undersurface) of the breasts, so these patients may not be ideal candidates for DITI. DITI is not currently recommended or approved as a substitute for screening mammography, and correlation of findings on DITI should be made with alternative imaging techniques.

- Deeper tissues need to be further investigated by combining different medical imaging modalities.
- Thermography is applicable, however it has the following disadvantages and limitations:
    - Its procedure is complex and requires subjective reading of the thermal image. Thus, it requires advances in image analysis and processing to improve the thermography to be more objective and simple with higher sensitivity and specificity.
    - It is limited by the temperature resolution capability of earlier imaging technology
    - Bulky equipment necessary to perform its procedures, and the general lack of computer analytical tools.

Thus, digitalized high-resolution imaging and sophisticated artificial intelligence-based neural network image analysis lead to a revolution in the thermography technology.

- Neural network framework has been established and it has great potential for future development of an intelligent breast cancer diagnosis system based thermal images. This will be useful to the teenagers and young adults who are unsuitable for mammography at a young age. An intelligent breast thermography-neural network will be able to give an accurate diagnosis of breast cancer and can make a positive impact on breast disease detection.

- Thermal image pre-processing, image transition, filtering, image derivative, image segmentation and classification are required.
- Extraction and selection of appropriate features is a crucial task in asymmetry analysis of breast thermograms.
- Soft computing and meta-heuristic optimization algorithms are highly recommended for thermal images analysis and processing.
- Since, it was concluded that thermography should be combined with other medical modalities, thus image fusion is recommended.
- With respect to quantification of side-to-side differences within a defined ROI, it is significant to use the medical analysis function of image fusion.
- More improvement in the sensitivity and specificity of thermography in diagnosing breast cancers is required.
- Automatic identification of object and surface boundary of breast thermal images is a difficult and challenging task.
- Researchers reported that the thermography accuracy in women whose thermal images are captured in a proper period of menstruation is better, where the woman's body temperature is stable in certain periods after menstruation.
- Watermarking for the thermal images is recommended to achieve secured transmission without loss of the image medical information content.
- Thermal images de-noising should be studied to achieve high detection and diagnosis accuracy.

Generally, several researches have been developed advanced techniques that can be used with thermal images as reported by (Tapia *et al.*, 2009; Shen, 2009; Cinque, 2012; Bersch *et al.*, 2013; Dey *et al.*, 2014; Araki *et al.*, 2014; Aguiar *et al.*, 2014; Fouad *et al.*, 2016).

CONCLUSION

Thermal imaging techniques including thermography have extensive applications in various domains. Due to their advantages including being non-invasive, radiation-free without surface contact, it proved its efficiency in the medical domain. Thus, thermal imaging techniques became complementary to anatomical investigations based on X-rays, ultrasound and three-dimensional scanning techniques such as CT and MRI. DITI medical applications are extensive, particularly in the fields of neurology, rheumatology, oncology, sports medicine, breast cancer, diabetes detection and physiotherapy. Thermal imaging systems are an economical easy-to-use tool for diagnosing, examining and monitoring patients quickly and accurately without contact. Combining high-speed computers technology and very accurate thermal imaging cameras, the heat from the human body is recorded and processed in the computer to produce an image map which can then be analyzed, printed and transmitted via email.

The main concept of the medicine thermal imaging is to detect areas of temperature changes associated with inflammation and areas of high metabolic activity. The captured thermal images are useful in diagnosing injury and pain, circulatory pathology, detecting tumors and in monitoring treatment. Thermal information can be illustrated in a pseudo colored image where each color refers to a specific range of temperature. Thermography is one of the thermal imaging tools that record the temperature distribution of a body using the infrared radiation emitted by the body surface. It has realized success in examining the body temperature extremities. Medical infrared thermography is utilization for breast abnormality detection.

Breast thermography is a physiological technique that supplies information based on the temperature changes in breast. Due to angiogenesis, and higher blood vessel and chemical activity of the precancerous tissues and the area around a cancerous tumor, higher temperature of these regions rather than a normal breast can be detected. Different methods can be applied to extract hot regions for detecting suspected ROI in the breast infrared images. Thus, breast thermography has the potential to detect early abnormal changes in the breast tissues by detecting the first sign of forming up cancer before mammography can detect. Breast thermography can be considered a risk assessment tool that can be used in addition to mammography and other tests or procedures. Thermography plays two potential roles in the clinical management of breast cancer as follows. First, it can be utilized as a way of screening for breast lesions, either malignant or benign. Second, it can be able to differentiate malignant from benign lesions that have been detected by other methods.

Widespread studies have been described in the current chapter including articles related to breast thermal images analysis and processing. Studies recommend thermography to be a complementary modality of breast disease detection. IR imaging can improve the breast cancer detection sensitivity by capturing abnormalities in early stages. All thermography studies are concerned with identifying thermal emissions that indicate potential risk markers only without suggesting diagnosis and/or treatment. Studies depicted that the earliest detection of breast cancer is realized when multiple tests are used together. These multimodal techniques including physical breast exams by the physician, mammography, MRI, ultrasound, thermography, and other tests assists breast cancer early detection and diagnosis.

Borders, shape and size of the images' hottest regions can help to determine features that used to detect abnormalities. Thus, image segmentation approaches can play a significant role to segment and extract the abnormal regions in the breast infrared thermal images. Further image processing using classification can be employed to classify the detected regions based on the temperature into normal and abnormal. Additional classification can be used to classify the abnormality classes into mild, moderate and sever cases. Consequently, image processing has an extensive role in thermal imaging applications. Development in the thermal camera as well as the in the thermal image processing algorithms will provide more reliable and accurate thermal imaging techniques.

REFERENCES

Aguiar, Y. P., Maria de Fátima, Q. V., Galy, E., & Santoni, C. (2014). Accounting for Individual and Situation Characteristics to Understand the User Behaviour when Interacting with Systems during Critical Situations. *International Journal of Ambient Computing and Intelligence*, *6*(2), 29–55. doi:10.4018/IJACI.2014070103

Ahmed, S. S., Dey, N., Ashour, A. S., Sifaki-Pistolla, D., Bălas-Timar, D., Balas, V. E., & Tavares, J. M. R. (2016). Effect of fuzzy partitioning in Crohn's disease classification: A neuro-fuzzy-based approach. *Medical & Biological Engineering & Computing*, 1–15. PMID:27106754

Anbar, M., Milescu, L., Naumov, A., Brown, C., Button, T., Carly, C., & AlDulaimi, K. (2001). Detection of cancerous breasts by dynamic area telethermometry. *Engineering in Medicine and Biology YK Magazine, IEEE*, *20*(5), 80–91. doi:10.1109/51.956823 PMID:11668900

Anonymous,. (2002). *Infrared training: Level 1 course manual.* North Billerica, MA: Infrared Training Centre.

Araki, T., Ikeda, N., Molinari, F., Dey, N., Acharjee, S. M., Saba, L., & Suri, J. S. et al. (2014). Effect of geometric-based coronary calcium volume as a feature along with its shape-based attributes for cardiological risk prediction from low contrast intravascular ultrasound. *Journal of Medical Imaging and Health Informatics*, *4*(2), 255–261. doi:10.1166/jmihi.2014.1251

Arora, N., Martins, D., Ruggerio, D., Tousimis, E., Swistel, A. J., Osborne, M. P., & Simmons, R. M. (2008). Effectiveness of a noninvasive digital infrared thermal imaging system in the detection of breast cancer. *American Journal of Surgery*, *196*(4), 523–526. doi:10.1016/j.amjsurg.2008.06.015 PMID:18809055

Ashour, A. S., Beagum, S., Dey, N., Ashour, A. S., Pistolla, D. S., Nguyen, G. N., & Shi, F. et al. (2017). Light microscopy image de-noising using optimized LPA-ICI filter. *Neural Computing & Applications*, 1–17.

Ashour, A. S., Samanta, S., Dey, N., Kausar, N., Abdessalemkaraa, W. B., & Hassanien, A. E. (2015). Computed Tomography Image Enhancement Using Cuckoo Search: A Log Transform Based Approach. *Journal of Signal and Information Processing*, *6*(03), 244–257. doi:10.4236/jsip.2015.63023

Bagavathiappan, S., Saravanan, T., Philip, J., Jayakumar, T., Raj, B., Karunanithi, R., … Jagadeesan, K. (2009). Infrared Thermal Imaging for Detection of Peripheral Vascular Disorders. *Journal of Medical Physics/Association of Medical Physicists of India*, *34*(1), 43.

Beagum, S., Dey, N., Ashour, A. S., Sifaki-Pistolla, D., & Balas, V. E. (2016). Nonparametric de-noising filter optimization using structure-based microscopic image classification. *Microscopy Research and Technique*. doi:10.1002/jemt.22811 PMID:27901295

Bersch, S., Azzi, D., Khusainov, R., & Achumba, I. E. (2013). Artificial immune systems for anomaly detection in ambient assisted living applications. *International Journal of Ambient Computing and Intelligence*, *5*(3), 1–15. doi:10.4018/ijaci.2013070101

Bezerra, L. A., Oliveira, M. M., Araújo, M. C., Viana, M. J., Santos, L. C., Santos, F. G.,... & Resmini, R. (2013). Infrared Imaging for Breast Cancer Detection with Proper Selection of Properties: From Acquisition Protocol to Numerical Simulation. *Multimodality Breast Imaging: Diagnosis and Treatment, E. YK Ng; U. Rajendra Acharya*, 285-332.

Bronzino, J. D. (2006). Infrared imaging of the breast–an overview. In Biomedical Engineering handbook–Medical Systems and Devices (3rd ed.). CRC.

Carmeliet, P., & Jain, R. K. (2000). Angiogenesis in cancer and other diseases. *Nature*, *407*(6801), 249–257. doi:10.1038/35025220 PMID:11001068

Cheriguene, S., Azizi, N., Zemmal, N., Dey, N., Djellali, H., & Farah, N. (2016). Optimized Tumor Breast Cancer Classification Using Combining Random Subspace and Static Classifiers Selection Paradigms. In *Applications of Intelligent Optimization in Biology and Medicine* (pp. 289–307). Springer International Publishing. doi:10.1007/978-3-319-21212-8_13

Cinque, M., Coronato, A., & Testa, A. (2012). Dependable services for mobile health monitoring systems. *International Journal of Ambient Computing and Intelligence*, *4*(1), 1–15. doi:10.4018/jaci.2012010101

Clark, M. R., McCann, D. M., & Forde, M. C. (2003). Application of infrared thermography to the non-destructive testing of concrete and masonry bridges. *NDT & E International*, *36*(4), 265–275. doi:10.1016/S0963-8695(02)00060-9

Colthup, N. (2012). *Introduction to infrared and Raman spectroscopy*. Elsevier.

Davis, A. P., & Lettington, A. H. (1988). Principles of thermal imaging. *Applications of Thermal Imaging*, 1-34.

Dey, N., & Ashour, A. S. (Eds.). (2016). *Classification and Clustering in Biomedical Signal Processing*. IGI Global. doi:10.4018/978-1-5225-0140-4

Dey, N., Ashour, A. S., Ashour, A. S., & Singh, A. (2015). Digital Analysis of Microscopic Images in Medicine. *Journal of Advanced Microscopy Research*, *10*(1), 1–13. doi:10.1166/jamr.2015.1229

Dey, N., Ashour, A. S., Chakraborty, S., Banerjee, S., Gospodinova, E., Gospodinov, M., & Hassanien, A. E. (2017). Watermarking in Biomedical Signal Processing. In *Intelligent Techniques in Signal Processing for Multimedia Security* (pp. 345–369). Springer International Publishing. doi:10.1007/978-3-319-44790-2_16

Dey, N., Dey, G., Chakraborty, S., & Chaudhuri, S. S. (2014). Feature analysis of blind watermarked electromyogram signal in wireless telemonitoring. In *Concepts and Trends in Healthcare Information Systems* (pp. 205–229). Springer International Publishing. doi:10.1007/978-3-319-06844-2_13

Diakides, N. A., & Bronzino, J. D. (2008). *Medical Infrared Imaging*. Taylor & Francis Group, LLC.

Fadlallah, S. A., Ashour, A. S., & Dey, N. (2016). Advanced Titanium Surfaces and Its Alloys for Orthopedic and Dental Applications Based on Digital SEM Imaging Analysis. *Advanced Surface Engineering Materials*, 517-560.

Fouad, K. M., Hassan, B. M., & Hassan, M. F. (2016). User Authentication based on Dynamic Keystroke Recognition. *International Journal of Ambient Computing and Intelligence*, *7*(2), 1–32. doi:10.4018/IJACI.2016070101

Fouad, K. M., Hassan, B. M., & Hassan, M. F. (2016). User Authentication based on Dynamic Keystroke Recognition. *International Journal of Ambient Computing and Intelligence*, *7*(2), 1–32. doi:10.4018/IJACI.2016070101

Gaussorgues, G., & Chomet, S. (2012). *Infrared thermography* (Vol. 5). Springer Science & Business Media.

Ghosh, A., Sarkar, A., Ashour, A. S., Bălas-Timar, D., Dey, N., & Balas, V. E. (2015). Grid Color Moment Features in Glaucoma Classification. *International Journal of Advanced Computer Science and Applications*, *6*(9). doi:10.14569/IJACSA.2015.060913

Golestani, N., EtehadTavakol, M., & Ng, E. Y. K. (2014). Level set method for segmentation of infrared breast thermograms. *Experimental and Clinical Sciences*, *13*, 241–251. PMID:26417258

Goswami, S., Dey, U., Roy, P., Ashour, A., & Dey, N. (2016). Medical Video Processing: Concept and Applications. *Feature Detectors and Motion Detection in Video Processing*, 1.

Hemalatha, S., & Anouncia, S. M. (2016). A Computational Model for Texture Analysis in Images with Fractional Differential Filter for Texture Detection. In *Biometrics: Concepts, Methodologies, Tools, and Applications: Concepts, Methodologies, Tools, and Applications*. Academic Press.

Heriansyah, R., & Abu-Bakar, S. A. R. (2009). Defect detection in thermal image for nondestructive evaluation of petrochemical equipments. *NDT & E International*, *42*(8), 729–740. doi:10.1016/j.ndteint.2009.06.008

Hildebrandt, C., Ammer, K., & Raschner, C. (2010). An Overview of Recent Application of Medical Infrared Thermography in Sports Medicine in Austria. *Sensors (Basel, Switzerland)*, *10*(5), 4700–4715. doi:10.3390/s100504700 PMID:22399901

Hore, S., Chakroborty, S., Ashour, A. S., Dey, N., Ashour, A. S., Sifaki-Pistolla, D., & Chaudhuri, S. R. et al. (2015). Finding Contours of Hippocampus Brain Cell Using Microscopic Image Analysis. *Journal of Advanced Microscopy Research*, *10*(2), 93–103. doi:10.1166/jamr.2015.1245

Hore, S., Chakroborty, S., Ashour, A. S., Dey, N., Ashour, A. S., Sifaki-Pistolla, D., & Chaudhuri, S. R. et al. (2015). Finding Contours of Hippocampus Brain Cell Using Microscopic Image Analysis. *Journal of Advanced Microscopy Research*, *10*(2), 93–103. doi:10.1166/jamr.2015.1245

Huang, C. L., Wu, Y. W., Hwang, C. L., Jong, Y. S., Chao, C. L., Chen, W. J., & Yang, W. S. et al. (2011). The application of infrared thermography in evaluation of patients at high risk for lower extremity peripheral arterial disease. *Journal of Vascular Surgery*, *54*(4), 1074–1080. doi:10.1016/j.jvs.2011.03.287 PMID:21784604

Infernuso, T., Loughin, C. A., Marino, D. J., Umbaugh, S. E., & Solt, P. S. (2010). Thermal Imaging of Normal and Cranial Cruciate Ligament-Deficient Stifles in Dogs. *Veterinary Surgery*, *39*(4), 410–417. doi:10.1111/j.1532-950X.2010.00677.x PMID:20459492

Ishimwe, R., Abutaleb, K., & Ahmed, F. (2014). Applications of Thermal Imaging in Agriculture—A Review. *Advances in Remote Sensing*, *3*(3), 128–140. doi:10.4236/ars.2014.33011

Izhar, L. I., & Petrou, M. (2012). Thermal imaging in medicine. *Advances in Imaging and Electron Physics*, *171*, 41–114. doi:10.1016/B978-0-12-394297-5.00002-7

Janssen, J. H., Ijsselsteijn, W. A., Westerink, J. H., Tacken, P., & de Vries, G. J. (2013). The tell-tale heart: Perceived emotional intensity of heartbeats. *International Journal of Synthetic Emotions*, *4*(1), 65–91. doi:10.4018/jse.2013010103

Jones, B. F., & Plassmann, P. (2002). Digital infrared thermal imaging of human skin. *Engineering in Medicine and Biology Magazine, IEEE*, *21*(6), 41–48. doi:10.1109/MEMB.2002.1175137 PMID:12613210

Kakuta, N., Yokoyama, S., & Mabuchi, K. (2002). Human thermal models for evaluating infrared images. *Engineering in Medicine and Biology Magazine, IEEE*, *21*(6), 65–72. doi:10.1109/MEMB.2002.1175140 PMID:12613213

Kapoor, P., Prasad, D. S., & Patni, S. (2012). Automatic Analysis of Breast Thermograms for tumor detection based on Biostatistical feature extraction and ANN. *International Journal of Emerging Trends in Engineering and Development*, *2*(7), 245–255.

Kapoor, P., & Prasad, S. V. A. V. (2010, February). Image processing for early diagnosis of breast cancer using infrared images. *2010 2nd International Conference on Computer and Automation Engineering (ICCAE),* (Vol. 3, pp. 564-566). IEEE.

Kim, Y. C., Bahk, J. H., Lee, S. C., & Lee, Y. W. (2003). Infrared thermographic imaging in the assessment of successful block on lumbar sympathetic ganglion. *Yonsei Medical Journal*, *44*(1), 119–124. doi:10.3349/ymj.2003.44.1.119 PMID:12619184

Kolarić, D., Lipić, T., Grubiaić, I., Gjenero, L., & Skala, K. (2011, September). Application of infrared thermal imaging in blade system temperature monitoring. *ELMAR, 2011 Proceedings* (pp. 309-312). IEEE.

Kotyk, T., Dey, N., Ashour, A. S., Balas-Timar, D., Chakraborty, S., Ashour, A. S., & Tavares, J. M. R. (2016). Measurement of glomerulus diameter and Bowmans space width of renal albino rats. *Computer Methods and Programs in Biomedicine, 126*, 143–153. doi:10.1016/j.cmpb.2015.10.023 PMID:26796351

Kumar, A., & Santhi, V. (2011). A review on geometric invariant digital image watermarking techniques. *International Journal of Computers and Applications, 12*(9), 31–36. doi:10.5120/1708-2275

Mahmoudzadeh, E., Montazeri, M. A., Zekri, M., & Sadri, S. (2015). Extended hidden Markov model for optimized segmentation of breast thermography images. *Infrared Physics & Technology, 72*, 19–28. doi:10.1016/j.infrared.2015.06.012

McDannold, N. J., King, R. L., Jolesz, F. A., & Hynynen, K. H. (2000). Usefulness of MR Imaging-Derived Thermometry and Dosimetry in Determining the Threshold for Tissue Damage Induced by Thermal Surgery in Rabbits. *Radiology, 216*(2), 517–523. doi:10.1148/radiology.216.2.r00au42517 PMID:10924580

Minkina, W., & Dudzik, S. (2009). *Infrared Thermography Errors and Uncertainties*. John Wiley & Sons, Ltd. doi:10.1002/9780470682234

Nandi, D., Ashour, A. S., Samanta, S., Chakraborty, S., Salem, M. A., & Dey, N. (2015). Principal component analysis in medical image processing: A study. *International Journal of Image Mining, 1*(1), 65–86. doi:10.1504/IJIM.2015.070024

Ng, E. K., Fok, S. C., Peh, Y. C., Ng, F. C., & Sim, L. S. J. (2002). Computerized detection of breast cancer with artificial intelligence and thermograms. *Journal of Medical Engineering & Technology, 26*(4), 152–157. doi:10.1080/03091900210146941 PMID:12396330

Ng, E.Y., Ng, F.C., & Sim, L.S. (2001). Statistical analysis of healthy and malignant breast thermography. *Journal of Medical Engineering & Technology, 25*(6), 253–263. doi:10.1080/03091900110086642 PMID:11780767

Ng, E. Y. K., & Chen, Y. (2006). Segmentation of breast thermogram: Improved boundary detection with modified snake algorithm. *Journal of Mechanics in Medicine and Biology, 6*(2), 123–136. doi:10.1142/S021951940600190X

Ng, E. Y. K., & Kee, E. C. (2008). Advanced integrated technique in breast cancer thermography. *Journal of Medical Engineering & Technology, 32*(2), 103–114. doi:10.1080/03091900600562040 PMID:17852648

Okada, Y., Kawamata, T., Kawashima, A., & Hori, T. (2007). Intraoperative Application Of Thermography In Extracranial-Intracranial Bypass Surgery. *Neurosurgery*, *60*(4), 362–365. PMID:17415175

Poljak-Blazi, M., Kolaric, D., Jaganjac, M., Zarkovic, K., Skala, K., & Zarkovic, N. (2009). Specific thermographic changes during Walker 256 carcinoma development: Differential infrared imaging of tumour, inflammation and haematoma. *Cancer Detection and Prevention*, *32*(5), 431–436. doi:10.1016/j.cdp.2009.01.002 PMID:19232842

President, I. H. (2011) International Consensus and Guidelines on Medical Thermography 2011 (ICGMT 2011) 2nd International Work Group for Medical Thermography Meeting (IWGMT 2011). *Neurosurgery*, 8.

Puri, C., Olson, L., Pavlidis, I., Levine, J., & Starren, J. (2005, April). StressCam: non-contact measurement of users' emotional states through thermal imaging. In CHI'05 extended abstracts on Human factors in computing systems (pp. 1725-1728). Academic Press.

Qi, H., & Diakides, N. A. (2009). Thermal infrared imaging in early breast cancer detection, Augmented Vision Perception in Infrared. *Advances in Pattern Recognition*, *II*, 139–152.

Ring, E. F. J. (2000). The discovery of infrared radiation in 1800. *Imaging Science Journal*, *48*(1), 1–8. doi:10.1080/13682199.2000.11784339

Ring, E. F. J., & Ammer, K. (2012). Infrared thermal imaging in medicine. *Physiological Measurement*, *33*(3), R33–R46. doi:10.1088/0967-3334/33/3/R33 PMID:22370242

Ring, F. (2010). Thermal imaging today and its relevance to diabetes. *Journal of Diabetes Science and Technology*, *4*(4), 857–862. doi:10.1177/193229681000400414 PMID:20663449

Saba, L., Dey, N., Ashour, A. S., Samanta, S., Nath, S. S., Chakraborty, S., & Suri, J. S. et al. (2016). Automated stratification of liver disease in ultrasound: An online accurate feature classification paradigm. *Computer Methods and Programs in Biomedicine*, *130*, 118–134. doi:10.1016/j.cmpb.2016.03.016 PMID:27208527

Saber, E., & Tekalp, A. M. (1998). Integration of color, edge, shape, and texture features for automatic region-based image annotation and retrieval. *Journal of Electronic Imaging*, *7*(3), 684–700. doi:10.1117/1.482605

Samanta, S., Choudhury, A., Dey, N., Ashour, A. S., & Balas, V. E. (2016). *Quantum inspired evolutionary algorithm for scaling factors optimization during manifold medical information embedding. In Quantum Inspired Computational intelligence: Research and Applications*. Elsevier.

Santhi, V., Thangavelu, A., & Arulmozhivarman, P. (2013). Adaptive visible watermarking in Hadamard domain for digital images. *International Journal of Information and Computer Security*, *5*(3), 202–223. doi:10.1504/IJICS.2013.055838

Saxena, R., Shah, K., Chawla, R., & Santhi, V. (2014). Biometric Watermarking for Copyright Protection of Digital Images. *International Journal of Applied Engineering Research*, *9*(24), 23681–23688.

Shen, X. (2009). Intrusive Evaluation of Ambient Displays. *International Journal of Ambient Computing and Intelligence*, *1*(4), 12–31. doi:10.4018/jaci.2009062202

Sivanandam, S., Anburajan, M., Venkatraman, B., Menaka, M., & Sharath, D. (2012). Medical thermography: A diagnostic approach for type 2 diabetes based on non-contact infrared thermal imaging. *Endocrine*, *42*(2), 343–351. doi:10.1007/s12020-012-9645-8 PMID:22411072

Smith, H. (2005). Screening for breast cancer. *Journal of the American Medical Association*, *293*(10), 1245–1256. doi:10.1001/jama.293.10.1245 PMID:15755947

Tapia, D. I., & Corchado, J. M. (2009). An ambient intelligence based multi-agent system for Alzheimer health care. *International Journal of Ambient Computing and Intelligence*, *1*(1), 15–26. doi:10.4018/jaci.2009010102

Tkáčová, M., Foffova, P., Živčák, J., & Hudak, R. (2010, January). The methodics of medical thermography in the diagnostics of the human body musculoskeletal system. *2010 IEEE 8th International Symposium in Applied Machine Intelligence and Informatics (SAMI),* (pp. 275-277). IEEE.

Trabelsi, I., & Bouhlel, M. S. (2017). Feature Selection for GUMI Kernel-Based SVM in Speech Emotion Recognition. In Artificial Intelligence: Concepts, Methodologies, Tools, and Applications (pp. 941-953). IGI Global.

Usuki, H., Ishimura, K., Hagiike, M., Okano, K., Izuishi, K., Karasawa, Y., & Maeta, H. et al. (2002). Thermographic examination for carcinoma. *Biomedical Thermology*, *21*(4), 1–7.

Vollmer, M., & Möllmann, K. P. (2010). *Infrared thermal imaging: fundamentals, research and applications*. John Wiley & Sons. doi:10.1002/9783527630868

Więcek, B., Zwolenik, S., & Jung, A. (1999, October). Advanced thermal, visual and radiological image processing for clinical diagnostics. In *Engineering in Medicine and Biology, 1999. 21st Annual Conference and the 1999 Annual Fall Meetring of the Biomedical Engineering Society BMES/EMBS Conference, 1999.Proceedings of the First Joint* (*Vol.* 2). IEEE. doi:10.1109/IEMBS.1999.804270

Winfree, W. P., & Heath, D. M. (1998, March). Thermal diffusivity imaging of aerospace materials and structures. In Aerospace/Defense Sensing and Controls (pp. 282-290). International Society for Optics and Photonics.

Yahara, T., Koga, T., Yoshida, S., Nakagawa, S., Deguchi, H., & Shirouzu, K. (2003). Relationship between microvessel density and thermographic hot areas in breast cancer. *Surgery Today*, *33*(4), 243–248. doi:10.1007/s005950300055 PMID:12707816

Zhang, X. H., Guimond, Y., & Bellec, Y. (2003). Production of complex chalcogenide glass optics by molding for thermal imaging. *Journal of Non-Crystalline Solids*, *326*, 519–523. doi:10.1016/S0022-3093(03)00464-2

KEY TERMS AND DEFINITIONS

Active Thermography: The type of thermography where an energy source is required to produce a thermal contrast between the feature of interest and the background. It is used with the applications that the inspected parts are usually in equilibrium with the surroundings.

Breast Cancer Screening: Checking a woman's breasts for cancer before there are signs or symptoms of the disease.

Classification: Defined as assigning a physical object or incident into one of a set of predefined categories. It is the analysis of the numerical properties of various image features and organizes data into categories.

Emissivity: A term that represents a material's ability to emit thermal radiation.

Feature Extraction: The transforming the existing features into a lower dimensional space.

Feature Selection: The selection of a subset of the existing features without a transformation in order to decrease the dimensions and reduce the processing time.

Features: Inherent properties of data, independent of coordinate frames.

Image Analysis: The extraction of meaningful information from images; mainly from digital images by means of digital image processing techniques.

Image Processing: The analysis and manipulation of a digitized image, especially in order to improve its quality.

Image Segmentation: The process of partitioning a digital image into multiple segments as a set of pixels or super-pixels. It assists to locate objects and boundaries in images.

Infrared Radiation (IR): The radiation that has a wavelength between 3.0 μm and 100 μm.

Magnetic Resonance Imaging: The use of nuclear magnetic resonance of protons to produce proton density images.

Mammogram: A medical imaging modality used to investigate the breast. It is an X-ray of the breast.

Medical Imaging: The technique by which medical images are produced, captured or created.

Medical Modalities: The devices/ instruments that used to capture (acquire) medical images for the organ under concern for further analysis and processing in order to support diagnosis.

Passive Thermography: The type of thermography where the features of interest are naturally at a higher or lower temperature than the background. It is commonly used by sports physicians and veterinarians to determine the inflammation areas of the body.

Stefan–Boltzmann Law: Also known as Stefan's law, which describes the power radiated from a black body in terms of its temperature. It states that "the total energy radiated per unit surface area of a black body across all wavelengths per unit time is directly proportional to the fourth power of the black body's thermodynamic temperature".

Thermal Camera: Also called infrared camera, thermal imaging camera, or thermographic camera, which defined as a device that forms an image using infrared radiation, similar to a common camera that forms an image in wavelengths as long as 14,000 nm (14 μm).

Thermal Images- are also known as the thermograms: which are the visual display of the amount of infrared energy emitted, transmitted, and reflected by an object.

Thermal Imaging Camera: A thermal camera that capable of performing algorithms to interpret the data and build an image. It shows approximation of the temperature at which the object is operating; the camera is actually using multiple sources of data based on the areas surrounding the object to determine that value rather than detecting the actual temperature.

Thermal Imaging: The approach of improving objects visibility in a dark environment or generally allows objects' infrared radiation detection. Then, create an image based on the temperature information of the object.

Thermal Infrared: Thermal radiation emitted by objects near the normal environmental temperature is infrared. IR radiation wavelengths; hotter objects, emit infrared at wavelengths shorter than thermal infrared.

Thermal Radiation: An electromagnetic radiation produced by the thermal motion of charged particles in matter. All matter/ objects with a temperature greater than absolute zero emit thermal radiation.

Thermography: A method of measuring the heat (thermo) emitted from a body. The science of thermography is the application of measured heat readings to locate abnormal function or pathology in the body. It studies the heat distribution in structures or regions to detect tumors for example. Thermography is used to determine the body areas that have irregular blood flow.

Chapter 5
Human Health Diagnosis System Based on Iris Features

Poonguzhali N
Pondicherry Engineering College, India

R. Hariharan
Pondicherry Engineering College, India

M. Ezhilarasan
Pondicherry Engineering College, India

N. Praveen Devaraajan
Pondicherry Engineering College, India

ABSTRACT

Iris feature has been used in authentication systems in many real time applications and is proved to provide high accuracy. Apart from authentication iris features can also be used for detecting pathological changes in human body and diagnose human health. The present study analyses the relationship between human iris anatomy and their health, as it is proved that changes in human health condition reflects the iris. Basically, in authentication system iris texture features are used for identification, in the proposed work iris texture and geometric features can also be deployed in diagnosing human health. The texture features present in the human iris are extracted using the mathematical statistical measure which is used to specify the characteristics of the texture of an image using gray-level co-occurrence matrix. The iris and pupil are extracted and correlated to the compactness features of the circle. Based on the comparison the system enables in prediction of abnormalities in the iris texture and identifies the affected person.

DOI: 10.4018/978-1-5225-2423-6.ch005

INTRODUCTION

A biometric system inculcates automatic recognition of a person on the basis of some characteristic or unique feature possessed by the individual, on whom some identification processes are carried out (Wildes 1997). There are various types of biometric system, which are based on hand geometry, handwriting, facial features, fingerprints, voice, retina, iris (Sanderson & Erbetta 2000). Iris biometric is said to be the feature that is highly unique and accurate, so the chances of any two individuals having the similar characteristic will be minimal and it will never change during one's life time. The current state-of-art of iris technologies in biometric are used for recognition system. This study is an attempt on iris biometric trait for the prediction of diseases or pathological organs in human body. In early days, iridology was a branch of science for iris diagnosis of human anatomy and used to diagnose disease based on manual investigation of iris texture pattern. Hence the system is automated using image processing techniques. The iris biometrics was generally used for authentication purpose, but in this paper, the approach is towards using the iris biometrics as a means for iris diagnosis. The iris is an annular part between the pupil and sclera, which is externally visible in an eye as shown in Figure 1.

Iris consists of two parts, the inner part is the pupillary zone and the outer part is the ciliary zone, which are divided by a zigzag pattern region called collarette (Adler, 1965). These characteristic features make it very useful for identifying the individuals (Daugman, 1994), (Boles & Boashash 1998). Image processing tech-

Figure 1. Iris regional boundaries

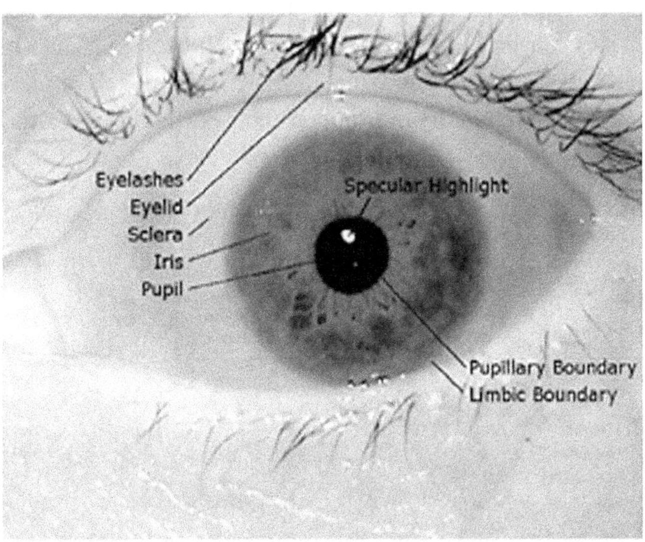

niques can be implemented to bring out the discreet iris pattern and to encode them into a biometric template. The biometric template is later stored in a database. This template of the biometric consists of quiet a few mathematical representation of the uniqueness of the iris, and allows comparisons to be done between templates. When an individual is to be identified by an iris biometric recognition system, the input human eye image is captured using sensors and a template is created for the iris region of the eye. This template is subjected to comparison with other templates, which are previously stored and either a matching is found, on identifying a person or a matching is note found. Iris recognitions is generally considered as the most unique, reliable and high accuracy form of biometric technology, compared to finger prints, voice, face and other form of recognitions. The iris biometric system is said to undergo various processing techniques of image processing such as segmentation, normalisation, encoding and matching.

In the embryonic stage of development, the first cells to form are the neurons (Wolff, 1976), (the basic cell of the brain and all the nervous system). The eye which starts as part of the brain, then get detached from the brain and is connected to the brain by the optic nerve, (bundle of millions of neurons), blood vessels and other tissues. Any changes that occur in the body will be reflected in the iris. The iris is made of muscles, nerve endings and other tissues. The nerve fibers in the iris as well as other tissues respond to changes in the body tissues by manifesting a reflex physiology that corresponds to the specific tissue changes and locations. This will manifest itself in changes in colour in certain parts of the iris. The weave of the muscular structure of the iris, can get retracted or atrophied (loss of thickness due to lack of blood & nerve supply) in certain areas, which will correspond or reflect the condition of certain organs in the body, as shown in Figure 2.

The proposed work is based on the texture pattern of the iris region in the eye, where the Bernard Jensen's iridology chart is being used to check for the possible pathological changes in the organ. The Bernard Jensen's chart (Jensen, 1980) is shown in Figure 3. Iridology is a branch of science, particularly health science which

Figure 2. Anatomy of iris: top view (left); cross-section view (right)

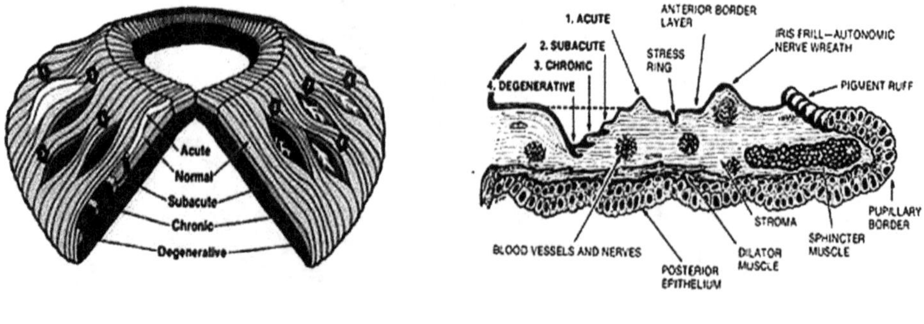

Figure 3. Iridology chart

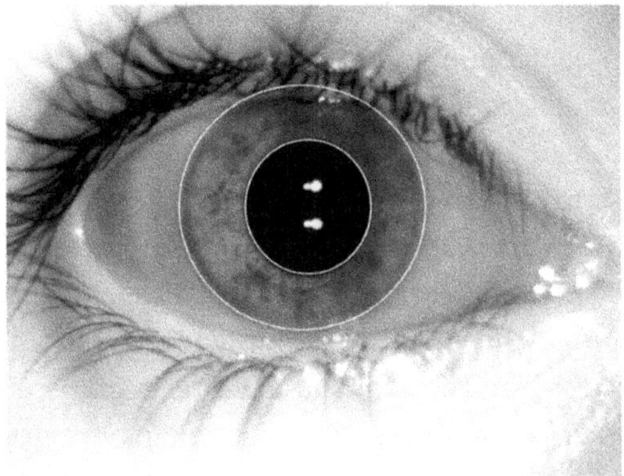

can describe about the disturbances in the organs inside a human body, it can identify which organ is affected and it can also reveal about the progress of such disturbances. In this science, it's about analysing the signs and fibre colours in the eye, which leads to the pathological changes in the organ. When the eye is observed under the magnifying instrument, one can clearly see that it is a splash of non-uniform colours, having various shapes and flaws, patterns, markings, etc., which gives a message about the organs and its functions. Similar to how computer stores information about various things, the iris also receives and stores the information about the whole human body. It is a receptor of all the signals from the body and imprints it in itself. The iris, which is also said to be a constituent of the nervous system, has thousands of nerve endings and impulses on it, which receive them from the rest of the body.

The paper is structured as follows. In Section 2, the related work is presented. The Section 3 reveals the image processing technique used for feature extraction. Section 4 deals with the inference obtained from the Bernard Jensen's Chart with regards to texture features. Section 5 depicts the experiment results on processing the real time database. Finally in Section 6, the conclusion for the proposed work in presented.

RELATED WORK

The patterns in iris are generally used for identification of individuals. There exist several methods for the analysis of iris patterns; the proposed system is concerned towards texture based analysis methods (Li Ma, Tieniu Tan, Yunhong Wang & Dexin Zhang, 2003). In history, the iridology was first formulated by Dr. Ignatz

Von Peczely. Iridology is a comfortable and convenient approach for patients towards medical diagnosis as it involves no pain and its precision is quiet high. In another work, the iris detection system has been built to facilitate the users, who are not familiar to the software. In this, (Lodin, 2009) has proposed to improvise the previously mentioned work to create a semi automated system. In (Ramlee & Ranjit, 2009), a system was developed to detect the cholesterol level present in the human body. Their work was based on recognize alimentary canal disease and nerve system. To detect chronic diseases such as bronchitis, erectile dysfunctions and brain dysfunction has been developed, in which the iris image in normalized using pseudo-polar coordinate system and the filter used here is Gaussian in vertical direction with which thinning algorithm is used to eliminate useless profiles by thinning the obtained binary image from the converted original image.

Diagnosis of diabetes mellitus has been proposed in (Wibawa & Purnomo, 2006), where procedures for finding the pancreas region in iris have been developed. The input iris image was transformed to gray level and filters were applied. Detection of iris was done using a novel case-based object recognition and case mining method. To obtain accurate iris image, a new method called water flow method have been incorporated in (Othman & Prabuwono, 2010). In this method new technique of segmentation in iris recognition was used to overcome the problem of noting down or considering the density of iris, its colour, the location of body organ in iris image and sign on iris image.

An eye detection system was proposed to provide the an reliable diagnosis clues about the patient based on the pupil. The algorithm detects the pupil and based on the change in the pupil size and pupillary response to light the system was deployed (Yan & Zhang, 2011). The gabor features extracted from iris code is used to diagnosis diabetic in an individual (Durga devi & Preethi, 2014) and was able to detect ophthalmic disease like corneal oedema, iridotomies and conjunctivitis. SVM technique had been proposed to detect diabetics in human from the iris image. The experimental results has been compared with the traditional iridology system and proved that existing iris recognition algorithm can be used to detect the disease (More & pergad 2015). An accurate diabetes detection method based on the broken tissues in the iris was primarily analysed. As a next step with the reference of iridology chart the change of color in the pancreases region has been detected and classification was performed using artificial neural network. (Jamal Firmat Banzi & Zhaojun Xue, 2015). A computer aided design is proposed to enable as an screening level for medical practinoer based on the ocular region of the human eye (Ramlee et al., 2015).

In (Haralick, 1973) has defined various textural features, based on his work. Specify an array of offsets to the gray co-matrix function to create multiple GLCMs, whose offsets can define pixel relationships of varying distance and direction. The statistics are manipulated from these GLCMs for the input images. A threshold value is fixed by taking the difference of the images in the database with input images.

TEXTURE FEATURE EXTRACTION

The proposed system is developed to ease the work of medical practitioners to automatically diagnose the patient, as medical images are quiet difficult to be analysed manually. The efforts of iris detection algorithms have been a great help to iridology practitioners to obtain accuracy in detecting changes in a person's iris. The modules in the proposed system are image acquisition, image pre-processing, feature extraction, feature encoding and disease prediction as shown in Figure 4.

The block diagram describes how the given input image is processed at each stage and how it is used for the next computational event. The first process is image acquisition where obtaining of eye images are discussed. The second part is the process of extraction of required region i.e. the iris from the whole obtained eye image. The third part is the conversion of the region of iris to required specifications for computational texture analysis. The final module is the process of analysing the texture features and the inference is obtained from the results of the analysis. In the data flow of this model, where the initial input is raw image and conversion of the raw image into various forms in the intermediate modules and the final arrival of inference of disease prediction as shown in Figure 5.

Image Acquisition

The image acquisition is the process to capture a pair of iris images from the subject using a specifically designed sensor. One of the major hurdles for practical applications is getting iris images, of extremely fair quality. As the iris is very small

Figure 4. Block diagram of the proposed work

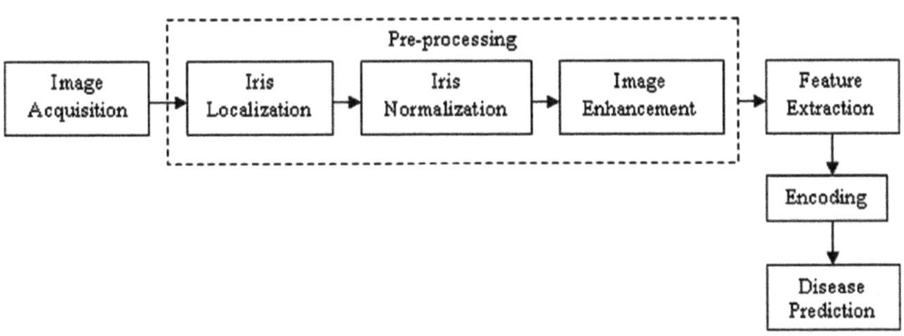

Figure 5. Architecture of the proposed system

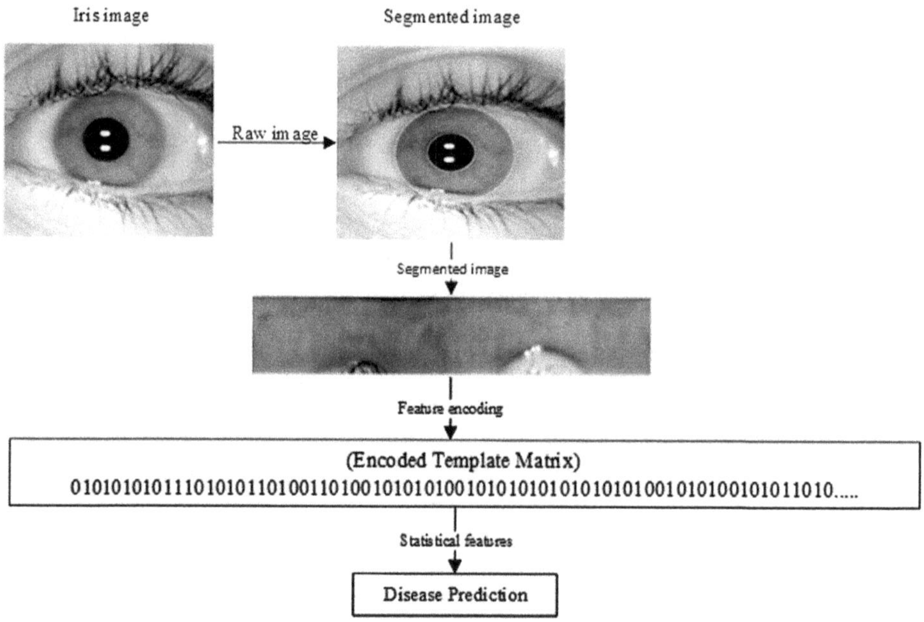

(diameter is about 1 cm), it shows numerous and variable texture features under near infrared lighting. The lighting, the positioning, and the physical capture of the system are the three main aspects one should consider when designing an image acquisition environment. Usually sequence of images is obtained rather than a single image, while capturing the eye images. Unfortunately, not all images captured are clear and sharp enough for further processing. The image may be out of focus or may contain many noticeable disturbing or overlapping lines, especially in regions close to the boundary, caused by eye motion, or severe occlusions by eyelids and eyelashes. Therefore, it is necessary to acquire a suitable image of high quality from an input sequence before it is used for processing in the forthcoming steps. Assessment of the image quality is a necessary factor as it affects the recognition accuracy of the obtained image. The subject has to be positioned in an appropriate place in the focal plane, if the subject moves during acquisition; a motion blurred image will result. When the subject opens his eye partially, the resulting image contains little useful information.

Image Pre-Processing

An iris image contains not only the region of interest (iris) but also some "unuseful" parts (e.g., eyelid, eyelashes and pupil). A change in the camera-to-eye distance may also result in variations in the size of the same iris. Furthermore, the brightness is not uniformly distributed because of non-uniform illumination. Therefore, before feature extraction, the original image needs to be pre-processed to localize iris, normalize iris, and reduce the influence of the factors mentioned above. Such pre-processing is detailed in the following subsections. The iris is an intermediate part between the 'sclera', which is said to be the outer boundary and the 'pupil', said to be the inner boundary. Both the outer and the inner boundary of an iris can be approximately taken as circle.

Iris Localization

The main aim of the process of localization is to segment the iris region from the rest of the image of eyes. There are different approaches in iris image pre-processing such as Daugman Integro-Differential Operator (Libor, 2003), (Widles & Asmuth, 1994) and Active Contours Model (Kass, Witkin & Terzopoulos, 1987). The localization process begins with the process of edge detection.

Canny Edge Detection

Edge detection of an image is defined as identifying points in a digital image at which the image brightness changes sharply or more formally, has discontinuities. The points at which image brightness changes sharply are typically organized into a set of curved line segments termed edges. These edges in this context refer to the edges of iris and pupil region. There are several methods in edge detections, one of the most suitable and efficient one, used in this work is the Canny's Edge Detection algorithm, has the following Steps:

1. **Smoothing:** Smooth the image with a Gaussian filter
2. **Finding Gradients:** Compute the gradient magnitude and orientation using finite-difference approximations for the partial derivatives.
3. **Non-Maximum Suppression:** It is to convert the blurred edges in the image of the gradient magnitudes to make sharp edges.
4. **Double Thresholding:** The edge-pixels remaining after the non-maximum suppression step are marked with their strength pixel-by-pixel.

5. **Edge Tracking by Hysteresis:** Strong edges are interpreted as "certain edges", and can directly be included in the final edge image. Weak edges are included if and only if they are connected to strong edges.

The maximum value of first derivative also corresponds to the minimum of the first derivative. In other words, both points with dramatic change of gray-scale (strong edge) and points with slight change of grayscale correspond to the second derivative zero-crossing point. Thus these two thresholds are used to detect strong edges and weak edges. Canny algorithm is not susceptible to noise interference enables its ability to detect true weak edges. It's an optimal edge detection algorithm. Figure 6 illustrates how Canny edge detection algorithm gives a binary threshold of image as output after edge detection.

Hough Transform

Parameters pertaining to simple geometric objects, present in an image such as circles, lines, etc., can be found out using a standard image processing algorithm called Hough Transform. The centre and radius of the pupil can be deduced from an iris region using circular Hough Transform. In doing this edge detection stage, Wildes et al. (1994) bias the obtained derivatives in the vertical orientation for finding the outer circular boundary of the iris and in the horizontal orientation for detecting the eyelids. The reason for this is that the eyelids are normally aligned in the horizontal orientation and also the eyelid map will affect the circular boundary in the iris edge map, if one uses all the gradient data.

Figure 6. Canny edge of iris image

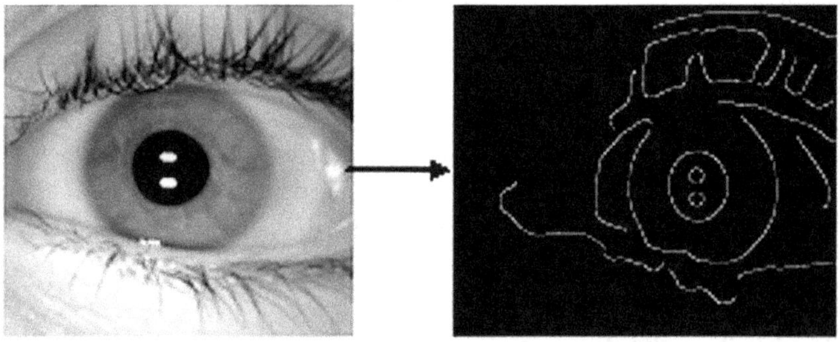

On considering only the vertical gradients for locating the iris boundary will minimise the influence of the eyelids when carrying out the circular Hough transform and not all the edge pixels are required. This makes it more efficient as well as accurate as only less edge points are used. The processes of Hough transform are illustrated, as shown below, in the Figure 7.

Iris Normalization

Various factors affect the size of the iris during acquisition. So compensating for iris deformation is a necessary factor for which Daugman found out a solution by projecting the iris image into a doubly dimensionless pseudo polar coordinate system other than in a Cartesian coordinate system. This normalises iris of different size, to be the same as it represents the parameters in a fixed interval. And in a similar manner the image can be unwrapped in counter clockwise direction, from a circular ring into rectangular block of fixed size m, which is denoted as,

$$I_n(X,Y) = I_o(x,y) \tag{1}$$

$$x = x_p(\theta) + \left(x_i(\theta) - x_p(\theta)\right)\frac{Y}{M} \tag{2}$$

$$y = y_p(\theta) + \left(y_i(\theta) - y_p(\theta)\right)\frac{Y}{M} \tag{3}$$

$$\theta = 2\pi\frac{X}{N} \tag{4}$$

Figure 7. Circular Hough transform

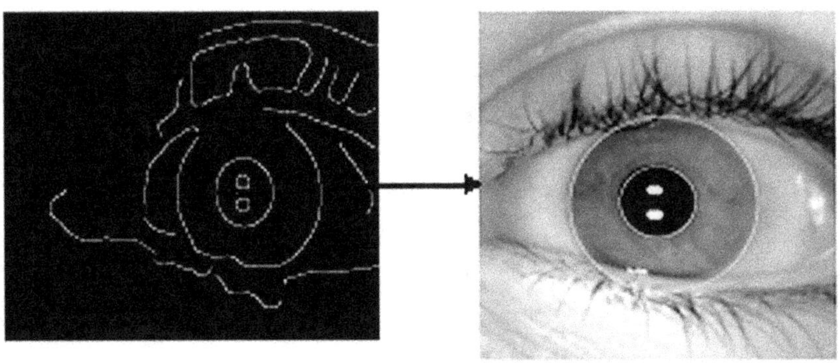

where I_n is an M x N (64 x 512 in our experiments) normalized image, and $\left(x_p(\theta), y_p(\theta)\right)$ are the coordinates of the inner and outer boundary points in the direction θ in the original image I_o.

This simplifies the subsequent processing of the iris image, which reduces the iris distortion to a great extent. In spatial domain, some particular filters such as Gabor filters are used to extract information of an image of certain geometric orientation and can be scaled.

The process of Daugman Rubber-sheet model is illustrated in Figure 8, where we transform the iris which appears like a circular ring, into rectangular block.

Image Enhancement

For texture analysis methods, Gabor filters are widely used and its elementary functions are oriented from complex sinusoidal functions of Gaussian modulations. New spatial filters can be defined according to the characteristics of iris texture. The difference between a defined filter and Gabor filter lies in the modulating sinusoidal function and their kernels as given below (even symmetric Gabor filters are used here),

$$G(x, y, f) = \frac{1}{2\pi\delta_x\delta_y}\exp\left[-\frac{1}{2}\left(\frac{x^2}{\delta_x^2}+\frac{y^2}{\delta_y^2}\right)\right]M_i(x, y, f); \text{ i=1, 2.} \quad (5)$$

$$M_1(x, y, f) = \cos\left[2\pi f\left(\sqrt{x^2+y^2}\right)\right], \quad (6)$$

Figure 8. Daugman rubber-sheet model

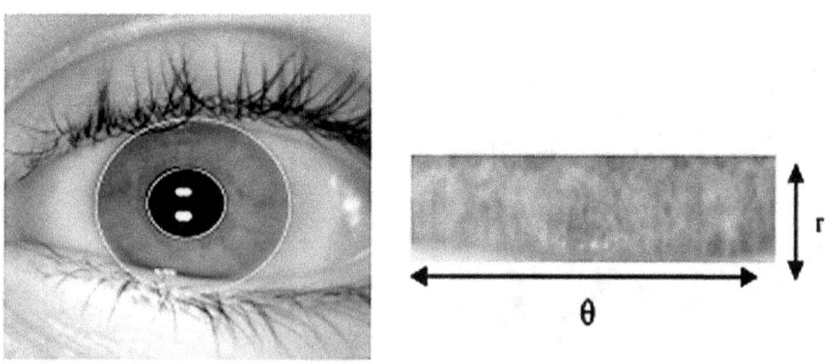

$$M_2\left(x, y, f\right) = \cos\left[2\pi f\left(x\cos\theta + y\sin\theta\right)\right]. \tag{7}$$

Due to positioning of light source, the normalized image has a low contrast and non-uniform brightness, and all these may have a negative impact on further processing in matching and feature extraction. It is first started by approximating the intensity of variations in the image so that a more well spread texture image can be obtained.

The mathematical mean of every 8 x 8 small blocks gives a rough estimate of the illumination in the background, which can be further expanded by bicubic interpolation of the normalized image. By subtracting the normalized image to compensate for various lighting conditions after which the enhancement could be done. After the above process the corrected image undergoes histogram equalization in every 16 x 16 regional blocks. This contributes for improving the contrast of the image.

Figure 9 shows the histogram equalization before enhancement and Figure 10 shows the histogram equalization after enhancement.

Figure 9. Iris enhanced rubber-sheet image and its histogram

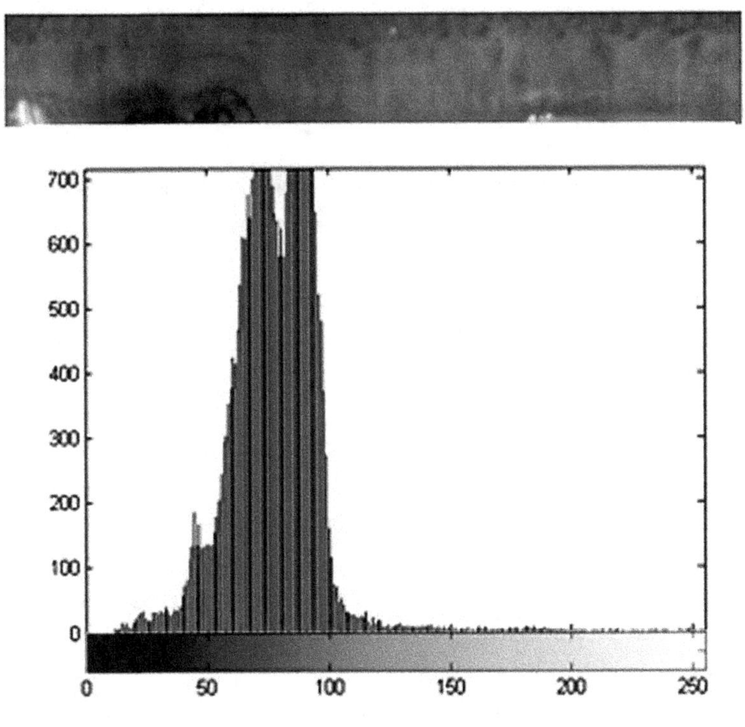

Figure 10. Image enhancement process

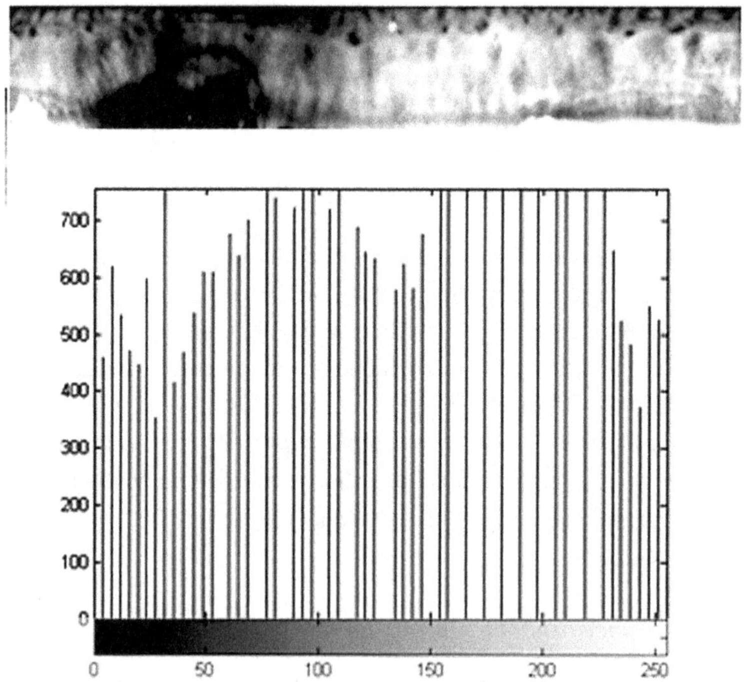

Iris Texture Feature Extraction

The iris has a particularly interesting structure and provides abundant texture information. So, it is desirable to explore representation methods which can capture local underlying information in an iris. From the viewpoint of texture analysis, local spatial patterns in an iris mainly involve frequency and orientation information. Generally, the iris details are spread along the radial direction in the original image corresponding to the vertical direction in the normalized image. As a result, the differences of orientation information among iris not seem to be significant. That is, frequency information should account for the major differences of iris from different people. We thus propose a scheme to capture such discriminating frequency information which reflects the local structure of the iris. In general, the majority of useful information of the iris is in a frequency band of about three octaves. Therefore, a bank of filters is constructed to reliably acquire such information in the spatial domain. As we know, coefficients of the filtered image effectively indicate the frequency distribution of an image.

This is implemented by refining the normalised iris pattern with 1D Log-Gabor wavelets. The 2D normalised patterns are spliced into a number of 1D signals and then these 1D are mixed with 1D Gabor wavelets. The rows in the 2D normalised pattern are considered as 1D signal. ach row corresponding to a circular ring on the iris region and instead of considering the radial direction the angular one is considered, which depicts to the columns of the normalised pattern because the maximum independence happens in the angular direction only. The average intensity of surrounding pixels is used instead of the intensity values at known noise areas in the normalised pattern in order to prevent the contribution of noise in the filtering output. After this, the output of filtering is phase quantised to four levels by the use of the filter, giving two bits of data for each phase. The output obtained here is a gray code, so that only one bit changes when going from one quadrant to another, which will in turn minimise the number of bits, not coming into the picture.

The process of encoding gives a bitwise template having information of number of bits, and an equivalent noise mask which implies to corrupt regions inside the iris pattern, and marks the bits in the template as corrupt. These regions where amplitude is zero are also marked in the noise mask, since the phase information will be meaningless. The total number of bits in the template will be the angular resolution times the radial resolution, times 2, times the number of filters used. Two statistic values are thus extracted from each small region in the filtered image to represent local texture information of the iris. A feature vector is an ordered collection of all features from the local regions.

Gray Level Co-Occurrence Matrix (GLCM)

In the proposed method the mathematical statistical measure, is used to specify the characteristics of the texture of an image using GLCM (gray-level co-occurrence matrix), also known as the gray-level spatial dependence matrix (Sarvesh, 2005). This method calculates the frequency of the occurrence of pairs of pixel, with specific values and in a specified spatial relationship and then by extracting the statistical measures from this matrix. Spatial relationship is known as the pixel of interest and the pixel to its horizontally adjacent (immediate) right, but one can specify the other relationships between the two pixels. Each and every element $\left(i, j\right)$ in the resultant GLCM is the ratio of sum of the number of times the pixel with 'i' value occurred in the spatial relationship to the pixel value 'j' in the input image.

At first, the dynamic scale range of the given image is brought down to reduce the number of intensity values. The size of the GLCM determines the number of gray levels. The certain properties about the spatial distribution of the gray levels in the texture image can be determined from the gray-level co-occurrence matrix. For instance, if many entries in the GLCM are intensified along the diagonal, then the texture is coarse in the specified offset. But, a single GLCM may not be enough to depict the textural features of the input image. For instance, gray co-matrix can create multiple GLCMs for a single input image, if a single horizontal offset might not be sensitive to texture with a vertical orientation. After the process of feature encoding, the process of computation and analysis of the obtained gray-scale co-occurrence matrix is done. As a process of encoding, using the Gabor filter, the image matrix has been converted as a grayscale co-occurrence matrix with which further computations are to be done.

In our approach the original masked image is resized to [256, 256] shown in Figure 11 below and then obtaining features by using GLCM.

The original encoded iris image is 64 x 512 pixels. In iris based disease prediction system the whole part of iris has its importance as each part points to a particular region of human body. In such a case, occlusions affect at a higher rate of efficiency in disease prediction. These images often occur in a captured sequence since the eye is in the state of continual motion and non-invasive image acquisition also requires users to adjust their position (hence, body motion). In the proposed study an analysis is made on the pupillary region of iris. This region of iris is re-

Figure 11. a) Normal iris image b) Iris image with filter applied c) Iris image after threshold

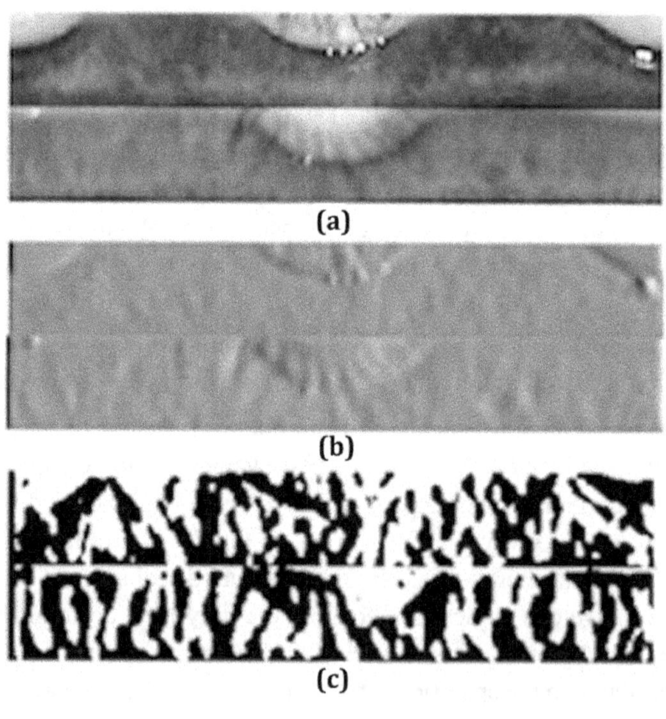

lated to gastro intestinal tract. Hence the first 8 rows of the grayscale co-occurrence matrix are taken as this particular part represents the pupillary region. The first process of computation is to extract the 8 x 512 matrix from the 64 x 512 co-occurrence matrix.

After the separation of this part, the texture feature analysis is made with the gray level co-occurrence matrix. The first order gray level histogram represents the probability of occurrence of gray levels in an image (Kramer & Aghdasi, 1999). A gray level co-occurrence matrix is considered to measure the second order histogram as it considers gray level distributions of pairs of pixels in each direction (Tuceryan and Jain, 1993).The gray level co-occurrence matrix is defined as $H\left(x,y,d\right)$, where H represents the matrix computing the frequency of occurrence of all pairs of gray level values $\left(i,j\right)$, with i and j in [0, 255], separated by a distance 'd' $\left(dx,dy\right)$ in direction 'θ'. That is, in a co-occurrence matrix the element $\left(i,j\right)$ represents the number of pixel pairs in the image separated by a distance 'd' in a direction such that the intensity of the first pixel is i and that of the second pixel is j. There is no criteria to select the distance 'd'. This results in a number of sparse co-occurrence matrices for different combinations of the distance value. These matrices do provide some amount of information on the spatial distribution of gray levels with respect to the displacement 'd' and direction 'θ'. For example if the diagonal of a co-occurrence matrix with displacement 'd' is concentrated then the texture is coarse in direction. But these matrices are not used directly to analyze texture. Many textural features can be computed from these matrices.

The statistical parameters of texture features that can be derived for GLCM are energy, entropy, contrast, homogeneity and correlation. These features can be used in isolation, or combinations classification.

In the proposed work, the following are statistical features used. The following are the formulas used to compute certain textural features from the co-occurrence matrix $M\left(i,j\right)$ where μ_x, μ_y, σ_x and σ_y are the means and standard deviations of $M\left(x\right)$ and $M\left(y\right)$ where,

$$M\left(x\right)=\sum_{i}\left(x,j\right) \tag{8}$$

$$M\left(y\right)=\sum_{i}\left(y,i\right) \tag{9}$$

1. **Energy:** Energy provides the sum of squared elements in the GLCM.

$$Energy = \sum_i \sum_j M^2 \left(i, j\right) \tag{10}$$

2. **Entropy:** Entropy is high when the elements of GLCM have relatively equal values. Low when the elements are closer to either 0 or 1.

$$Entropy = \sum_i \sum_j M \left(i, j\right) \log M \left(i, j\right) \tag{11}$$

3. **Contrast:** Contrast measures the local variations in the gray-level co-occurrence matrix.

$$Contrast = \sum_i \sum_j M \left(i, j\right)^2 M \left(i, j\right) \tag{12}$$

4. **Homogeneity:** Homogeneity measures the closeness of the distribution of elements in the GLCM to the GLCM diagonal.

$$Homogeneity = \sum_i \sum_j \frac{M \left(i, j\right)}{1 + \left|i - j\right|} \tag{13}$$

5. **Correlation:** Correlation metrics use the intensity-based methods to compare intensity patterns in images.

$$Correlation = \frac{\sum_i \sum_j \left(i - \mu_x\right)\left(j - \mu_y\right) M \left(i, j\right)}{\sigma_x \sigma_y} \tag{14}$$

INFERENCES OF TEXTURE FEATURE ANALYSIS BASED ON BERNARD JENSEN'S CHART

By analysing the texture features of the iris templates of the images, the first process of computation was made with the whole dataset available. There were several factors which were influencing the process of computation. As this process is a diagnosis based approach, real datasets are needed to check the efficiency of the system.

The first process was to obtain the data values of the texture features of the iris pairs, followed my special examination of the abnormal iris which were marked during the process of data collection. Before the application of texture features, the exhaustive study over Bernard Jensen's chart was done for computation of the texture features. The first eight rows of the iris encoded image denote the pupillary region of iris which has been inferred from proper examination of Bernard Jensen's Chart.

According to the chart the pupillary are is correlated with some of the gastro intestinal diseases. The following abnormalities are based on the chart:

- **Central Toxemia:** Orange/brown around the pupil. This condition prohibits efficient digestion and ability to absorb nutrients as toxic environment leads to lethargy. This is caused by junk food, medications especially antibiotics killing of friendly bacteria, negative emotions, anger, and depression. The liver, which filters the accumulated toxins, will be under stress. Flatulence and lethargy are the common symptoms.
- **Hyperacidity:** Silverfish/white ring immediately around the pupil that comes from consumption of acid-forming food/stress. Often leads to hypo acidity and the symptoms may include indigestion, heartburn and ulcers.
- **Hypo-Acidity:** Brown/black ring, in this case there is a difficulty in absorbing acid-forming food and a simple diet is required to put as little stresses possible on the digestive process in the stomach. The most common symptom includes indigestion.
- **Spastic Colon:** Happens when the colon has lost its tone and miss-shapen causing erratic bowel movements. It builds over the years caused by stress, diet of refined white flour and symptoms may include inability to maintain the peristaltic movement.
- **Tight Colon:** Tension in the walls, the colon squeezes and narrows the area through which the feces moves leading to constipation. Causes may include refined/spiced food. Constipation is the most common symptom.
- **Ballooned Colon:** Loss of tone and strength leading to sluggish bowel movement. The hormonal glands which lie directly outside the ANW (Autonomic Nervous Wryth) are also affected. Symptoms include constipation, flatulence and bloatedness.
- **Pancreas:** In case of under-functioning sugars and protein-rich foods will add strain.

Bowel pockets/Diverticulae may be caused by old fecal matter stuck in the inner wall of the intestines because of sluggish bowel movements. Signs include black/dark brown holes or pockets inside the ANW. Impactions in the ascending colon may create a reaction in the liver, breast or lungs. The most common symptom includes irritable bowels. The main digestive organs are Liver/Gall bladder/Pancreas. Toxicity is caused by high fat diet sugars, alcohol, drugs, medicines, anger frustration and other toxicity in the body leading to under-functioning liver (incapable to filtrate the toxins). Gall bladder is affected by high saturated fats, chocolates, cheeses, meat creams etc.

EXPERIMENTAL RESULTS AND ANALYSIS

The proposed work is experimented with real time database. In real time dataset, iris scanners were used where around eyes of 150 subjects of age group 19-25 years were scanned (300 iris). The subjects were engineering under graduate students. During image acquisition the student were orally enquired about their health condition and was noted. Iris of subjects who reported about their problems in health was set as the test cases and the rest of the data were the training set. The data base consists of 150 subjects of which 110 was test set and 40 was training set. The iris image is acquired using a Moblie-Eyes™ sensor which is a product of L-1 Identify Solution. The images are captured under near infra-red (NIR) illumination. It captures 752 x 480 wide SVGA images. The image capture rate is 30 frames per second. The device is a dual-iris enrolment i.e., both the left and right eyes are auto captured simultaneously. It operates in bright sunlight as well as in dark operating environment. The scanned image is stored in.PNG file format and the resolution of the image is set as 640 x 480. The sensors operate on Iris Analysis Software SDK which has image segmentation module, encoding module, image assessment scoring module and focus measure module. The scanners are kept at about 5 cm distance from the subject and were taken under normal room lightning. The subjects were instructed to wide open their eyes in order to avoid occlusions. This is also taken care by the image assessment scoring module present in the SDK which stores or captures image only if it is above certain assessment score. The experiment was done in two ways, first the individual iris (300 iris) were considered and next it was experimented on both the iris (left and right iris) for 150 image pairs.

Experiment 1: Analysis with Individual Iris Images

By analysis, the values from the abnormal iris were used to fix the threshold values based on the statistical values of mean, standard deviation, energy, entropy, contrast, correlation. Among these five values of the abnormal iris, the values deviate drastically than that of a normal iris. These values were analysed and then the obtained threshold limits were applied over the real time data set. The mean and standard deviation distribution of the each individual value of the features are shown in Figure 12 and Figure 13 respectively.

The energy and entropy distribution of the each individual value of the statistical features are shown in Figure 14 and Figure 15 respectively.

The contrast distribution of the each individual iris is shown in Figure 16. The sample data sets for all the statistical values of texture features in experiment 1 is shown in Table 1.

Figure 12. Distribution of mean values

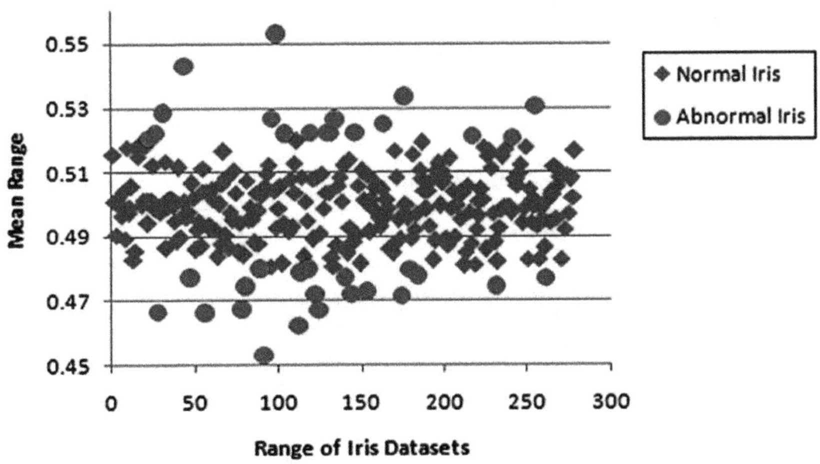

Figure 13. Distribution of standard deviation values

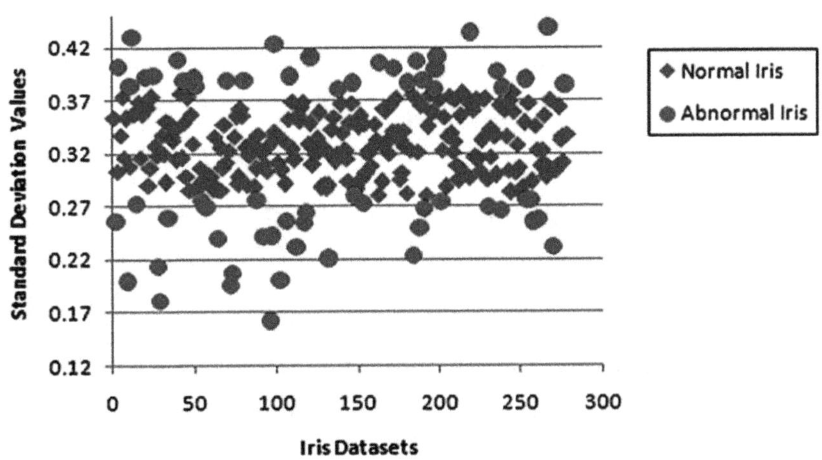

Figure 14. Distribution of energy values

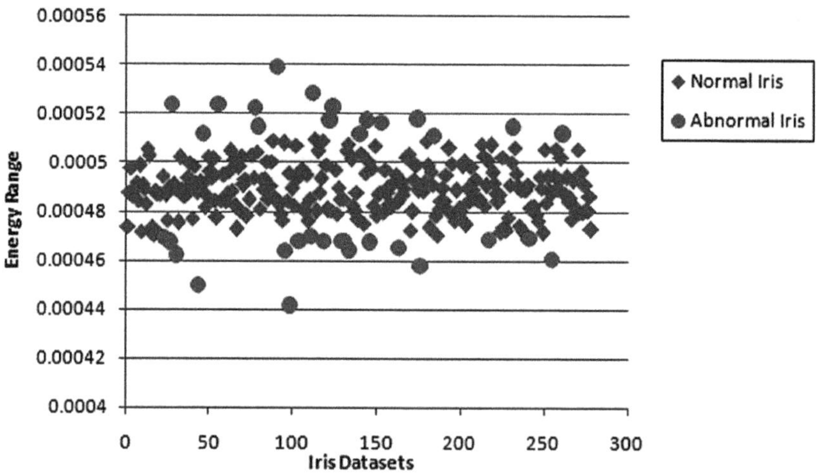

Figure 15. Distribution of entropy values

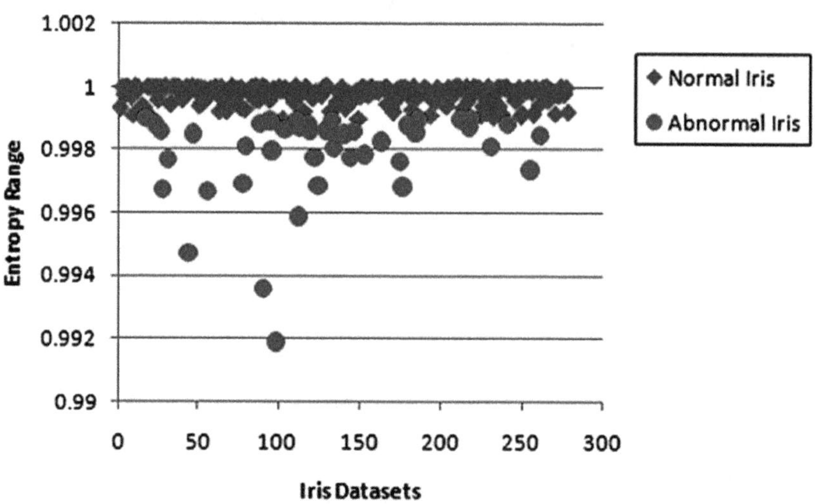

Experiment 2: Consistency between the Pair of Iris

Another property of iris is that the consistency in the texture pattern between the left and the right iris. The Bernard Jensen's Chart of left iris and the right iris is reflex to each other. But the pupilary region in both the charts represents the gastrointestinal organs. Hence an easy comparison can be made between a person's left and right iris. The consistency here refers to the consistent pattern between the pair of

Figure 16. Distribution of contrast values

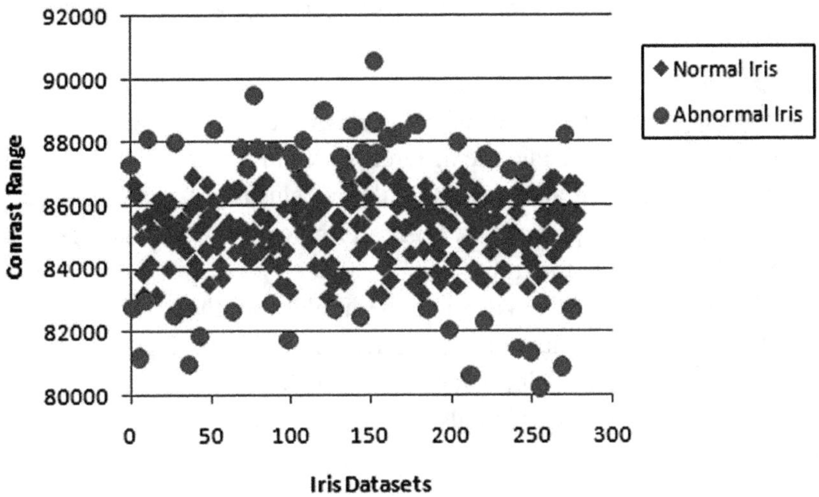

Table 1. Sample statistical values of texture features for Experiment 1

Sl. No.	Mean	Standard Deviation	Energy	Entropy	Contrast
1	0.499268	0.389489	0.000489	0.999998	86330.25
2	0.500977	0.275758	0.000487	0.999997	83716.92
3	0.530273	0.367724	0.00046	0.997354	80215.47
4	0.493164	0.275936	0.000495	0.999865	82851.42
5	0.482666	0.292676	0.000506	0.999133	85545.44
6	0.498779	0.25492	0.000489	0.999996	85689.54
7	0.495361	0.346297	0.000493	0.999938	84876.24
8	0.486328	0.323467	0.000502	0.999461	86489.3
9	0.476807	0.257302	0.000512	0.998447	85068.43
10	0.501465	0.307336	0.000487	0.999994	86850.27
11	0.494141	0.322882	0.000494	0.999901	84391.61
12	0.503906	0.355359	0.000484	0.999956	86838.16
13	0.50293	0.310352	0.000485	0.999975	85734.85
14	0.511719	0.297915	0.000477	0.999604	85879.35
15	0.495361	0.43832	0.000493	0.999938	83578.53

iris but not similarities in the iris. By comparing the values of the pair of iris, the values does not change much among the pair of iris i.e., even though they are not the same, their intensity and texture pattern are nearly constant to one another for a normal pair of iris.

By comparing the mean values of the left as well as the right iris, their difference in mean was computed and the difference was in a range of 0 to.0025, which is very meagre. This once again gives us the conclusion of a strong evidence of concurrency between the pair of iris. Figure 17 gives the graphical representation of the means difference in the iris.

By analysing the standard deviation in the similar way, the value range is around 0 to 0.0007 which is a very less change in deviation of the values between the iris pair, as shown in Figure 18. Whenever an abnormality is seen, there is a change in intensity at either of the iris which also results in the easy detection via the comparison among the pair of iris.

The values of energy also change in the similar way like mean and standard deviation with the range of 0 to 0.00003, as given in Figure 19. The only limitation which we can pose here is that this way of comparison of values between the pair of iris can only be made to Gastro-Intestinal organs where the chart is same representing the same region for both the iris. Some sample data sets for statistical values of texture features of left iris for experiment 2 is shown in Table 2.

The sample data set for statistical values of texture features of right iris for experiment 2 is shown in Table 3.

Figure 17. Distribution of difference of means between left and right iris

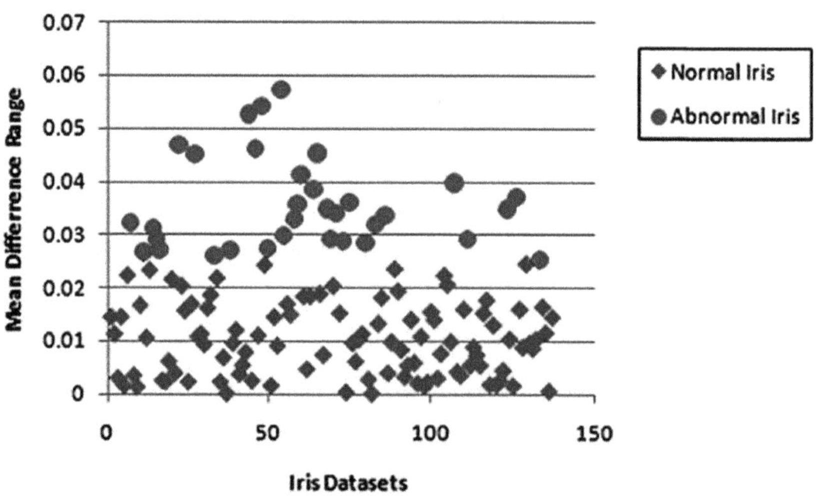

Figure 18. Distribution of difference in standard deviation of left and right iris

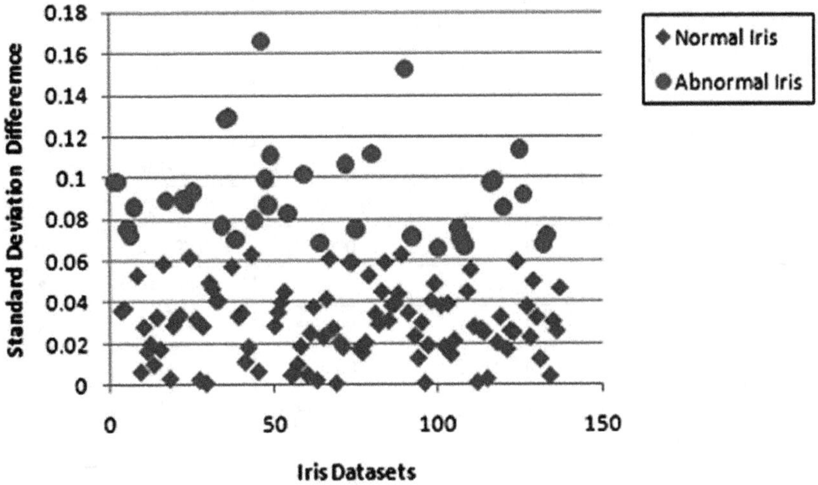

Figure 19. Distribution of difference in energy of left and right iris

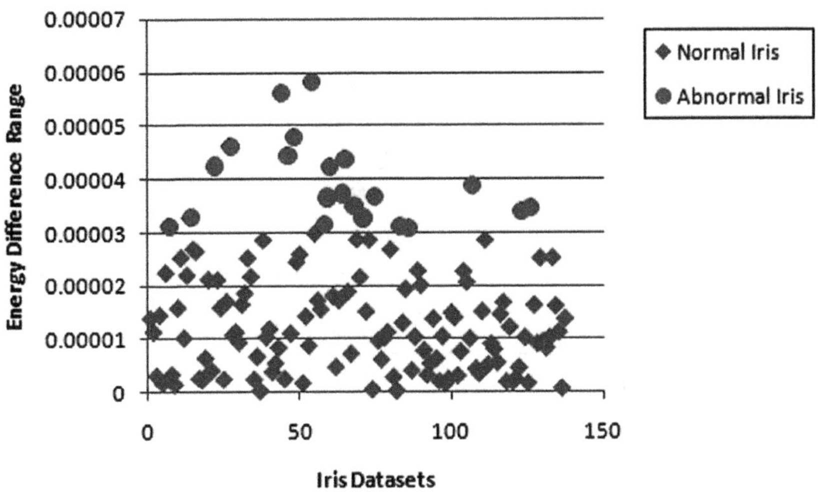

From the above analysis the changes in human health causes abnormality in the iris. A $E\ factor$ is initially analyzed for abnormal iris with which the normality range is being fixed. The $E\ factor$ is defined as:

$$E\ factor = \frac{m}{m_m} + \frac{sd}{sd_m} + \frac{enr}{enr_m} + \frac{ent}{ent_m} + \frac{cont}{cont_m} \quad (15)$$

Table 2. Sample statistical values of texture features of left iris

Left Iris					
Sl. No.	Mean	Standard Deviation	Energy	Entropy	Contrast
1	0.515381	0.354158	0.000474	0.999317	87277.42
2	0.490479	0.303509	0.000498	0.999738	86644.06
3	0.499512	0.33807	0.000489	0.999999	85508.62
4	0.50415	0.316832	0.000484	0.99995	85016.66
6	0.497314	0.308315	0.000491	0.999979	82998.73
7	0.505371	0.429532	0.000483	0.999917	88093.11
8	0.485596	0.358636	0.000503	0.999401	85203.66
9	0.515137	0.369298	0.000474	0.999339	83123.07
11	0.499756	0.357831	0.000489	1.000000	85108.66
12	0.518066	0.392354	0.000471	0.999058	85800.86
14	0.493896	0.290697	0.000494	0.999893	85124.67
15	0.501465	0.373285	0.000487	0.999994	84002.29

Table 3. Sample statistical values of texture features of right iris

Right Iris					
Sl. No.	Mean	Standard Deviation	Energy	Entropy	Contrast
1	0.500732	0.256266	0.000488	0.999989	86281.49
2	0.501953	0.40152	0.000486	0.999961	81157.85
3	0.496338	0.373777	0.000492	0.999682	83120.64
4	0.489502	0.353588	0.000499	0.999997	85655.74
6	0.499023	0.383762	0.000489	0.999157	84178.14
7	0.48291	0.357372	0.000506	0.999083	84934.19
8	0.517822	0.272521	0.000471	0.99898	85841.68
9	0.518799	0.316382	0.000471	0.999996	86190.77
11	0.501221	0.364176	0.000487	0.999996	85564.09
12	0.501221	0.364483	0.000487	0.998786	85974.02
14	0.520508	0.306824	0.000469	0.99957	86150.83
15	0.512207	0.392953	0.000477	0.998575	84884.54

The E *factor* is a value which gives a range, within which the normal and abnormal iris are classified for prediction and the value of E plays a predominant role in prediction of abnormal iris.

$$m_m = \sum_{i=1}^{n} m_i \tag{16}$$

$$sd_m = \sum_{i=1}^{n} sd_i \tag{17}$$

$$enr_m = \sum_{i=1}^{n} enr_i \tag{18}$$

$$ent_m = \sum_{i=1}^{n} ent_i \tag{19}$$

$$cont_m = \sum_{i=1}^{n} cont_i \tag{20}$$

Thus, the texture feature is a set of patterns rather than a set of values and these values were normalized into a single value of texture pattern, with which the disease prediction was done. In the data set about 40 were identified to have health issues related to Gastro-intestinal. In the 40 abnormal iris, 32 iris were abnormal when experimented with the proposed system. Hence, the proposed system gives a better result in predicting the human health with iris texture features.

CONCLUSION

The proposed work is designed as an assisting tool for medical practitioners. The system takes human iris as input. The image processing techniques such as iris localization, normalization, enhancement, feature extraction and encoding were applied for the input. Later the texture patterns nearer to the pupil region were extracted. The texture patterns in the iris region were experimented with real time database which reveals the human health condition. The statistical features for the texture pattern are computed and used for predicting the abnormalities. The results based on the prediction of the proposed system were also approved by medical practitioner.

REFERENCES

Adler, F. (1965), Physiology of the Eye: Clinical Application (4th ed.). London: The C.V. Mosby Company.

Banzi & Xue. (2015). An Automated Tool for Non-contact, Real Time Early Detection of Diabetes by Computer Vision. *International Journal of Machine Learning and Computing*, *5*(3), 225–229. doi:10.7763/IJMLC.2015.V5.511

Boles, W., & Boashash, B. (1998). A Human Identification Technique Using Images of the Iris and Wavelet Transform. *IEEE Transactions on Signal Processing*, *46*(4), 1185–1188. doi:10.1109/78.668573

Daugman, J. (1994). *Biometric Personal Identification System Based on Iris Analysis.* United States Patent no. 5291560.

Devi & Preethi. (2014). Disease Identification in Iris Using Gabor Filter. *International Journal of Engineering And Computer Science*, *3*(4), 5396–5399.

Haralick, , & Shanmugam, & Dinstein. (1973). Textural features for image classification. *IEEE Transactions on Systems, Man and Cybernetics, 3* (6), 610-621. doi:10.1109/TSMC.1973.4309314

Jensen, B. (1980). *Iridology Simplified* (2nd ed.). Escondido.

Kass, M., Witkin, A., & Terzopoulos, D. (1987). Snakes: Active Contour Models. *International Journal of Computer Vision*, 321–331.

Kramer, D., & Aghdasi, F. (1999). Texture analysis techniques for the classification of micro calcifications in digitised mammograms. *IEEE AFRICON*, *1*, 395–400.

Lodin & Demea. (2009). Design of an iris-based medical diagnosis system. *International Symposium on Signals, Circuits and Systems*, 1(4), 9-10.

Ma, L., Tan, T., Wang, Y., & Zhang, D. (2003). Personal identification based on iris texture analysis. *IEEE Transactions on Pattern Analysis and Machine Intelligence*, *25*(12), 1519–1533. doi:10.1109/TPAMI.2003.1251145

Masek, L. (2003). *Recognition of Human Iris Patterns for Biometric Identification.* The University of Western Australia.

More & Pergad. (2015). On A Methodology For Detecting Diabetic Presence From Iris Image Analysis. *International Journal of Engineering Sciences & Research Technology*, *4*(7), 2277–9655.

Othman, Z., & Prabuwono, A. S. (2010). Preliminary study on iris recognition system: Tissues of body organs in iridology. IEEE Conference on Biomedical Engineering and Sciences, 115-119.

Ramlee, R. A., Ramli, A. R., & Sulaiman, H. A. (2015). A Review on Diseases Manifestation by Ocular Diseases using Computer Aided Diagnosis (CAD). *IACSIT International Journal of Engineering and Technology*, *7*(4), 1343–1348.

Ramlee, R. A., & Ranjit, S. (2009). Using Iris Recognition Algorithm, Detecting Cholesterol Presence. *International Conference on Information Management and Engineering*, 714(717), 3-5. doi:10.1109/ICIME.2009.61

Sanderson, S., & Erbetta, J. (2000). Authentication for secure environments based on iris scanning technology. *IEEE Colloquium on Visual Biometrics*, 1-7. doi:10.1049/ic:20000468

Sarvesh, M. (2005). *Analyses and synthesis of iris images* (M.S. thesis). West Virginia University, Morgantown, WV.

Wibawa & Purnomo. (2006). Early Detection on the Condition of Pancreas Organ as the Cause of Diabetes Mellitus by Real Time Iris Image Processing. *IEEE Asia Pacific Conference on Circuits and Systems*, 1008(1010), 4-7.

Wildes, R. P. (1997). Iris Recognition: An Emerging Biometric Technology. *Proceedings of the IEEE*, *85*(9), 1348–1363. doi:10.1109/5.628669

Wildes, R. P., & Asmuth, J. C. (1994). A System for Automated Iris Recognition. *Second IEEE Workshop on Applications of Computer Vision*, 121-128.

Wolff, E. (1976). *Anatomy of the Eye and Orbit* (7th ed.). H. K. Lewis & Co. LTD.

Yan, H., & Zhang, Y. (2011). Medical Eye Image Detection System. *ICAIC, Part V, CCIS*, *228*, 619–627.

Section 3
Application of Thermal Imaging in Industries

Temperature measurement is an essential requirement for a large number of smart applications in industries. Due to the advancement in hardware technology, infrared thermal cameras could capture images with high spatial resolution. Similarly advancements in image processing techniques allow fast and accurate processing of thermal images. By combining advancements in both thermal imaging techniques and image processing techniques, many real world problems in military and industrial domains could be addressed with less effort. Keeping this in view, in this section an insight into thermal imaging in smart application is presented. In addition, usage of thermal images in security application and its enhancement methods are presented in detail. Moreover, optical analysis of solar concentrator with an image process is presented, to improve thermal efficiency of the solar concentrator.

Chapter 6
Thermal Imaging in Smart Applications

Aqeel ur Rehman
Hamdard University, Pakistan

Iqbal Uddin Khan
Hamdard University, Pakistan

Tariq Javid
Hamdard University, Pakistan

Ahmar Murtaza
Hamdard University, Pakistan

ABSTRACT

Temperature measurement is an essential requirement for a large number of smart applications in medicine, agriculture, environment, and security domains, to name but a few. Conventionally, temperature measurements are mostly performed using thermometers, thermocouples, thermistors, and resistance temperature detectors. Most of these instruments require physical contact with the object to measure temperature at specific points. Infrared thermography has revolutionized the concept of temperature measurement. Infrared thermal imaging (IRTI) can provide the temperature mapping without a physical contact with the object of interest from a reasonable distance. A typical IRTI system comprises of a thermal camera equipped with infrared detector, a signal processing unit and an image acquisition system, usually in the form of an embedded system. Such cameras are utilized for applications like fault detection, irrigation management, motion detection, etc. This chapter briefly introduces use of thermal imaging in medicine, agriculture, environment, smart home/cities and security applications.

DOI: 10.4018/978-1-5225-2423-6.ch006

BACKGROUND

Infrared Technology

By means of our eyes, we perceive the ecosphere in visible light. However visible light seals simply a minor fragment of the radiation spectrum, the invisible light shelters most of the remaining spectral kind. The radiation of invisible light conveys greatly further material.

Just for a swift review from history, probing for different photosensitive substances, in 1800 William Herschel accidentally discovered radiation of a different wavelength which is now known as infrared radiation. He initially darkened the top of a very sensitive mercury based thermometer. In William Herschel's thermometer, a glass prism was added to direct the sun rays on his working table where he completed his preparation for measurement. With this setup, he long-established the temperature ranges of different colors of the visible electromagnetic spectrum. Gently stirring the mercury bulb of the darkened thermometer over the various colors of the spectrum, he surprisingly observed that the temperature is rising from violet towards red. He observed that the temperature kept rising more rapidly in the area after the red band of the spectrum. After multiple attempts, he concluded that the maximum temperature is after the red band of visible spectrum. These days maximum temperature area after red band is named as "Infrared Area" (GmbH).

The Electromagnetic Radiation Band

A band in the technical term is the concentration of a combination of electromagnetic waves as the function of the wavelength or band. The electromagnetic radiation band encompasses wavelength's area of about 10^{23} and differs from region to region in origin in establishment and area of usage for that specific radiated wavelength. The complete categories of electromagnetic radiation tail the same physics' laws

Figure 1. A glance of electromagnetic spectrum (GmbH)

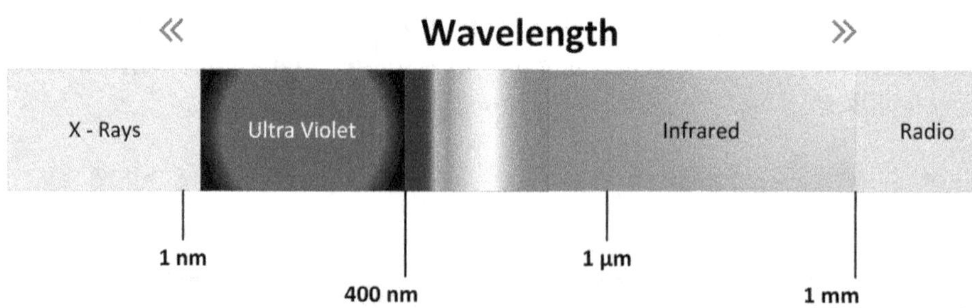

of scattering; change of course, reflection and polarity. Their spreading out fleetness resembles as the speed of light under ordinary circumstances. The product of multiplying wavelength (**λ**) with the frequency (**f**) is continuous (GmbH; Hunt):

$$\lambda \cdot f = c$$

The infrared radiation provides shelters a narrow but workable fragment within the range of electromagnetic continuum: The workable range starts at the 0.78 μm and finishes around 1 mm of wavelength. The wavelengths starting from 0.7 μm and up to 14 μm are significant for infrared based temperature measuring process. Above the mentioned wavelengths the energy level within the band is so stumpy, that the detectors which are currently available are not considerably sensitive to identify their presence (Hunt).

History of Infrared Technology

Afterward observing the pattern, Herschel prepares himself to measure the temperature fair further than the red slice of the spectrum in an area where no sunlight was noticeable. To his revelation, he establishes that this region had the utmost temperature of entire spectrum.

In 1880, the American astronomer Samuel Langley used a bolometer to sense body heat from a cow at approximate height of 100 feet. A bolometer measures the variation in electrical resistance associated to the variation in temperature, rather than determining voltage alteration. Sir John Herschel son of Sir William Herschel, used a device named an Evaporograph, produced the first infrared image in 1840. The thermal image caused from the differential evaporation of a thin film of oil and was observed by light reflecting off the oil film (Corporation, 2009).

Initial varieties of thermal imagers were known as photo-conducting sensors. From 1916 to 1918, the American discoverer Theodore Case investigated with photo-conducting sensors to produce a signal over uninterrupted interface with photons rather than via heating. The result was a more rapid, more sensitive photo-conducting sensor. Through the 1940s and 1950s, thermal imaging knowhow was expanded to meet a growing number of military solicitations. German scientists exposed that by cooling the photo-conducting sensor, whole performance is improved (Corporation, 2009).

Not before 1960s, thermal imaging was not utilized for military uses. Even if early thermal imaging systems were somewhat ungainly, gentle to acquire data, and had very low or plain resolution, they were mostly used for industrialized uses such as the examination of outsized electrical transmission and distribution systems. Persistently advances in the 1970s for military uses produced the first transportable systems that could be used for such applications as building diagnostics and the non-destructive analysis of materials.

In the 1970s, thermal imaging systems were much mature, long-lasting and dependable however the features and resolution of the obtained images was very low as compared to modern thermal imaging equipment. By way to the 1980s, thermal imaging was being broadly used for medical dedications, in typical industry, and for building examinations. Thermal imaging systems were regulated to deliver fully radiometric images, so that radiometric temperatures could be measured at any point of captured thermal image (Hellier, 2003). A radiometric image is a thermal image that contains temperature measurement calculations for various points within the image.

As the technology advances and fabrication of sensitive sensors emerges, consistent thermal imager coolers were developed to substitute the compressed or liquefied gas that was used to maintain temperature of thermal imagers. Comparatively less expensive, tube-based, Pyro Electric Vidicon (PEV) thermal imaging systems were also industrialized and broadly manufactured. Even though not radiometric, PEV thermal imaging systems were lightweight, portable, and operable without cooling.

At the end of 80s era, an innovative device now a days it is known as Focal Plane Array (FPA) was out by the military research center into the open market for sale and local use (Corporation, 2009). The Focal Plane Array (FPA) is device used for image sensing. This device comprises of a grid array (typically in rectangular shape) of infrared sensors at the focal plane of a lens.

This was an essential development over the generic scanned detectors, as the result there was enhancement of quality and its spatial resolution of resulting image (Rogalski, 2003). Arrays on recently developed thermal imager, distinctively have pixels that varies from 16 × 16 up to 640 × 480 pixels. For remembering, a pixel, is said to be the tiniest independent element of an FPA that is used in array to detect infrared's availability and strength. For any dedicated purposes, customized arrays can be used with pixels of 1000 × 1000 resolution. The first quantity denotes the number of vertical count (columns) of that specific array and whereas the second quantity represents the number of horizontal count (rows) of array as well as the image to be displayed on the screen. For example, an array of 640 × 480 is equal to 307,200 total pixels (640 vertical elements × 480 horizontal elements = 307,200 pixels).

The expansion and advancement of FPA technology has paced up the usage of thermal sensors and imagers since the year 2000. The thermal imager which is able to sense infrared radiation of wavelength which is between 8 μm and 15 μm is said to be Long Range Thermal Imager. For remembering, a micrometer (μm) is a unit

Figure 2. Structure of FPA
K, Dat, & K, 2013; Liu et al., 2013.

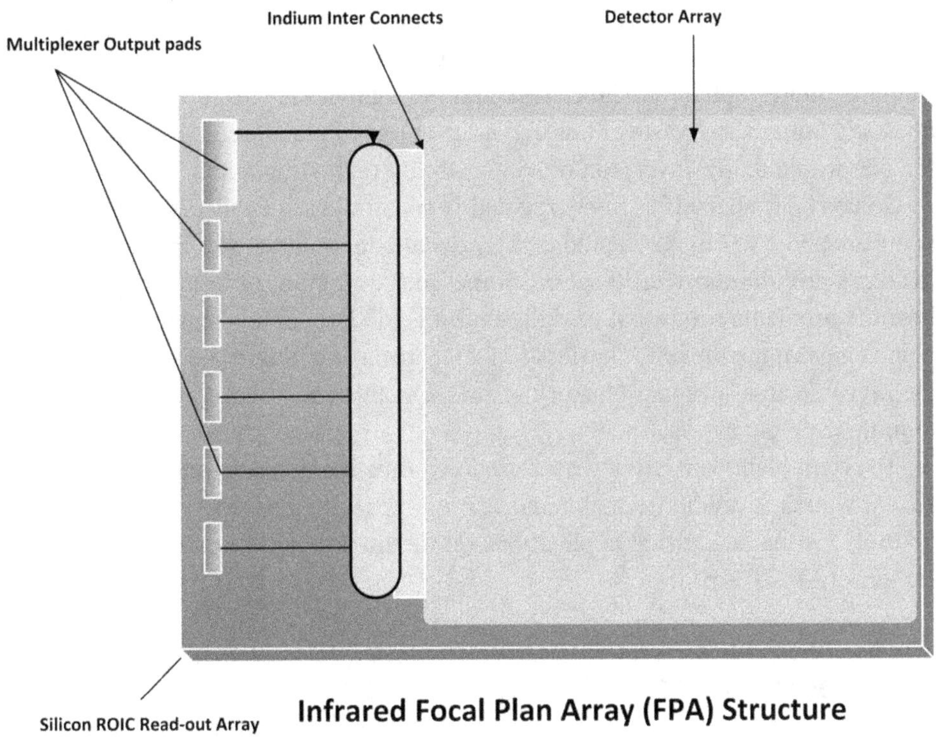

of measurement which is equal to thousandth part of a millimeter (0.001 m) which is itself the thousandth part of a meter. Moving further, a thermal imager that is able to detect infrared radiation of wavelength which is between 2.5 μm and 6 μm is said to be Mid Wave Thermal Imager. Equally, long and mid wave thermal imager systems are nowadays accessible in entirely radiometric varieties, frequently using image fusion plus thermal sensitivities of 0.05°C or less *(K et al., 2013;Liu et al., 2013).*

The budget of these systems has released from higher to lower price by more than a factor of ten over the past few years and quality has affectedly increased. Additionally, the use of computer aided tools and engineering for image processing has grown enormously. Approximately, all commercially offered recent infrared systems employ software to smoothen the analysis and report compilation. Reports can be quickly generated and can be sent over the Internet or can be well-kept-up in a common format, such as a PDF, and recorded on various types of digital storage devices (K et al., 2013; Rogalski, 2003).

Why Use Thermal Imaging?

Manufacturing more rapid, improved, more professionally and at a lesser cost. To accomplish aforesaid goals, manufacturing plants of each and every field need to be running uninterrupted round the clock and throughout the year. So there should be no costly failures, no waste of time, so as in charge of plant predictive maintenance the one would really have a ton of responsibility on his shoulders.

So consider yourself as in-charge and If you could see explicitly when components are about to fail, you could exactly decide the accurate slot in available time to take appropriate corrective action. Sometimes we can say unfortunately, the most horrible problems persist out of sight until it is too late. In this regard, thermal imaging cameras are the only flawless tool for forecasting failures because they make the invisible areas of plant / factory visible. On a thermal image, problems appear to jump right out at you.

To retain plants operating with zero down time many manufacturing industries have joined their predictive maintenance programs with the utmost prized diagnostic tools for manufacturing applications on the market place i.e. thermal imaging cameras (AB, 2011b).

INTRODUCTION TO THERMAL IMAGING

Thermal imaging is not anything very common till recent years due to the high cost of thermal imaging sensors and devices. This is now going to change rapidly as new detectors, sensors, new substitutes, constituents, fabrication materials and techniques and other mass production improvements in sensor's manufacturing are motivating the measurements and falling of per unit cost making them more adaptable. Thermal cameras are now the part of multiple domains of business such as the avionics industry, shipping industry, security / surveillance, firefighting and law's administration.

Any thermal imaging or thermography camera converge the electromagnetic radiations which are molded into an image same as other cameras however, an orthodox cameras works within the boundary of visible light spectrum, the range of visible wavelength is between 400 and 700 nanometers (0.4 – 0.7 μm) approximately. The design of a thermal imaging camera is to sense radiations having superior wavelengths as compared to visible band, up to 14,000 nanometers (14 μm) approximately. Radiations of this slot within the electromagnetic spectrum are stated as infrared, or simply IR, which further can be distributed into a number of sub categories according to the variations of wavelength with in IR band.

Wavelength of about 0.7–1.5 μm is known as near-infrared light, which is fair further than human's visible wavelength. Cameras or sensors are developed and manufactured to make a useful use of this sub infrared band. A commonly known and used cameras (for both Day and Night usage) use an IR cut filter throughout daylight to filter out IR radiations, so it will not create distortion in the colors of captured images and keep it closer to what shades and colors are observed by the human eye. While, when the same camera is working in the night mode, the IR cut filter is not involved. As it is understood that the human eyes are blind to see infrared light, the working camera shows the image in monochrome mode (Black and White). Nearer to read infrared light band correspondingly requires some sort of light source – whichever may be moonlight, or a man made, such as street lights or a dedicated IR lamp (Communications, 2009).

After visible light spectrum, the Infrared region started and as we move on words in infra-red region towards radio waves the Infrared region is further categorized in the following categories, wavelengths stated below are in micrometer (μm):

- Short Wave Infrared (SWIR), approx. 1 – 3.
- Mid Wave Infrared (MWIR), approx. 3 – 5.
- Long Wave Infrared (LWIR), approx. 8 – 12.
- Very Long Wave Infrared (VLWIR), approx. 12 – 25.
- Far Wave Infrared (FWIR), approx. 25 μm – 1 000 (1 mm).

It is easily noticed that the range between 5 μm to 8 μm is not named in categories above. This range is not adoptable by any thermal imaging solutions, due to high spectral absorption of spectrum of said range. (Lawrence, Diego, & Eveland, 2005).

Microwaves have wavelength more than 1mm and after that wavelength of radio waves starts. Radio waves have wavelength of 1 meter or grater. While on the other side of microwave, there are waves of shorter wavelength as compare to visible ones named as ultraviolet, x-rays, and gamma rays (Communications, 2009; Lawrence et al., 2005).

What Makes It Work?

What a human eye receives in form of image, it is able to be defined as the light reflected back by different entities. Absence of reflection means no light in the surroundings consequently the human eye become sightless under said type of circumstances. In contrast, thermal images are not reliant on any visible light source and its related spectrum. As an alternative, required images are obtained as if we work in the infrared spectrum. It delivers efficient result flawlessly well no matter how much darkness is there as the experiments have proved that ambient light's presence is effect less in the result of IR camera.

The fact which makes it promising is that no matter it is an organic or inorganic compound / substance, it radiates and infrared radiation is present due to their body temperature levels. It is also in notice that every living organism have temperature above than 0° K which is equal to -273° C or -459° F, moreover we can mention coldest objects, like an ice or a steel post on roadside in winter seasons, also emits fair amount of thermal radiations (Roy Alves, 2010).

The emissivity is the capability to radiate out captivated energy. All types of materials have emissivity (e) depending on their nature and chemical composition. Their values range from 0 to 1 depending on their properties. Other values later than 1 applies only on an object which is theoretically said to be a Black body. For understanding it can be said that the grayish and blackish materials could have emissivity up to 1. In other words, materials or body which is more reflective has lesser value than black or grey. As an example, consider an exceptionally polished silver and brass have emissivity of around 0.02 and 0.03 accordingly. Same as iron having emissivity around 0.14 – 0.035 only if it is well polished, but approximately 0.61 if it is rusted and shaded as red. Similarly any conventional glass product, having ability to block thermal radiation effectively, holds an emissivity likely up to 0.92 (Instrument; Mikrons Instruments Company. Inc).

The thermal radiation of any object is not only dependent on it surface's smoothness but also on its temperature, warmer the object is the more thermal radiation it will emits. As we know that it is not visible to us by a naked eye, but we can easily sense it, let's have an example, as we approach toward a camp fire the greater the temperature of specific section, the richer thermal images we can capture.

Moreover, as compared to long wave spectrum area, in mid wave infrared area very few materials will unveil the thing in emissivity. As we consider imaging purposes these variations in material's emissivity are for the utmost chunk and as for lesser meaning sensitivity of a camera can be defined as its aptitude towards differentiating of shades with respect to the temperature variations.

Although thermal images are infrequently linked with high toned and strong colors that might look unusual, bearing in mind that "Is there a camera that might work outside the visible spectrum?" To reply, the answer to question is "The reproduction of colors is done digitally, known as pseudo colors". Every shade or tone characterizes as a different value of temperature, typically red and white for upper temperature ranges as compared to green, violet and blue for cooler tones. As we all knows that the human eye is superior at differentiating multiple shades of a single color than dissimilar tones of grey.

Sensors which are specially designed for thermal imaging purpose knows how to be approximately distributed into different kinds: The Cooled thermal imagers are characteristically designed to function within the boundary of mid wave infrared (MWIR) band whereas the uncooled thermal imagers are designed to function in the long wave infrared (LWIR) range (Schioppa, 2014).

Cooled Detectors

Cooled infrared detectors are designed to perform in an enclosure which is vacuum sealed and temperatures are maintained as from 60 to 100° K, subjected to the level of performance and area of usage. Such exceedingly down temperatures are maintained and achieved with the help of coolants like cryogenic. Thermally prompted noise are balanced and moderated via cooling so it is necessary, as high temperature increases the sensors became vulnerable to the hazard of being flooded or "blinded" by their unnecessary thermal radiation of a particular wavelength. Such type of sensing equipment is not only bulky but compatibly expensive due to additional hardware, they are also considered more energy consumer up to some extent as compared to uncooled detectors.

Even though this technology is not only costly but a challenging job to keep it maintained, but consider that it has paybacks. This type of detectors work within the mid wave of spectral band, which makes availability of the enhanced spatial resolution as the wavelengths are far shorter that make them to offer more thermal contrast as compared to the long-wave band (as higher the frequency larger the data rate in radio frequency). Henceforth, cooled detectors designed to sense the smaller temperature variations and in result yield crisp as well as high resolution images.

Additional benefit is that better sensitivity also permits the use of higher Focal-number lenses. Therefore, cooled detectors are a better option for long range detection, i.e. 5 km – 16 km (3 – 10 mi) (Schioppa, 2014; Sharma, 2004).

Cool but Uncooled

In an uncooled thermal camera, sensors used are not dependent on any cryogenic type of cooling mechanism. These types of thermal sensors are cooled by maintaining their temperature approximately equal to the ambient temperature, which requires less complex temperature controlling systems. Sensors develop to run and sense long wave infrared band are designed on the principle discussed above.

Mostly bolometer technology bases designs are available. Characteristically it is a tiny resistor having high temperature-reliant properties on its silicon section, which is further thermally shielded and read electronically. Temperature variations of sight will develop fluctuations in the bolometer which are further processed to convert in electrical signals and attached processor then processes those electrical signals into an image. The sensitivity level of infrared radiation of any camera is said to be the ability to differentiate among very small dissimilar temperature alterations in any particular scene which is known as Noise Equivalent Temperature Difference (NETD). Now the advanced versions of these types of bolometers have achieved their NETD value as low as 20 mK. (Lim, Manzur, & Kar, 2014; Pillans, 2013; Schioppa, 2014).

Uncooled thermal image sensors are lesser in size and constructed with less moving parts. Not only this do make them lesser in cost than their cooled corresponding item but also permits for much longer service intermissions. Whereas the cooled infrared or thermal cameras typically need a rebuild for the cryocooler every 8K to 10K duty hours while an uncooled camera can remain in working mode continuously for years (Pillans, 2013).

Raining or Sun Shine

Thermal cameras are designed so that they do not only perform flawless in entire darkness, but also, they have matchless performance under challenging atmospheric situations such as hazy, dusty, rainy, snowy and smoky. But still, there are few physical limitations for the thermal cameras to deliver performance.

Humidity, dewdrops or minor dust elements are always present in the air. They will certainly obstruct the transference of thermal emission from a particular entity, creating it tougher for camera to detect from long distance. Accordingly, snowy, hazy, and rain fall will affect the performance of thermal imager. Water has ability to confine the emitted thermal emission whereas presence of moisture in the air flat out temperature variations among diverse items in the observable span. In the contrast, any thermal camera will yield improved and richer images throughout the winter seasons by means of clear skies and blameless climatic situations than under varying situation of atmospheric situations during the summer season when the humidity retains its higher ratio in climate.

Nonetheless, although these margins considered, when detection's result is concerned, thermal imaging cameras are much superior to the orthodox cameras underneath a widespread variety of harsh conditions of weather (Pillans, 2013; Roy Alves, 2010).

New Scope for Thermal Cameras

Targeting to improve a ubiquitously existing thermal imager, at this point researchers has established an innovative perception of less cost thermal imager that could be embedded within a palm top device like cellular phone, which is now known as MTIS (Mobile Infrared Image System) (Lee, Chen, & Liu, 2015). As now most of the hand-held devices are equipped with the availability of wireless connectivity and easy approachability to any location and scene, the MTIS may be known as an innovative instrument for thermal image usage in near future.

Proceeding forward in the said domain, a measurement console is developed by researchers that deliver the vital utility for thermal imaging as well as for temperature measurement. Within the console, a several number of thermal sensing modules (Thermal Infrared Module, TIM) are matched and synced in a hub to map dissimilar positions. The entire design wireless, the network is created on a Wi-Fi router, where the cellular phone acts as a console in the control unit. Whereas on the mobile phone, there will be a special application to accomplish the remote monitoring and image display. As soon as any of the available cellular phone clients imitate and establish connection with the console, the Infrared Imaging Application software (IRAPP) will displays the assigned IP address of the responded TIM to point out the monitoring position. Afterwards, the TIM obtains the emission of thermal radiations from the measured aim and then reports the thermal data back to the client. For information, software application (IRAPP) uses the User Datagram Protocol (UDP) for data receiving from TIM to keep less over head of handshaking and acknowledgement, and uses data do the relevant calculations, and as a final point shows the thermal image.

An Excellent Complement

As we are moving forward in technology, thermal imaging continuously becoming comparatively economical and a vital component of Internet Protocol oriented surveillance schemes, a complete range of usages turn into both conceivable and carefully feasible. Thermal cameras are the only match in various state of affairs wherever orthodox cameras are or maybe inadequate or deficient (Lee et al., 2015).

They are, obviously, incomparable in circumstances of entire darkness. They could be a choice in zones that are very challenging to brighten up effectively, as an example a harbor (sea boarder), or whichever additional vast expanse of open sea. In the same way, non-natural (other than sunlight) lights are the carriers of risk factor, as it might be illuminating the location of positioned cameras, permitting parties to duck or destroy them, on the other hand they can also generate or project unnecessary shades or multiple shadows in which a trespasser can dodge the detection and surveillance mechanism used in any specific area. So, the usage of cameras which do not depend on any visible light source can be the number one solution in several dissimilar traffic circumstances, no matter whether it is deployed in a railway tunnel or on the airports' air strips or on any regular street. To conclude, the thermal imaging cameras cannot be blinded by any source of bright light or even laser beam.

On the whole, thermal imaging network connectivity cameras flawlessly balances and finishes a network based video surveillance system, constructing well defined objects, people, and instances those are observed 24/7 and 365 days continuously (AB, 2011a; Eastman, 2006).

A REFERENCE SYSTEM ARCHITECTURE

The thermal imaging cameras are complex set of multiple systems. Those sub-systems impose their features to a captured frame or image so that output can appear in appropriate form for further processing as per requirement of deployment. It is not simple as it appears in application domain. When we are dealing with a special or selected wavelength, we require multilayers of filters and stages of features' additions and improvement factors. After these processes, we can have an appropriate frame or image so that it can be utilized by higher end systems and processors to deliver required results (Rogalski, & Chrzanowski, 2014). Moreover, these features require more hardware acceleration and control as well as they also requires stable power and heat control as in thermal imaging we are dealing with the wavelengths having higher energy levels as compared to visible band. In Figure 3, a basic architecture is illustrated, mentioning the core blocks of any thermal imaging cameras.

The Figure 3 shows basic camera architecture with ten different hardware blocks which is commonly known as a camera head. The thermal camera head consists of an optics' section which is responsible for accumulating incoming beams, imaging, on demand or on requirement zooming, enabling focusing to a certain object, along with required optical filters and their assembles. All afore-discussed section is known as Interchangeable optics. In further discussion, each block will be elaborated within basic architecture of camera.

Internal Imaging Optics

Internal imaging optics' section is applied for amplification and enchantment of limited band of electromagnetic spectrum. It increases the intensity of the received wavelengths and it is also known as image intensifiers. These intensifiers are characteristically used where the imaging is required with low light levels along with time constrains, in other words, where exposure time of any object is very less in very less light, like insects and parasites in garden or farms mostly appear while sun is about to rise or almost to set. In the said scenarios, image intensifiers simply amplify the received photo-electrons before the detecting arrays and preventing captured frame or image from degrading due to the readout noise of sensing array. Moreover, image intensifier considerably improves the incoming image to be processed for

Figure 3. Basic thermal camera system architecture
K et al., 2013; Liu et al., 2013.

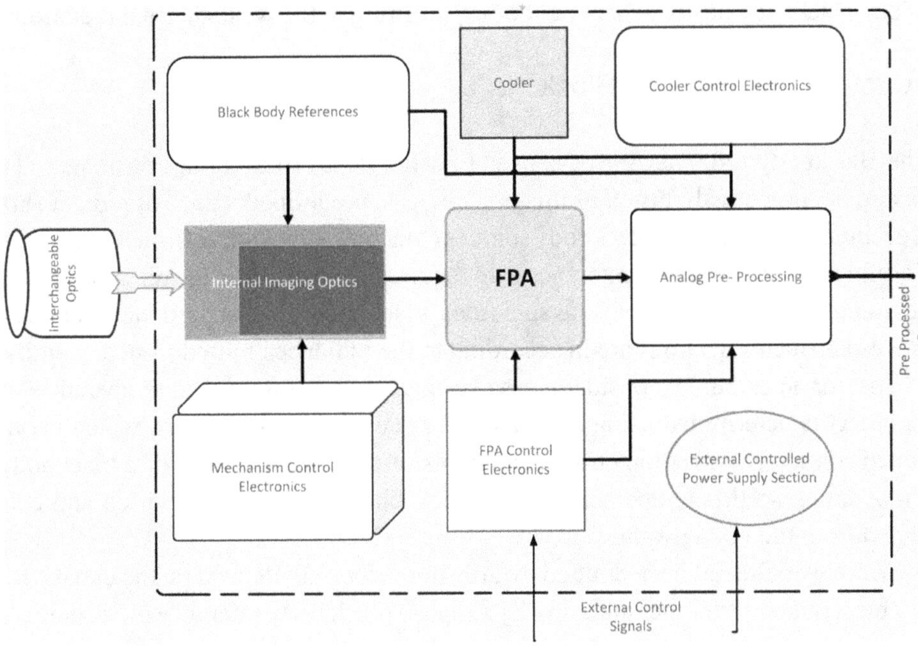

night visioning. The image intensifiers use a photo cathode within its assembly which enhances photosensitivity and give broader spectral response as compared to conventional systems and a human eye. There are three types of image intensifiers; one is electrostatically, second is magnetically and third is proximately focused.

Black Body Reference

What Is a Black Body?

Any surface mostly smooth, which approximately reflect all incident rays back, mostly are metallic surfaces are known as reflecting surfaces or reflecting objects. Adding more, when a non-smooth object or surface reflects all incidents rays in all direction equally is termed as white object or white body. In contrast with white or reflection surfaces or objects, there are surfaces having ability to absorb incident rays completely or having ability not to reflect back complete wavelength of incident ray. So a surface having ability to absorb complete or partial wavelength of incident ray is known as Black Body or Black object. In addition, G. Kirchhoff states that any object having ability to capture all incoming spectral band without reflection from its surface is considered as black body (Planck, 1959). Furthermore, for understanding any black body or object must absorb heat. Explicitly the absorbing property of any blackbody destroys heat radiations, resulting in the weak thermal radiations.

Function of Reference Block

The Blackbody reference block is used to calibrate thermal frames or images. To ensure the proper calibration of imaging devices, predefined values of radiant flux are required mostly from blackbody sources (Machin, Simpson, & Broussely, 2009). Careful and well observed references are required to ensure that the reference block has radiance parameters as close as some ideal blackbodies so that the images captured and established variations are in according to the radiance emitted from any entity.

Just for an example, consider a passive single band system. The temperature of an object is determined using a predefined system calibration chart, which is obtained from various radiometric calculations of the output signals of a blackbody temperature, so that temperature of each non blackbody can be verified and corrected from the known values.

Moreover, the radiance emitted by grey body does not depend on the emissivity of object but on its temperature. In this scenario black body reference block enforce correction over received frame.

Cooler

The use of semiconductors in thermal imaging field requires cooling to maintain their results. Temperature could vary from room temperature to 4°K according to the technology used. Most of the detectors having cooling operation can operate from 60°K to 100°K range (Loose, Beletic, Garnett, & Xu, 2007). Without proper cooling mechanism within it, the thermal cameras could be blinded or face radiation flood by its own radiation. Furthermore, cooling systems required more power to consume as well as consume time to process received wavelengths.

Cooler Control Electronics

As discussed earlier, cooling is the vital process to obtain reliable and accurate images in all circumstances. The cooler is not able to run unless it is guided and controlled by its electronic circuitry. The function of cooler control electronics circuitry is to maintain cooling process in varying condition of sensor, as sensor might get quickly warmer according to its deployed area's condition. Its circuitry also maintains the speed of cooling process to maintain acceptable amount of temperature of sensor.

It is a feedback control system which continually monitors the sensor's temperature and sets speed of cooling process according to the temperature.

Mechanism of Control Electronics

This block is responsible for focusing the image or frame. The image intensifiers are able to set focus as on requirement to maintain results (Ishimwe, Abutaleb, & Ahmed, 2014). Image intensifiers are of three types that are electrostatically focused, mechanically focused and proximately focused. As discussed earlier all three types of image intensifiers require complex circuitry for feedback type control to set focus on any incoming image.

Focal Plane Array (FPA)

Focal plane array (FPA) is a grid of pixels mostly in rectangular shape and the smallest element used to sense incoming spectrum and generate its results is known as pixel (Fundamentals of Remote Sensing, 2016). The FPA is measured by the number of pixels present with it according to rows and columns e.g. 320 x 240 or 2048 x 2048 FPAs.

FPA Control Electronics

This section controls the clock according to the system interface for visioning and generating biasing current upon requirement to set image intensity which will be read out by the analog preprocessing unit. Both of the mentioned functions are controlled by an embedded microcontroller so that it can hold the clock or biasing current till the requirement is changed from external interface units or preprocessing unit.

Analog Preprocessing

This unit is responsible for reading out data on each pixel and converting it to a composite output which can be further usable for other interfaced devices and processors. It is consisting of array controllers, memory blocks and Digital to analog converters.

External Controlled Power Supply

The externally controlled power supply system is externally monitored and controlled. It involves continues monitoring of applied and required power and makes available only required amount of power so that attached hardware can perform optimally and avoid unnecessary heating of camera head.

SMART THERMAL IMAGING APPLICATIONS

Thermal Imaging in Medicine

As per Ring (2004), "The association between disease and human body temperature is as old as medicine itself." John Herschel, the son of William Herschel known for the discovery of infrared radiation, created the first thermal image and called it a Thermogram. Early imaging systems using infrared were developed in 1940s and available for use in 1959. The first color thermographs were created in late 1960s whereas 1970s marked beginning of quantitative thermography.

According to Ring and Ammer (2012), "An infrared thermogram is an image of temperature distribution of the target." Focal plane array infrared cameras become available in the 1990s. They have provided greater spatial resolution and higher temporal resolution to measure the human body skin temperature. The International Standards Organization (ISO) has specified the requirements for qualification and assessment of personnel to use thermography for condition monitoring and diagnostics of machines in the part 7 of standard ISO 18436:2014. Additionally, ISO standard documents ISO 18434:2008, ISO 10878:2013, and ISO 18436:2014 describe general procedures, vocabulary, and condition monitoring and diagnostics of machines. Use of a rainbow color palette is preferred in medicine which has red color for hot and blue or black color for cold. Body positions and regions of interest in thermal images have significant influence on measurements accuracy and precision.

Applications of thermal imaging in medicine are listed below:

- Complex regional pain syndrome,
- Fever screening,
- Inflammatory arthritis,
- Malignant diseases,
- Osteoarthritis,
- Oral Health screening,
- Peripheral circulation,
- Soft tissue rheumatism:
    - Muscle spasm and injury,
    - Enthesiopathies,
    - Tennis elbow,
    - Fibromyalgia.

In addition to applications of thermal imaging in above areas of medicine, a number of other applications have also emerged. These include burns, renal dialysis, and physiological reactions.

Campbell and Thomas (2013) have described the use of energized instrumentation during surgery. Energy-based instrumentations are mostly based on ultrasound, laser, or radio frequency and can facilitate vessel sealing and dissection procedures. Most electrosurgical damages are not recognized during surgeries. Use of thermal imaging system makes it possible to minimize these damages and improves the surgical practice. A normal human body temperature is at 310 K having peak output around to 10 μm and in between 2 to 20 μm range most of thermal radiation lies. Most of the thermal imagers manufactured today by leading manufactures detectors based on indium-antimonide and gallium-arsenide materials to detect radiations in bands 3-5 μm and 8-9 μm, respectively. Both are appropriate for quantitative imaging of hotter surfaces occur by use of energized surgical instrumentation.

The success of medical infrared thermographic intervention combined with laser therapy is based on following factors:

- Appropriate conditions during the monitoring process,
- Evaluation of activity and development of standards and protocols,
- Problem or condition to be monitored,
- Setup and correct operation of infrared system.

Giordano and Abramson (2013) argue on the use of infrared based techniques or technology in medicine must be relevant and consistent with moral obligations and ethical integrity. The explicit knowledge of capacity and limitations will ensure appropriate use of this technology in clinical medicine.

The review by Matos, Neves, Norte, Rosa, Reis, & Vilaça-Alves (2015) investigates the application of thermal imaging to monitor skin temperature while cryotherapy is applied on clinical injuries and exercise recovery. Authors showed the use of thermal imaging is superior to other methods to monitor human body skin temperature.

When facial nerves damages due to any reason, causes weakness of muscles on any side of the face, which can be observed on face of any suffering individual as he might not be able to close eye completely of effected side or his mouth might see a bit dropped. These type of appearances damages the actual facial appearance and accent production while expressing any view or in discussion (Liu, Dong, & An, 2015). Thermal imaging can be used to analyze the severity of facial paralysis. By using thermal imaging, doctor can observe the temperature variation throughout the human face along with the comparison of temperature distribution characteristics of damaged region and non-damaged facial regions, which gives two data sets of each under observed facial structure. Using thermal imaging systems, the region of interest is identified and then these characteristics are extracted from facial asymmetry, like temperature difference, area ratio and temperature distribution and using neural network these features are utilized to construct a facial nerve function.

Thermal imaging has a strong role for analyzing facial expressions and emotions without having physical contact with the person under examination. Facial thermograms are used for parameter collection (Costea, Dumitrescu, & Mine, 2016). Thermal analysis is not dependent on amount of light, reflection or cannot be degraded via shadows, as they can be captured perfectly in a dark surrounding.

Thermal imaging systems can play a vital role in healthcare fields including medicine (Cao, & Cui 2016). The radiation wave can help to diagnose about physical condition of any human. Tumor can be detected using thermal imaging. A normal or healthy person would have symmetric distribution around region of interest whereas if there is tumor present the temperature distribution would be asymmetric.

Fever is a well-known symptom leading a person to suffer from influenza and other similar dieses. It can be also observed that when human body's immune system responds to dieses, aimed to destroy effecting microbiological organisms. Now some contactless fever detection methods are available using thermal imaging, in which foreheads are analyzed after face detection. Forehead is used to analyze as it is the common area to sense the fever.

Thermal vision can also be utilized for diagnosis of cancer and paralysis in human. Early cancer detection in oral or other body parts can also be diagnosed using this technology. Thermal imaging has potential to diagnose early stages of cancer without biopsy, as biopsy is found to be very hard for various patients and also give some false results.

Thermal Imaging in Smart Agriculture

The application of thermal imaging in agriculture is well established due to knowledge of plant shape, internal structure of leaves, and presence of water (Ishimwe, Abutaleb, & Ahmed, 2014). Potential use of thermography in agriculture includes irrigation scheduling, monitoring, stress detection, and yield estimation. Calibration of images in accordance to weather conditions is important for monitoring vegetation growth. Spectral reflectance characteristics of vegetation depend on leaf pigmentation, structure, thickness, and amount of water in the leaf. Green vegetation has low reflectance in visible regions of spectrum and reflectance increases in near infrared region. Yield estimation is a challenging task and much research is needed in this area.

The reflectance characteristics can be quantified by part of incident energy, E_r, reflected by earth surface feature (Lillesand, Kieffer, & Chipman, 2015). This measure is called spectral reflectance, denoted by ρ_λ, expressed as a percentage, and defined mathematically by

$$\rho_\lambda = \frac{E_R(\lambda)}{E_I(\lambda)} = \frac{\textit{Energy of wavelength } \lambda \textit{ reflected from the object}}{\textit{Energy of wavelength } \lambda \textit{ incident upon the object}} \times 100\%$$

where E_R denotes reflected energy.

A spectral reflectance curve is a graphical representation of the spectral reflectance of an object which is specified as a function of wavelength. Figure 4 shows generalized spectral reflectance curves in both infrared and visible regions. An overlap of spectral reflectance curves for both coniferous (needle-bearing) and deciduous trees are obvious in visible range. Distinct regions in near infrared range make identification of trees easy.

An example is given by Javid (2014) to demonstrate the use of infrared image with visible range by means of digital image processing to obtain enhanced vegetation index information. Spectral Python module (Spectral Python) is used on the image chip from Moderate-resolution Imaging Spectroradiometer (MODIS) containing 220 images in it. MODIS imaging instrument has a swath width of 2,330 km. It makes images of the entire Earth surface every one to two days and generates two vegetation index (VI) products for land regions. These are the standard Normalized Vegetation Index (NVDI) and the Enhanced Vegetation Index (EVI). The MODIS VI products have ground cell resolutions at 250 m, 500 m, and 1000 m; and 0.05-degree spatial resolution. The EVI is given by

$$EVI = Gx \frac{\rho_{NIR} - \rho_{red}}{\rho_{NIR} - C_1 x \rho_{red} - C_2 x \rho_{blue} + L}$$

where ρ_x are surface reflectance, L is the canopy background adjustment, C_1 and C_2 are the coefficients, and G is a gain factor; for EVI algorithm, $C_1 = 6$, $C_2 = 7.5$, $G = 2.5$, and $L = 1$.

The measure of temperature is an important physical parameter in the domain of plants (Gupta, 2014). Temperature measurements using thermography depend on leaf orientation, surface properties, IR absorption, and more. Use of thermal imaging to analyze early stages of plant development is a recent research area. Thermal imaging has shown useful to detect biophysical and biochemical changes during imbibition and germination. This information is useful to predict quiescent seed survival. The study reproduced results from (Kranner et al., 2010; Belin et al., 2011) with the need of physiological interpretation of observed thermal discontinuity.

Figure 4. Generalized spectral reflectance envelops for deciduous and coniferous trees
Lillesand, Kieffer, & Chipman, 2015.

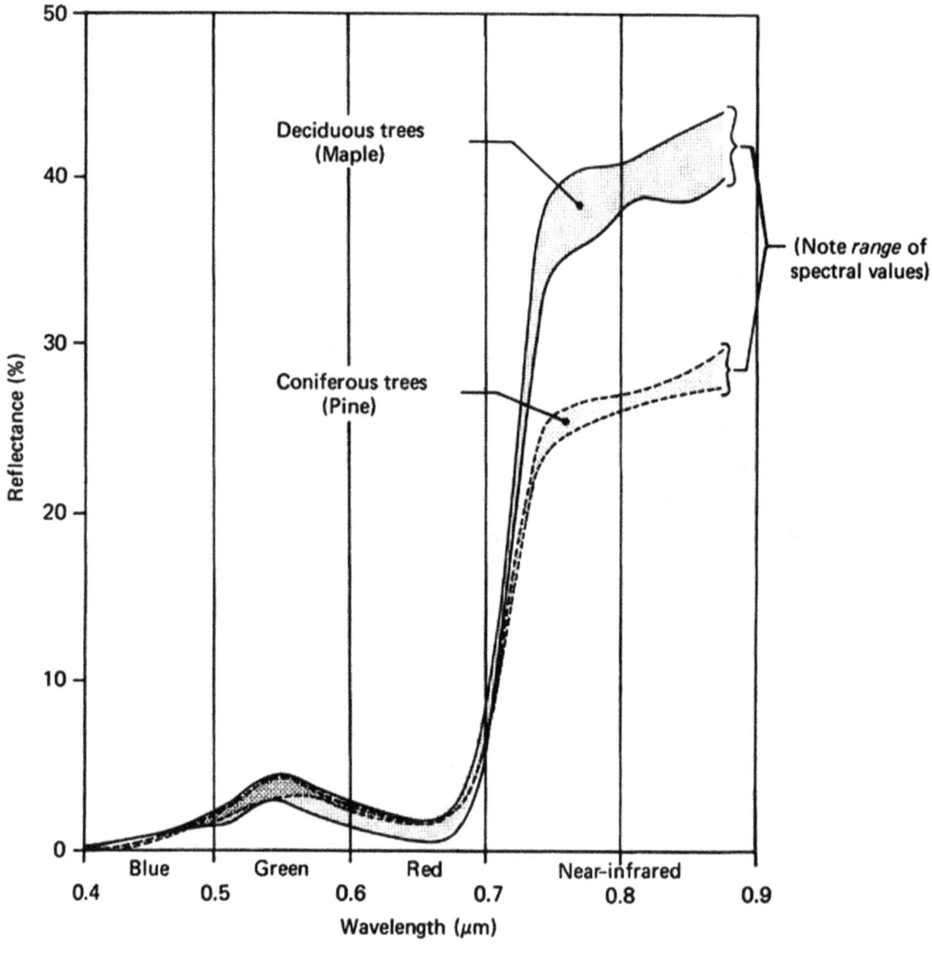

Powdery mildew is the most harmful disease causes decrease in wheat yield. As per the study by Awad, Abdullah, Bayoumi, Abd-Elsalam, & Hassanien (2014) thermal spectral sensing of normal and affected leaves is used to detect powdery mildew.

The measure of temperature is an important physical parameter in the domain of plants (Gupta, 2014). Temperature measurements using thermography depend on leaf orientation, surface properties, IR absorption, and more. Use of thermal imaging to analyze early stages of plant development is a recent research area. Thermal imaging has shown useful to detect biophysical and biochemical changes during imbibition and germination. This information is useful to predict quiescent seed survival. The study reproduced results from (Kranner et al., 2010; Belin et al., 2011) with the need of physiological interpretation of observed thermal discontinuity.

Thermal Imaging in Environment

Torgersen, Faux, Mcintosh, Poage, and Norton, D. (2001) has developed a remote sensing method to measure the stream temperature and has evaluated the factors which influence the accuracy. Thermal remote sensing is highly effective as compared to conventional methods which include use of in-stream data records. Steam temperature patterns are important due their influence on aquatic ecosystems. These measurements were taken with sensors in wavebands 3-5 and 8-14 µm. Factors influence temperature measurements include atmospheric absorption, emissivity, surface characteristics, and thermal infrared reflection. For flight planning and image data acquisition, meteorological data on air temperature, cloud cover, relative humidity, and surface winds were collected from multiple resources. Thermal and visible image pairs were registered within the Geographic Information System (GIS) and sample points were selected.

Water temperature measurement is important to know freshwater ecosystem status (Johnson, Wilby, & Toone, 2014). The Loughborough University TEmperature Network (LUTEN) has an array of thirty-six air and water temperature monitoring sites for prediction of water temperature for given catchment properties, river reach, and air temperature. To keep a record of water temperature measurements is also a mandatory requirement in many countries. Relationship between air temperature and water temperature is nonlinear and best described by use of a logistic function (Mohseni, Stefan, & Erickson, 1998). The LUTEN sites have spacing of 1.7 km on average and are monitoring both air temperature and water temperature.

Water temperature measurement is important to know freshwater ecosystem status (Johnson, Wilby, & Toone, 2014). The Loughborough University TEmperature Network (LUTEN) has an array of thirty-six air and water temperature monitoring sites for prediction of water temperature for given catchment properties, river reach, and air temperature. To keep a record of water temperature measurements is also a mandatory requirement in many countries. Relationship between air temperature and water temperature is nonlinear and best described by use of a logistic function (Mohseni, Stefan, & Erickson, 1998). The LUTEN sites have spacing of 1.7 km on average and are monitoring both air temperature and water temperature.

Thermal Imaging in Security

Willer, Saraji, Khorsandi, Geiser, and Schade (2006) reviewed infrared laser based non-intrusive detection of gases with emphasis on evanescent-field spectroscopy and direct absorption spectroscopy and the ability is demonstration to analyze volcanic gases. Further, article proposed detection and distinction between explosives by use of mid-infrared laser approach.

The research work by Liu, Kuyken, Roelkens, Baets, Osgood, and Green (2012) proposed spectral translation of mid-infrared signals to shorter wavelengths for detection by wide-bandgap photo-detectors that could eliminate cooling requirements. They used four-wave mixing in silicon nanophotonic wires to translate a signal at 2,440 nm to the telecom band at 1,620 nm. The silicon nanophotonic integrated circuit used in mid-infrared systems in spectral range 2-8 µm and has applications in industrial and environmental monitoring, medical diagnostics, and threat detection.

IR thermal imaging has a lot of applications in this field (Vollmer, & Llmann, 2010). It is used extensively to protect critical infrastructure, for example, airports, petrochemical installation, power grid, public transport, and warehouses. It is used to create thermal fences around a facility using IR cameras for perimeter monitoring application. Surveillance cameras market has a number of options to select. These include either cooled medium-wave detectors or uncooled microbolometer long-wave detectors; lenses, field of view, and image sizes. These cameras are usually equipped with a lens heater to avoid ice and water films on the lens. Factors influencing IR camera range are given below:

- Spectral range, type of detector (cooled or uncooled), and sensitivity,
- Lens and size of object,
- Temperature of the target and the background,
- Atmospheric transmission for ideal clear sky conditions,
- Humidity of the air,
- Aerosol contribution in the atmosphere: haze,
- Additional atmospheric properties: fog, rain.

Separating passing individuals from transit area of any bus or train station is also an important area for researchers. Speedily increasing number of individuals in every metropolitan area also increases security risk especially for important and critical areas other than public mass transit. The systems deployed in those specific areas tends to allow only a single individual to pass through them, but still the thread of piggybacking stands. By using the thermal imaging devices and equipment, results have improved as they are fair and accurate to detect and alert the concerned about trespasser (Siegmund, Handtke, & Kaehm, 2016). To identify any human in system, there are various techniques which are applied. Using pattern recognition technique, human head is identified and its dimension, shape and positions are analyzed by system. The thermal imaging systems are equipped with algorithms to have an adaptive adjustment to analyzed temperature range. The finalized image contains colors, which might vary from other systems as they depend on the use of that specific system's area and temperature of mantrap used. For reference, image areas having higher temperature are visualized by lighter colors like yellow or white and areas having lower temperature than mean are visualized via dark color like blue.

Use of video cameras is also increasing to monitor constantly for sensitive areas. Around the globe security surveillance and monitoring is done via closed-circuit television cameras to provide live video feed to a single or cluster of monitors. But the constraint in those monitors is the continuous monitoring by an individual person or a dedicated team to verify any intruding activity. Now thermal imagers are available to work in a distributed network for more intelligent remote monitoring and assisting teams to ensure protection of civilians and critical assets. The thermal video surveillance equipment is used in a predefined zone to monitor any intrusion in real time. In thermal video streaming, there are various approaches for intrusion detection (Li et al., 2015). Most of the systems use Improves Particle Filtering system and Pan – Tilt – Zoom (PTZ) thermal imagers simultaneously utilized over the Field of View (FOV). As the intruding object enters within any FOV, the thermal imager is activated instantaneously according to the spatial relationship developed by pose estimation.

Thermal Imaging in Smart Homes and Cities

As the technology is advancing, the concept of smart is setting to be the part of each and every domain of human life e.g. Smart cities and Smart homes. The smart city is a concept in which city will able to manage its resources according to the availability and efficiency e.g. the headend would have control over opening and exit of city resources, like water and waste management, metropolitan transit control and electrical power management. The concept of smart city and smart homes carry the factor of continuously evolving for which researchers are working to bring its deployment from academic domain. Smart city and homes would be able to accommodate upcoming technologies, upgrades its existing systems and operations. Furthermore, any smart city should have ability to get engaged with citizens and services provided to and for them in their social and private living to entertain their needs. It requires a very vast and emergent infrastructure along with all possible technologies, with the property of sustainability according to the economy. In the near future, metropolitan cities are required to emerge along with abilities to keep up mobility, environment sustainability as well as efficient infrastructure.

Thermal imaging can play a vital role in such Smart concepts to detect and analyze human activities from a distance means the method of noninvasive activity detection. The activity detection can be done within any smart city's platform along with advanced communication technologies as thermal cameras work fairly without depending on the light and can capture radiated infrared waves from any scene. Pixel based detection or object based detection methods can be utilized to count or track people and detect objects as people and objects having different temperature in day and night as compared to their surroundings.

A person tracking is an important application which can be done in ambient assisted environments by deployment of thermal imaging sensors (Mandel, & Autexier, 2016). For instance, a wheelchair can be sent to a goal to pick up a person if the person's initial location is known or it can assist any disable person on wheel chair to cross a road.

By using thermal imaging tools, the variation of temperature through and around any specific building or home can be visualized to find the structural damage or fault (Goodhew, Boomsma, Pahl, & Goodhew, 2015). Thermal imaging systems are capable to visualize the apparent surface temperature of any wall or ceiling of house and can also prevail the radiations from any of its surface, this might help to analyze any specific building structure or damage and related faults or defects so that the escape of energy can be protected and might help to enhance energy conservation.

Kitchen is the most important and active part of any home. Food cooking process is now been advanced and people are more health conscious that engages them in kitchen with different and interesting activities. As the interaction of humans with kitchen increases, it also increases the risk of hazards during cooking. The kitchen is equipped with stoves and ovens that make kitchen the second place where most accidents occur within any residential place (Yared, Abdulrazak, Tessier, & Mabilleau 2015). The accidents might happen due to careless using of equipment or over heating or cooking of ingredients, for example over cooking in a cooker might lead to cooker burst and will surely harm humans near it. By using thermal imaging above mentioned risks can be reduced. In smart home the kitchen can be put under surveillance by thermal imaging system, and temperature of cooking oil, oven or any pot with respect to the type of food is cooking, can be constantly observed. This might be able to avoid accidents by timely turning off or alerting the concerned persons. In addition, it can also observer the fire and can alert or manage unnecessary fire and smoke in kitchen.

Child monitoring and care have a vital role in growth of every child and its personality. It not only provides entertainment but also help to enhance its physical and psychological growth (Murphy et al., 2015). It is not possible for parents or guardians to sit beside any playing child all the time. Smart homes may provide assistance in such requirements by using thermal sensors/cameras by monitoring child's activity and alert if any fall or abnormal position of child is detected.

Table 1. Wavelength and spatial resolution comparison chart according to application

S. No.	Wavelength Range (µm)	Spatial Resolution	Applications				
			Medical	Agriculture	Environment	Security	Smart Home/ Cities
1	0.52 - 0.72	1 Km	×	✓	×	×	✓
2	0.75 - 1.4	5mm	✓	×	×	✓	✓
3	3.78 - 4.03	4 Km	×	✓	✓	×	✓
4	6.47 - 7.02	4 Km	×	✓	✓	×	✓
5	10.2 - 11.2	4 Km	×	✓	✓	✓	✓
6	11.5 - 12.5	4 Km	×	✓	✓	✓	✓
7	12.6 – 13.4	1 Km	×	×	✓	×	✓

CONCLUSION

Thermal imaging is a non-invasive technology providing many advantages and wide variety of applications in various domains. Under this chapter the background of thermal imaging, introduction of the technology, reference architecture and applications of thermal imaging in various important domains are presented. The wide range of applications of thermal imaging technology is from medicine to agriculture, environment to security and smart home to smart cities. As per the study of applications of thermal imaging in smart environments, it is observed that different wave lengths and spatial resolutions are required for different applications. The details are presented in Table 1.

Literature review corroborates that the thermal imaging technology with support of different other technologies will revolutionize the world.

REFERENCES

AB, F. S. (2011a). Thermal Imaging Guide Book for Building and Renewable Applications. FLIR System AB.

AB, F. S. (2011b). *Thermal Imaging Guide Book for Industrial Applications*. Retrieved from: http://www.flirmedia.com/MMC/THG/Brochures/T820264/T820264_EN.pdf

Cao, M., & Cui, W. (2016). Study on the Measurement Technique of the Alignment Variation of Medical Infrared Thermal Imaging System. *International Journal of Smart Home*, *10*(1), 81–88. doi:10.14257/ijsh.2016.10.1.09

Communications, A. (2009). *Some like it hot: Thermal cameras in surveillance* (White Paper). Retrieved from http://textlab.io/doc/538217/some-like-it-hot--thermal-cameras-in-surveillance

Corporation, F. (2009). *Introduction to Thermography Principles.* Author.

Costea, I., Dumitrescu, C., Minea, M., & Banica, A. (2016). Analysis of facial expressions using thermal imaging. In *Electronics Technology (ISSE), 2016 39th International Spring Seminar on* (pp. 306-309). doi:10.1109/ISSE.2016.7563210

Eastman, J. R. (2006). Introduction to Remote Sensing and Image Processing. In *IDRISI Guide to GIS and Image Processing* (Vol. 1, pp. 17 - 34). DRISI Production.

Fundamentals of Remote Sensing, Natural Resources Canada. (n.d.). Retrieved from http://www.nrcan.gc.ca/node/9309

Gmb, H. O. (n.d.). *Basic Principles of Non-Contact Temperature Measurement* (IR Basics Brochure-E2014-01-A). Retrieved from www.optris.de

Goodhew, J., Boomsma, C., Pahl, S., & Goodhew, S. (2015). It's fffffffrrreeezsing,.... thought thermal imaging might shed some light as to where the heats going!'Can visualisation shed a light on heating. *Proceedings from ECEEE.*

Hellier, C. J. (2003). *Thermal Infrared Testing Handbook of Nondestructive Evaluation.* The McGraw-Hill Companies, Inc.

Hunt, J. L. (n.d.). *Radiations in Environment.* Creative Commons. San Francisco. Instrument, M. Table of Total Emissivity.

Ishimwe, R., Abutaleb, K., & Ahmed, F. (2014). Applications of Thermal Imaging in Agriculture—A Review. *Advances in Remote Sensing.*, *3*(03), 128–140. doi:10.4236/ars.2014.33011

K, N., Dat, R., & K, A. (2013). Advances in Infrared Detector Array Technology. In P. S. Pyshkin (Ed.), *Optoelectronics - Advanced Materials and Devices*. In Tech.

Lee, F. F., Chen, F., & Liu, J. (2015). Infrared thermal imaging system on a mobile phone. *Sensors (Basel), 15*(5), 10166–10179. doi:10.3390/s150510166 PMID:25942639

Li, Z., Yang, J., Han, P., Ran, Y., & Chai, Z. (2015). Intrusion detection and cooperative tracking using PTZ network thermal imagers. *International Journal of Distributed Sensor Networks, 2015*, 1.

Lim, G., Manzur, T., & Kar, A. (2014). Noise sources and improved performance of a mid-wave infrared uncooled silicon carbide optical photodetector. *Applied Optics*, *53*(36), 8410–8423. doi:10.1364/AO.53.008410 PMID:25608189

Liu, R., Kong, Y., Jiao, B., Li, Z., Shang, H., Lu, D., & Zhang, Q. et al. (2013). A substrate-free optical readout focal plane array with a heat sink structure. *Journal of Semiconductors*, *34*(2), 024005. doi:10.1088/1674-4926/34/2/024005

Liu, X., Dong, S., An, M., Bai, L., & Luan, J. (2015). Quantitative assessment of facial paralysis using infrared thermal imaging. In *2015 8th International Conference on Biomedical Engineering and Informatics (BMEI)* (pp. 106-110). doi:10.1109/BMEI.2015.7401482

Loose, M., Beletic, J., Garnett, J., & Xu, M. (2007). High-performance focal plane arrays based on the HAWAII-2RG/4G and the SIDECAR ASIC. In Optical Engineering+ Applications (pp. 66900C-66900C). International Society for Optics and Photonics.

Machin, G., Simpson, R., & Broussely, M. (2009). Calibration and validation of thermal imagers. *Quantitative InfraRed Thermography Journal*, *6*(2), 133–147. doi:10.3166/qirt.6.133-147

Mandel, C., & Autexier, S. (2016, May). People Tracking in Ambient Assisted Living Environments Using Low-Cost Thermal Image Cameras. In *International Conference on Smart Homes and Health Telematics* (pp. 14-26). Springer International Publishing. doi:10.1007/978-3-319-39601-9_2

Murphy, F. E. (2015). i4Toys: Video technology in toys for improved access to play, entertainment, and education. In *2015 IEEE International Symposium on Technology and Society (ISTAS)* (pp. 1-6). doi:10.1109/ISTAS.2015.7439423

Pillans, L. A. (2013). *Performance Evaluation of an Uncooled Infrared Array Camera (Ph.D)*. London: London University.

Planck, M. (1959). *The theory of heat radiation: Authorized translation by Morton Masius*. New York: Dover Publications.

Rogalski, A. (2003). *Infrared detectors: status and trends* (00796727). Retrieved from http://www.sciencedirect.com/science/article/pii/S0079672702000241

Rogalski, A., & Chrzanowski, K. (2014). Infrared Devices and techniques (revision). *Metrology and Measurement Systems*, *21*(4), 565–618. doi:10.2478/mms-2014-0057

Roy Alves. (2010). *Thermal technology: a hot new trend in video surveillance.* Johannesburg: Axis Communications.

Schioppa, E. J. (2014). *The color of X-rays: Spectral X-ray computed tomography using energy sensitive pixel detectors.* Academic Press.

Sharma, U. C. (2004). *Infrared Detctors.* M. Tech Credit Seminar Report, Electronic systems Group, EE Dept, IIT.

Siegmund, D., Handtke, D., & Kaehm, O. (2016, May). Verifying isolation in a man-trap portal via thermal imaging. In *Systems, Signals and Image Processing (IWSSIP), 2016 International Conference on* (pp. 1-4). doi:10.1109/IWSSIP.2016.7502726

Wolff, Socolinsky, & Eveland. (2005). Face Recognition in the Thermal Infrared. In I. P. Bir Bhanu (Ed.), Computer Vision Beyond the Visible Spectrum (pp. 167-191). Springer-Verlag London Limited.

Yared, R., Abdulrazak, B., Tessier, T., & Mabilleau, P. (2015, July). Cooking risk analysis to enhance safety of elderly people in smart kitchen. In *Proceedings of the 8th ACM International Conference on Pervasive Technologies Related to Assistive Environments* (p. 12). doi:10.1145/2769493.2769516

Chapter 7
Optical Analysis of Solar Concentrators Using Photogrammetry Process

Safa Skouri
Research and Technology Center of Energy, Tunisia

Salwa Bouadila
Research and Technology Center of Energy, Tunisia

ABSTRACT

As the optical efficiency of solar concentrators has a high impact on its thermal performance. However a qualification method determining the geometrical accuracy of a solar concentrator system is necessary. The purpose of this chapter is to gives an optical analysis of solar concentrator with an imaging process in order to improve the thermal efficiency of the solar concentrator. In this order measurement techniques used to determine geometric errors of the solar concentrating system have been described. Intercept factor, slope error and displacement error have been identified and analyzed. Examples of the intercept factor for concentrator reflector along with optical efficiency has been developed and determined related to the experimental results given by photogrammetry measurement technique.

DOI: 10.4018/978-1-5225-2423-6.ch007

SECTION I: INTRODUCTION

Today the fossil energy situation indicated several weakness aspects due to its limited and pollutant character. Therefore, the investment in the renewable energy sector especially solar energy seems to be one of the potential solutions.

Several under-developed regions around the world receive large amounts of sunlight. Alonso-Montesinos and Batlles (2015) mentioned that Northern Africa and Central Asia receive as much as 7.5 kWh/m^2/day. There is great opportunity to use solar power to provide basic energy needs in these regions. The concentrated solar power (CSP) is considered as the most prominent solar energy technology. There are four types of CSP technologies solar Parabolic trough concentrator, Linear Fresnel mirror, Solar towers (Heliostat field collectors) and Parabolic dish concentrator.

Many solar concentrator applications and projects exist in the world. El Fadar (2015) presents a modeling and optimization investigation of a solar driven adsorption cooling system working with activated carbon–methanol pair they found that there is an optimal range of number of fins varying from 15 to 20 that allows obtaining an optimal performance; the corresponding range of fin spacing is found to be between 4 and 5.2 cm.

Bouvier (2015) presented an experimental campaign that has been conducted on a solar parabolic trough collector feeding a micro-CHP system (micro-cogeneration). The originality of the solar field is the DSG (direct steam generation) using a two axis tracker and a small size (46.5 m2) collector On the studied sunny day, the system produces saturated steam during 8 h with a quality higher than 0.6 for a flow rate of 33 kg/h

(Dascomb, 2009) An economic parabolic dish concentrating system was built at the Sustainable Energy Science and Engineering Center (SESEC) at Florida State University in Tallahassee, Florida. The goal of the project was to provide 6.67 kW of thermal energy. This is the amount of energy required to produce 1 kW of electricity with a conventional micro steam turbine.

Wua SY et al. evaluated the thermal-electric conversion performance of parabolic dish/AMTEC solar thermal power system. The overall conversion efficiency of the system could reach 20.6% with an output power of 18.54 kW. Blanco J et al. described ''Solar detoxification technology in the treatment of persistent non-biodegradable chlorinated industrial water contaminants'' project,

In order to develop a commercial solar water treatment technology based on compound parabolic collectors (CPC). Kleih (1991) realized a test facility for parabolic dish systems with sterling motor in their focus. Performance ranges from 5 to 25 kWel (depending on concentrator geometry, insolation, and engine). This author described the video-camera measuring system HIMAP used in this work to qualify concentrating modules within the test facility.

Brazil is looking for innovative alternatives to supply the country's increasing electricity demand and provide flexibility to cope with higher shares of variable renewable energy (VRE). Concentrated solar power (CSP) can help solving this double challenge. Three energy planning tools, namely MESSAGE-Brazil, TIMES-TiPs-B and REMIX-CEM-B, have been combined to analyze the opportunities that CSP plants offer to the power system and to the wider energy system of the country. Rafael Soria et al show that CSP can be a cost-effective option under stringent mitigation scenarios. CSP can provide firm energy and dispatchable capacity in the Northeast region of Brazil, optimally complementing wind and PV generations.

José M. Rodríguez et al (2016) developed a techno-economic assessment of thermal energy storage solutions for a 1 MWe CSP-ORC power plant they found that Thermocline storage enables a large reduction in the cost of the storage system (33%).

To obtain an optimal focused solar flux at the CSP receiver. It is necessary to develop an accurate measurement method to facilitate the installation and operation of the solar concentrator system.

The present chapter is an optical analysis of solar concentrator with an imaging process in order to improve the thermal efficiency of the solar concentrator. This paper is organized as follows.

In the first step the solar concentrator power technologies (SCP) and their drawbacks are given. In the second step the measurement methods of the solar concentrators systems are described specially the photogrammetric technique that is used to analyze the slope error and the displacement error to determining the geometric deformation of a SCP system. In the last examples results of intercept factor and optical efficiency estimation were analyzed. The main remarks of this work are reported in the conclusion

SECTION II: BACKGROUND

To obtain an optimal focused solar flux at the CSP receiver. It is necessary to develop an accurate measurement method to facilitate the installation and operation of the solar concentrator system. Shuai Y et al. (2008) mentioned that the optical error of the reflector affects the system optical performance. It is relevant to identify the influence of the optical errors on the direction of the reflected ray to calculate the optical performance of the system. Traditionally, the optical analysis of concentrated irradiations has been carried using ray-tracing programs described by Jiang S. et al (2010) and Ho C.K (2008).

Bendt P et al. (1981) simulate various optical behavior including different parameters such as reflector material, contour errors, tracking errors and with different shapes of receivers. Shuai Y. et al (2008) and Zanganeh G. et al. (2012) used Monte

Carlo ray tracing method to determine the thermal performance of a solar parabolic concentrator (SPC) based on optic and geometric studies and validated experimentally. Zhiqiang et al. (2012) presented a procedure to design a facet concentrator; the optimum size and position of each facet are determined using Monte Carlo ray tracing analysis to achieve the most concentrated flux.

Sellami et al. (2012) developed a new geometry of a 3-D solar concentrator compact, which collect the maximum of solar energy. They obtained an optical efficiency around 70% for different dimensions of the Square Elliptical Hyperboloid.

Optical Analysis of 3-D static elliptical hyperboloid concentrator, using the ray tracing technique has been presented by Saleh Ali et al. (2013). They found that effective concentration ratio for 3-D EHC for different aspect ratios increases as the concentrator height increases and it is followed by a decrease of the acceptance angle. The optimum aspect ratio is equal to 5.

The surface shape and the optical quality of the solar concentrators have a great influence on its thermal efficiency that needs an accurate and practical tool to measure the concentrator optical aspect. Generally, three types of measurement methods are used to define and analyze the optical errors and therefore to determine the intercept factor: C), Deflectometry and Photogrammetry. Jun Xiao et al (2012) reviewed and compared many measurement methods of the solar concentrators systems.

Khaled A., (2012) März T. et al. (2011), Weber C et al. (2014) and Ulmer S et al. (2007) have adapted the deflectometry technique to enhance the optical shape quality of the reflectors-for different solar concentrators. Ulmer S. et al. (2007) have been compared and validated by measuring a typical mirror panel under optimal conditions. Additionally, a flat water surface has been measured as an absolute reference object using deflectometry.

Photogrammetry is a geometric analysis tool of solar concentrators. This measurement technique determines the slope deviation compared to the ideal shape of the concentrator. applied photogrammetry technique to determine the real shape and the intercept factor for both solar dish and parabolic through. Prahl C. et al. (2013), described the development and assessment of a measurement technique for the qualification of parabolic trough collector modules based on the distant observer method called TARMES (Trough Absorber Reflection Measurement System). Instead of a stationary camera at ground level taking pictures of a turning collector, the new approach called QFly makes use of an airborne camera vehicle which allows a completely automated and fast measurement of large numbers of collectors under relevant operating conditions. Pottlera K. et al. (2014) mentionned that a Target values for the intercept factor are in the range of 96-99% for typical operating conditions. These values can only be reached if appropriate quality specifications are fulfilled. Specification values are suggested and their implications on intercept factors are discussed based on the well-proven statistical calculation model from

Bendt and Rabl (1981). A variety of measurement techniques which can be applied for measurement and control of the geometry parameters are discussed by Pottlera K. et al. (2014).

At the Plataforma Solar de Almería, a set of methods and algorithms based on close-range photogrammetry has been developed on the last years. These methods allow us to evaluate the geometrical quality of solar concentrators. The application of these techniques to different concentrators prototypes, permits us to determine their geometrical, optical and energy behavior as well as to detect improvements on their design. Fernández-Reche J. et al. (2012) gives the analysis performed to a small parabolic solar concentrator designed for industrial process heat applications. Lee H. explained the fundamental difference between the rigorous relation and the conventional relation between surface slope error and reflected ray error and he illustrated The implementation of the rigorous relation in the Monte Carlo ray-tracing. Its effect on the concentrated solar flux distribution is investigated with an exemplary case of dish concentrators

King P et al (2014) describes the development of an inexpensive, highly portable photogrammetry technique for measuring the position and form of large mirror segments for solar collectors. The accuracy of the technique has been validated using a large Coordinate Measuring Machine (CMM) with results showing a measurement capability of better than 100 μm.

García Cortés S. et al. (2012) mentionned that a work carried out to determine the real shape and the intercept factor of a new prototype of parabolic solar collector. Convergent photogrammetry with off-the-shelf equipment was used to obtain a 3D point cloud that is simultaneously oriented in space and adjusted to a parabolic cylinder in order to calculate the deviations from the ideal shape. The normal vectors at each point in the adjusted surface are calculated and used to determine the intercept factor. Deviations between adjusted shape and the theoretical one suggest mounting errors for some mirror facets, resulting in a global intercept factor slightly below the commonly accepted limit for this type of solar collector.

SECTION III: SOLAR CONCENTRATOR TECHNOLOGY AND THEIR DRAWBACKS

There are four categories of concentration collectors (Figure 1), each of which is discussed below:

- Parabolic trough concentrator,
- Linear Fresnel mirror,

- Solar towers (Heliostat field collectors),
- Parabolic dish concentrator.

Parabolic Trough Concentrator

The collector of this type of concentrator is made by bending a sheet of reflective material into a parabolic shape, A metal black tube, covered with a glass tube to reduce heat losses, is placed along the focal line of the receiver.

It is sufficient to use a single axis tracking of the sun thus producing long collector modules. The collector can be orientated in an east–west direction, tracking the sun from north to south. Over the period of one year, a horizontal north–south trough field usually collects slightly more energy than a horizontal east–west collector. However, the north–south field collects a lot of energy in summer and much less in winter. The east–west field collects more energy in the winter than a north–south field and less in summer, providing a more constant annual output. Therefore, the choice of orientation depends on the application and whether more energy is needed during summer or during winter.

PTCs can effectively produce heat at temperatures between 50 and 400 °C and they are the most mature solar technology to generate heat at temperatures up to 400 °C for solar thermal electricity generation or process heat applications. However, use of oil-based heat transfer media restricts operating temperatures today to 400°C, resulting in only moderate steam qualities.

The Parabolic trough concentrator is used for various applications such as the steam production and water desalination.

The Parabolic trough concentrator is the most concentrated solar power technology (CSP) developed. We can cited some representative plants of this technology: The central Solar Energy Generating Systems (SEGS) in California, Acciona Solar One near Boulder City in Nevada and the center Plataforma Solar de Almeria in Spain SSPSDCS.

Linear Fresnel Mirror

This technology relies on an array of linear mirror strips that concentrate light on to a fixed receiver mounted on a linear tower. The linear Fresnel reflectors field can be imagined as a broken-up parabolic trough reflector. The main advantage of this type of system is that it uses flat or elastically curved reflectors which are cheaper compared to parabolic glass reflectors. Additionally, these are mounted close to the ground, thus minimizing structural requirements.

In order to avoid the shading and blocking between adjacent reflectors, advanced design is adopted as shown in the following chart: However, linear Fresnel reflectors are less efficient than troughs in converting solar energy to electricity and it is more difficult to incorporate storage capacity into their design.

Panna Lal Singh et al. made a comparative study of the linear Fresnel mirror performance equipped with four kinds of absorbers to optimize the best absorber.

Solar Tower (Heliostat Field Collector)

Solar Tower can be used for extremely high inputs of radiant energy to reflect their incident direct solar radiation onto a common target. This is called the heliostat field or central receiver collector. By using slightly concave mirror segments on the heliostats, large amounts of thermal energy can be directed into the cavity of a steam generator to produce steam at high temperature and pressure. The concentrated heat energy absorbed by the receiver is transferred to a circulating fluid that can be stored and later used to produce power. Central receivers have several advantages:

- They collect solar energy optically and transfer it to a single receiver, minimizing thermal-energy transport requirements.
- They typically achieve concentration ratios of 300 –1500 and are highly efficient, both in collecting energy and in converting it to electricity.
- They can conveniently store thermal energy.
- They are quite large (generally more than 10 MW) and thus benefit from economies of scale.

The tower plants are less common than parabolic trough systems, they perform better and better energy storage capacity. As an example of this technology we can mention: Central Solar Two in Daggett, California and Planta Solar 10 (PS10) in Sanlucar la Mayor, Spain. Various solar power tower existing in the world.

Solar Tower Stack

In this concept, the solar radii are not concentrated. The air is heated by a solar collector area formed of a transparent cover and serving as a greenhouse. Warm air is lighter, it escapes through a large central fireplace. The temperature difference between the lower part and the upper part of the chimney resulting in a constant movement of air (phenomenon of natural convection). This airflow then allows turbines located at the entrance of the chimney generate electricity. The main advantage of this system is that it can operate intermittently without using the sunlight during the day and the heat stored in the ground at night.

Parabolic Dish Concentrator

A Parabolic Dish concentrator is a point-focus collector that tracks the sun in two axes, concentrating solar energy onto a receiver located at the focal point of the dish. The dish structure must track fully the sun to reflect the beam into the thermal receiver. Parabolic dish reflector (PDR) The receiver absorbs the radiant solar energy, converting it into thermal energy in a circulating fluid. The thermal energy can then either be converted into electricity using an engine-generator coupled directly to the receiver, or it can be transported through pipes to a central power-conversion system. Parabolic dish systems can achieve temperatures in excess of 1500°C. Parabolic dishes have several important advantages:

- Because they are always pointing at the sun, they are the most efficient collector systems.
- Typically, they have a concentration ratio in the range of 600–2000 and are highly efficient at thermal-energy absorption and power conversion systems.
- They have modular collector and receiver units that can either function independently or as part of a larger system of dishes.

Parabolic-dish systems that generate electricity from a central power converter collect the absorbed sunlight from individual receivers and deliver it via a heat-transfer fluid to the power-conversion systems. The need to circulate heat transfer fluid throughout the collector.

The sensors dishes Stirling Energy Systems (SES) and Science Applications International Corporation (SAIC) of UNLV and Big Dish Central Australian National University in Canberra, Australia is representative of this technology.

Solar Concentrator Technologies Drawbacks

- The diffuse radiation is poorly used in high concentrations systems, only direct radiation is concentrated.
- Optical losses can be very significant due to multiple reflections and refractions and geometric characteristics of the system.
- The high concentration systems have a narrow field. So it is necessary to point to capture the direct radiation through referral mechanisms.
- High concentration systems require precise optical assemblies and therefore expensive.

We must maintain the optical qualities of the system (against oxidation, dust...).

Figure 1. Solar concentrating technologies

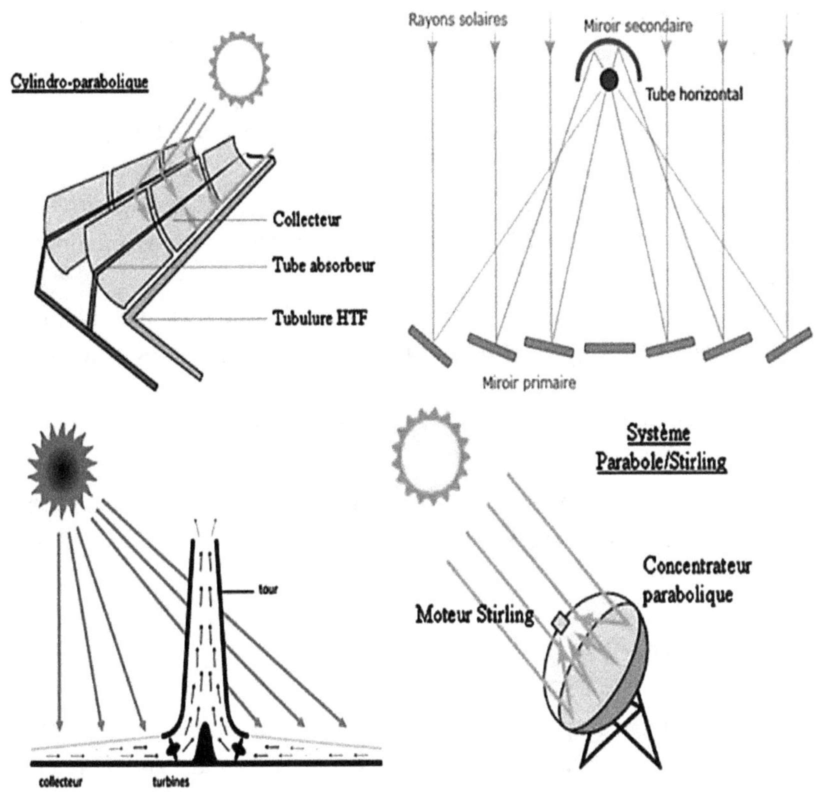

- The absorber is subjected to large temperature variations during cloudy pass.

These problems can be solved either by the choice of materials used in the construction of the sensor or by the management.

SECTION IV: METHODS OF SURFACE SHAPE MEASUREMENT

Deflectometry

Deflectometry (Figure 2) is an optical measurement technique originating in the car manufacturing industry, it is an optical characterization technique of reflective surfaces, and this technique is used to determine the geometrical errors of the reflectors.

The basic components of this characterization technique are: a light source with regular pattern, a projection screen, a DSLR camera (Digital Single Lens Reflex), an object surface to be characterized and image processing software.The main steps of this technique are:

1. A light source generates a regular pattern fringe system (eg projector, sun, etc) to a diffusing screen.
2. The screen displays a pattern consisting of a periodic pattern (horizontal example). This screen will play the role of "ruler" to know the orderly light source points.
3. The surface of the object to be treated (example: the reflector) is illuminated by the system of fringes whose surface imperfections show deformations of the image.
4. A camera which detects the line patterns of the images reflected on the surface to be characterized.
5. This software processes the images taken by calculating the local normal vector of the reflecting surface.

VSHOT Method

In 1996, Jones et al. developed a prototype Video Scanning Hartmann Optical Test (VSHOT) which is an evolutionary step beyond the original SHOT system. It is a laser ray trace device utilizing a video system to measure the optical quality of point focus solar concentrating mirrors, providing higher speed testing and requiring less operator oversight. A schematic of the VSHOT is shown in Figure 3. The distance from Mirror Specimen to target is slightly greater or less than twice its focal length.

And a laser beam is controlled by a 2-axis scanner. A laser beam strikes on the mirror and reflects back to the target. The CCD camera takes the image and the data digitized by a computer video board, and then centric location of the reflected spot is calculated. This method measures the slope of a reflector at many locations and compares that information to the mathematical simulation. Because the VSHOT was based on direct measurement of slope and provided a quantitative map of optical errors, it is proved to be suitable to characterize the optics of not only the dish-type solar concentrators but also any large reflector with a focal length over diameter ratio greater than 0.45, and RMS optical error in the 0.1–10 mrad range. And the VSHOT hardware, software, and operation are also described.

The outdoor testing appears feasible and a number of suggestions to improve the VSHOT capabilities in outdoor testing have been provided.

Figure 2. Deflectrometry main steps

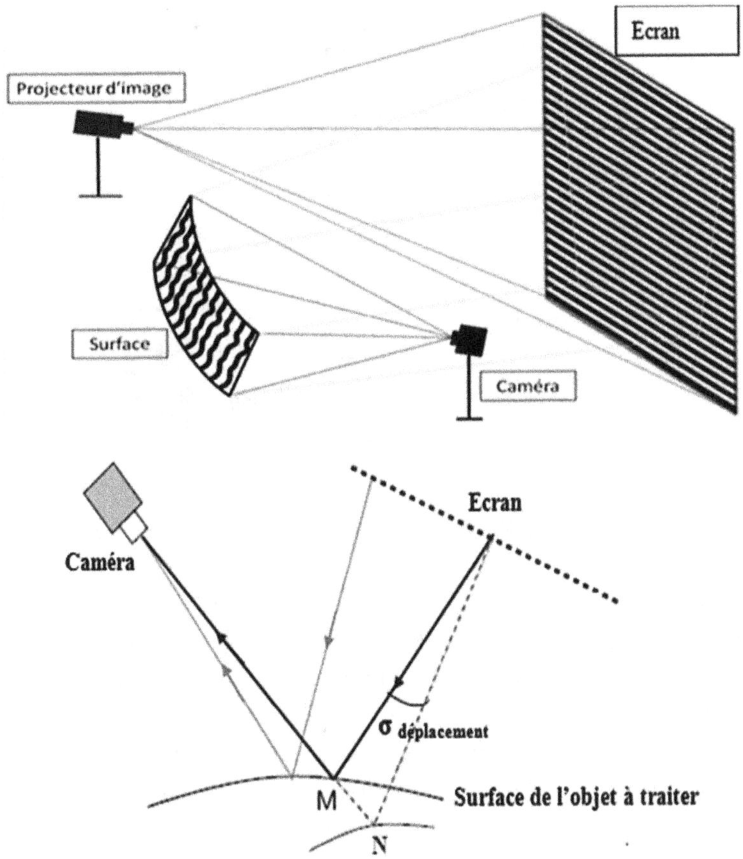

The Photogrammetry Measurement Technique

A geometric study of the experimental SPC has been done using the photogrammetry measurement technique. The Photogrammetry measurement technique determines the geometric properties of objects from photographic images Figure 4. Photogrammetry can deliver results with a level of precision going from 0,1mm ± 0,1mm/m to 0,05mm ± 0,005mm/m. This technique has been carried out on the solar collector structure and on its reflective surface in order to determine the shape and surface deformations of the SPC reflector, as well as the actual geometrical position of the receiver (absorber).

This measurement technique uses a high resolution DSLR Camera (Nikon D300) with a set of coded and non-coded optical Targets. Targets cover the majority of the SPC area that needs to be measured and mechanism axis. Also a set of scale bars and

Figure 3. VSHOT method

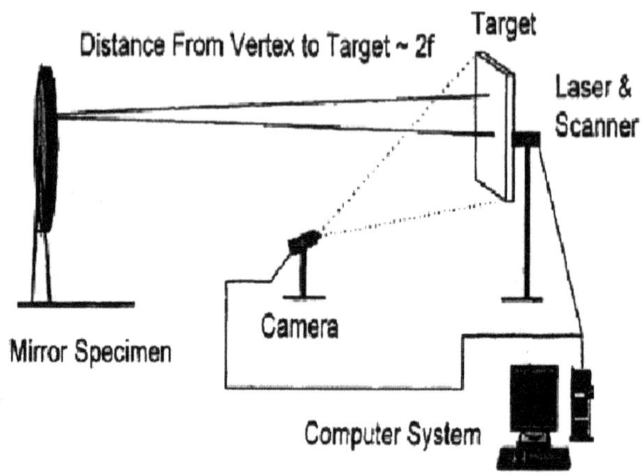

a reference cross are added to the photographed system in order to setting a referential graduation to the measured points. Adapters Targets are used to measure different features such as holes, ridges or other specific points of reference. Subsequently, the object must be photographed from different angles, distances and in different directions. The captured images are uploaded to the computer and then processed with an image treatment system software "Aicon 3D". The system calculates automatically the 3D coordinates of all the points photographed. The calculation is based on the principle of using triangulation between the spaces of the images.

Generally there are two types of Targets with two reflective colors. The first type are coded Targets, they have a specific geometry shape in order to be identified as known Targets with specific code after taking images. The second type is non-coded Targets which will have a code defined by the software algorithm.

For a low illumination, reflective color is used with the camera flash light during the photo shooting. When there is sufficient light, white color Targets are used without the flash light.

In order to setting a referential graduation to the photographed points, reference cross and scale bars are used. In advanced step, the treatment software will recognize the reference objects and basing on their dimensions it will automatically give the real coordinates for each photographed Target.

The DSLR camera used in this application is a Nikon D300, the capture mode of the camera is set to monochromatic colors (black and white). All other parameters like exposure, aperture, focal length and ISO level are set depending on the light conditions and the dimension of the photographed object to ensure better image quality with illuminated Target images. The used camera lens was a Nikon MF20 mm. An image check software 'DPA CHECK' was used to guaranty the required image quality of the Targets.

The basic of the treatment software 'AICON 3D'is the micro-triangulation. It will determine the position of the high contrast objects: illuminated point or dot with a black background.

As a first step, the taken photos are loaded to the software, then the camera parameters the reference cross dimensions, the scale bars and the Targets are injected to the software database. In the second step the software treats the photographed images t and delivers all the calculated positions of the camera even as the calculated points coordinates. Finally, the software produce the 3-D coordinates of all the recognized points (Targets) and a three dimensional graphical model of the real system shape.

In order to setting a referential graduation to the photographed points, reference cross and scale bars are used. In advanced step, the treatment software will recognize the reference objects and basing on their dimensions it will automatically gives the real coordinates for each photographed Target.

The DSLR camera used in this application is a Nikon D300, the capture mode of the camera is set to monochromatic colors (black and white). All other parameters like exposure, aperture, focal length and ISO level are set depending on the light conditions and the dimension of the photographed object to ensure better image quality with illuminated Target images. The used camera lens was a Nikon MF20 mm. An image check software 'DPA CHECK' was used to guaranty the required image quality of the Targets.

The basic of the treatment software 'AICON 3D'is the micro-triangulation. It will determine the position of the high contrast objects: illuminated point or dot with a black background.

As a first step, the taken photos are loaded to the software, then the camera parameters the reference cross dimensions, the scale bars and the Targets are injected to the software database. In the second step the software treats the photographed images t and delivers all the calculated positions of the camera even as the calculated points coordinates. Finally, the software produce the 3-D coordinates of all the recognized points (Targets) and a three dimensional graphical model of the real system shape.

Photogrammetry Parameters

Rim Angle

The rim angle of a solar parabolic concentrator is the angle between the axis of the parabola and the line connecting the parabola edge to its focal point. The angle dimension differs from one system to another depending on the focal distance and

Figure 4. Photogrammetry method description

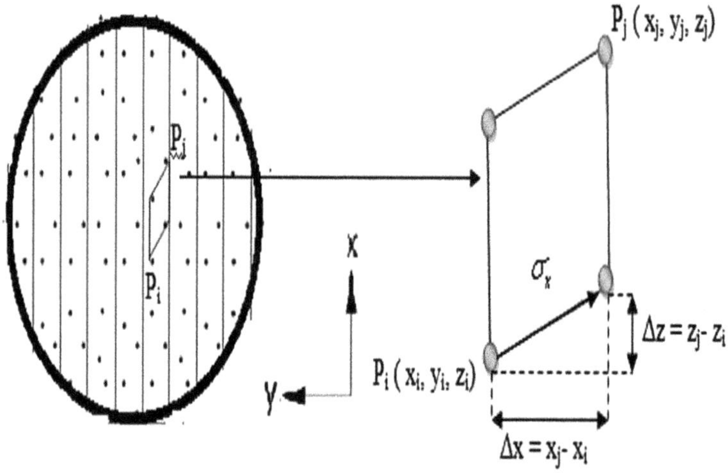

the parabola diameter. The point A in the reflector edge is defined by the following coordinates $(0, y_A, z_A)$ in the $R(O, \vec{x}, \vec{y}, \vec{z})$ coordinate system.

The equation of the parabola is defined as follow:

$$y_A = \left[4.f \cdot z_A\right]^{1/2}$$

The depth of the parabola is determined by the following expression.

$$h = D_0^2 / (16 \cdot f)$$

The rim angle ψ of the solar parabolic concentrator is defined by the following relation

$$\tan\left(\psi\right) = \frac{y_A}{(f-h)} = \frac{\left(4.f.z_A\right)^{1/2}}{\left(f - \left(\frac{D_0^2}{16.f}\right)\right)} = \frac{\left(4.f.h\right)^{1/2}}{\left(f - \left(\frac{D_0^2}{16.f}\right)\right)} = \frac{\frac{D_0}{2}}{\left(f - \left(\frac{D_0^2}{16.f}\right)\right)}$$

$$= \frac{1}{\left(\frac{2.f}{D_0} - \left(\frac{D_0}{8.f}\right)\right)} = \frac{\frac{f}{D_0}}{\frac{f}{D_0} \cdot \left(\frac{2.f}{D_0} - \left(\frac{D_0}{8.f}\right)\right)} \quad (1)$$

f is the focal length and D_0 is the aperture diameter of the concentrator.

$$\tan\psi = \frac{(f / D_0)}{2(f / D_0)^2 - (1/8)} \tag{2}$$

Total Optical Error

In practice, there are optical errors associated with the operation of the collector that will affect the intercept factor in addition to the fact that solar rays are incident from different directions covering the solar disk.

σ_{tot} is the standard deviation of the total errors it can be written as:

$$\sigma_{tot}^2 = \sigma_{optical}^2 + \sigma_{sun}^2 \tag{3}$$

$$\sigma_{optical}^2 = \sigma_{slope}^2 + \sigma_{specular}^2 + \sigma_{displacement}^2 + \sigma_{tracking}^2 \tag{4}$$

where $\sigma_{optical}$, σ_{slope}, $\sigma_{specular}$, $\sigma_{displacement}$, $\sigma_{tracking}$, and σ_{sun} are standard deviations of the total optical errors, slope error, specular error, receiver displacement error tracking errors and sun error, respectively.

Sun Error

The error of the sun is the distribution of energy directed toward the receiver, also known as a sun represents a constant value equal to 2.6 mrad.

Tracking Error

The tracking error is associated with the accuracy of the tracking drive. It is the misalignment of the receiver relative to the focal line of the collector. It depends on the tracking system used.

Tracking systems are required for collectors to follow the sun in order to concentrate the direct solar radiation onto the small receiver area. High concentration ratio collectors cannot work without a tracking system. Various forms of tracking mechanisms, varying from simple to complex, have been proposed. They can be divided into two broad categories –mechanical and electrical/electronic systems. The electronic systems generally exhibit improved reliability and tracking accuracy.

Specular Error

The specular error is the non-specularity of the reflecting surface of a solar concentrator. Knowing that the specular reflection is called when the incident ray gives birth to a single reflected ray. Ideally, the energy of the incident ray is totally reflected in the reflected beam.

Slope Error

The slope error is the deviation of the slope of the parabolic reflector surface from the required parabolic shape. To estimate this error we need to determine the level difference $\Delta_{z,level}$ which is the difference between the ideal coordinate along the $(O,\vec{z})$ axis of a point P_i of the parabola and the measured coordinate along the $(O,\vec{z})$ axis of the item which is determined experimentally using photogrammetry technique.

The level difference is defined as follow:

$$\Delta_{z,level} = z_i^{cal} - z_i^{mes} \tag{5}$$

Then the measured slope along the $(O,\vec{x})$ axis σ_x^{mes} of the surface element containing each two adjacent points (P_i and P_j), was calculated according to the following equation:

$$\sigma_x^{mes} = arctg\left(\frac{z_j^{mes} - z_i^{mes}}{x_j^{mes} - x_i^{mes}}\right) \sim \frac{z_j^{mes} - z_i^{mes}}{x_j^{mes} - x_i^{mes}} \tag{6}$$

(z_j^{mes}, x_j^{mes}) and (z_i^{mes}, x_i^{mes}) are the measured coordinates along $(O,\vec{z})$ and $(O,\vec{x})$ for two adjacent points P_i and P_j, respectively.

The calculated bending angle along the $(O,\vec{x})$ axis σ_x^{cal} for every two adjacent points (P_i and P_j) is defined as follows:

$$\sigma_x^{cal} = arctg\left(\frac{z_j^{cal} - z_i^{cal}}{x_j^{cal} - x_i^{cal}}\right) \sim \frac{z_j^{cal} - z_i^{cal}}{x_j^{cal} - x_i^{cal}} \tag{7}$$

(z_j^{cal}, x_j^{cal}) and (z_i^{cal}, x_i^{cal}) are the calculated coordinates along $(O, \vec{z})$ and $(O, \vec{x})$ for two adjacent points P_i and P_j, respectively.

The elementary slope error $\Delta\sigma_{slope}$ is the difference between both calculated and measured bending angles for every two adjacent points. It is calculated according to the following equation:

$$\Delta\sigma_{slope} = \sigma_x^{cal} - \sigma_x^{mes} \tag{8}$$

The slope error σ_{slope} is the average value of the elementary slope error $\Delta\sigma_{slope}$ on the whole parabola shape and is determined by the following expression:

$$\sigma_{slope} = \left(\sum_1^n \left(\Delta\sigma_{slope} \right)^2 / n \right)^{1/2} \tag{9}$$

n is the total number of directions.

Displacement Error

The receiver displacement error is defined as the angular deviation of the location of the receiver centerline from the design focus as seen from local points (*Pi*) on the reflector. The methodology of calculation of the local displacement error is illustrated in Figure 5. The standard deviation $\sigma_{displacement}$ were performed using the following equation:

$$\sigma_{displacement} = \left(\sum_1^n \left(\Delta\sigma_{displacement} \right)^2 / n \right)^{1/2} \tag{10}$$

n is the number of points taken along the x axis to calculate the local slope errors and $\Delta\sigma_{displacement}$ is the local displacement errors which depends on both m_1 and h_1 distances:

$$\Delta\sigma_{displacement} = \tan^{-1}\left(m_1 / h_1 \right) \tag{11}$$

Figure 5. Displacement error

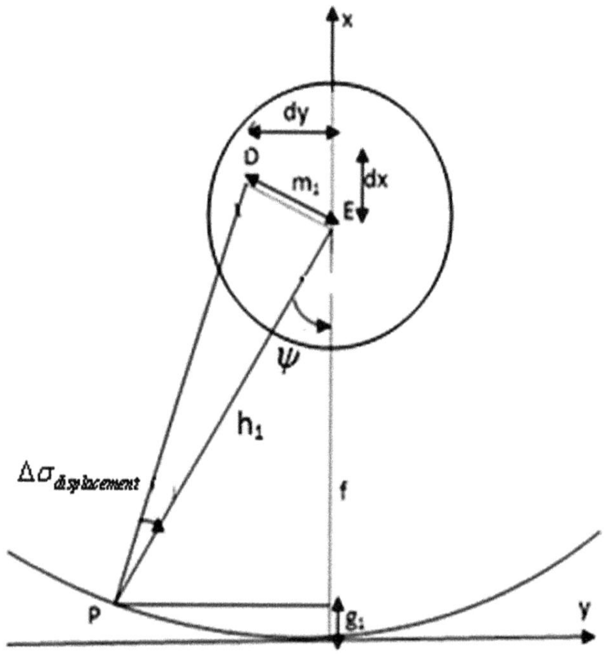

Calculation of the h_1 Distance

h_1 is the distance between the point P on the side of the concentrator and the E point of the absorber along the axis $(O,\vec{x})$.

$$h_1 = (f - g_1) / \cos(\Delta\sigma_{displacement}) \tag{12}$$

f is the focal length and g_1 is the distance between the reflector center and the projection of the point P on the focal axis. $\Delta\sigma_{displacement}$ is the angle between the ray received by the reflector and reflected back to the absorber (PE). According to the parabola equation $g_1$distance is defined as follow:

$$g_1 = (x_p^2 + y_p^2) / 4.f \tag{13}$$

$$\Delta\sigma_{displacement} = \tan^{-1}\left(y_p / (f - g_1)\right)$$

Calculation of the m_1 Distance

m_1 is the distance between the point E and the point D which is the intersection of the reflected beam with the absorber. m_1 is defined by the following expression (Fig 6).

$$m_1 = d\cos(b)$$

where

$$b = (a + \Delta\sigma_{displacement}) - \pi / 2$$

and

$$a = \tan^{-1}\left(d_y / d_x\right)$$

dy and *dx* are determined

$$\begin{Bmatrix} dy = y_E - y_D \\ dx = x_E - x_D \end{Bmatrix} \text{and } d = \sqrt{d_y^2 + d_x^2}$$

To determine the displacement error it is necessary to estimate the elementary variations *dx* and *dy* of the absorber.

Three non-coded targets where used to estimate the location of the receiver with respect to the vertex of the parabolic shape and to estimate the average deviation of the receiver centerline from the ideal location of the focus which is designed to be at x= y= 0 mm. The targets were fixed on the outer surface of the receiver. Thus, on the z-axis, the measured points were compared to an imaginary point that would be located on the outer surface of the receiver if the receiver was perfectly located along the focus.

Calculation of the Total Optical Error

In practice, there are optical errors associated with the operation of the collector that will affect the intercept factor in addition the solar rays are incident from different directions covering the solar disk. The total optical error of the SPC is calculated by Eq (3).

Figure 6. Calculation of m1and h1 distances

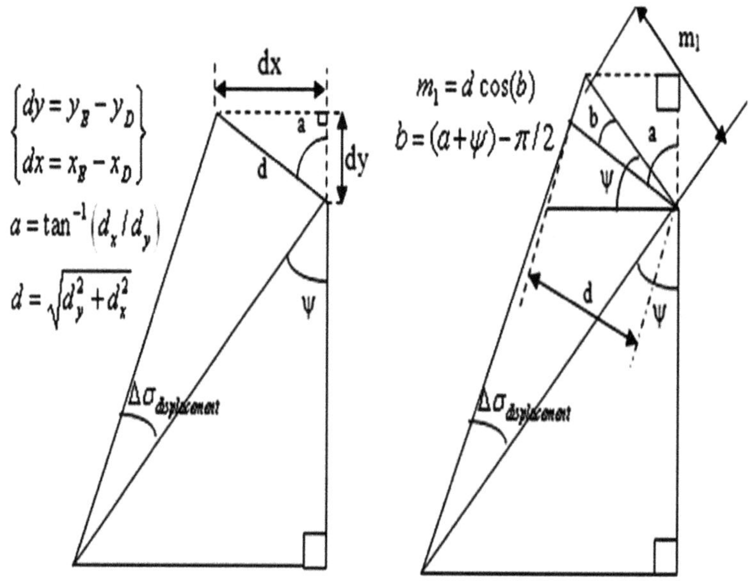

SPC tracking error is defined according to the used sun tracking system and the sun error represents a constant value. We considered specular error negligible. After determining the displacement error and the slope error using photogrammetry method. The average values of the different optical errors are summarized.

Intercept Factor for Parabolic Reflector with a Spherical Receiver

The intercept factor of solar parabolic concentrator system is defined as the ratio of the energy intercepted by the receiver to the energy reflected by the focusing device.

Erf is the error function.

We found the following expression of the intercept factor:

$$\gamma = 1 + \frac{\exp\left(-\frac{k^2}{2}\right)}{\tan^2\left(\frac{\psi_R}{2}\right)} - \frac{\exp\left(-\frac{k^2}{2}\cos^4\left(\frac{\psi_R}{2}\right)\right)}{\sin^2\left(\frac{\psi_R}{2}\right)} - \sqrt{\frac{\pi}{2}} \cdot \frac{k}{\tan^2(\psi_R/2)}\left[Erf\left(\frac{k}{\sqrt{2}}\cos^2\left(\frac{\psi_R}{2}\right)\right) - Erf\left(\frac{k}{\sqrt{2}}\right)\right] \quad (14)$$

The intercept factor for parabolic reflector with a sphere receiver under Gaussian distribution depends on the rim angle ψ_R and the $k = \theta_0 / \sigma_{total} \sim R_a / f.\sigma_{total}$ ratio, which requires the determination of this error.

Figure 7 shows the intercept factor of the parabolic reflector with a spherical receiver under Gaussian distribution for different k ratio and four rim angles. It is calculated by MATLAB program according to the Eq (14). It can be seen that the intercept factor increases when the k increases for the same rim angle. The intercept factor decreases when the rim angle increases for the same k. The intercept factor of a point of the receiver is only related to the k ratio. It is about 99% when the k=5. But the k for the point of reflection will decrease when the polar angle of the point increases, so the intercept factor decreases when the polar angle of the point increases. As the intercept factor of the system is the average of all points of the reflector, it will decrease when the rim angle of the system increases. The intercept factor corresponding to the rim angle of our experimental device ($\psi = 0.44.\pi$)

Figure 7. Intercept factor of the parabolic reflector with a spherical receiver under Gaussian distribution for different k ratio and four rim angles variations

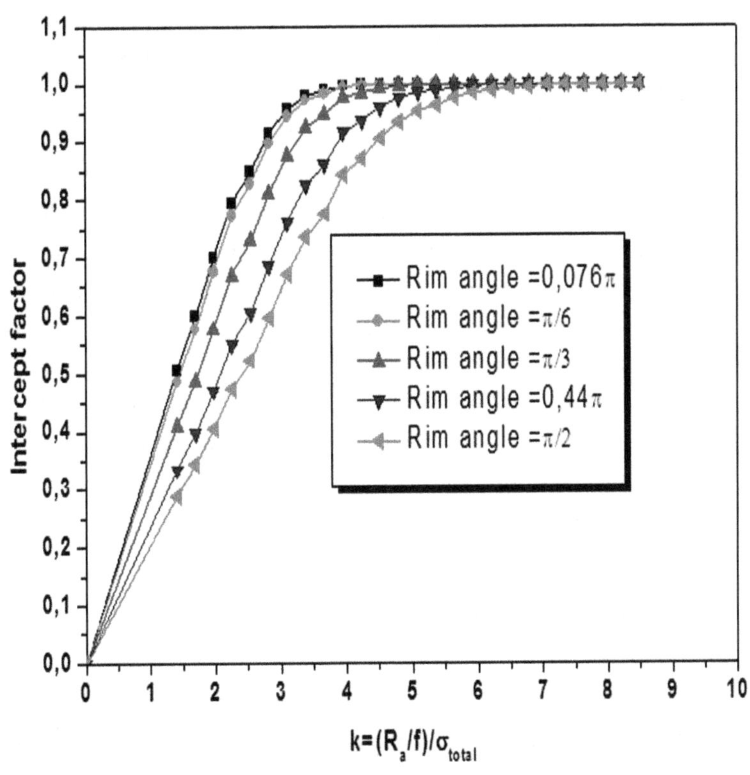

maintain a constant value equal to 95% for k = 5 which corresponds to a total optical error equal to 0.026 rad. Therefore for a lower total optical error, intercept factor is equivalent to 95%.

Optical Efficiency

The following equation form can be used to perform an approximated optical efficiency analysis [35]:

$$\eta_{op} = \gamma\rho\tau\alpha\cos\beta \tag{15}$$

ρ and $\tau\alpha$ are dish surface reflectance and transmittance–absorptance product, respectively. γ is the intercept factor of the receiver, β is the incidence angle of solar beam into the dish which is equal to zero degree, therefore the Eq. (1) can be written as:

$$\eta_{op} = \gamma\ \rho\ \tau\ \alpha \tag{16}$$

SPC optical efficiency is given by Eq.(15) it depends of the reflector material which is characterized by transmittance–absorptance product $\tau\alpha$ and specific reflectivity ρ in other hand it is related to the intercept factor γ.

Figure 8 illustrates the optical efficiency variation with $k = R_a / (f.\sigma_{total})$ for different rim angles. The optical efficiency increase when the $k = R_a / (f.\sigma_{total})$ increase for the same rim angle. The ratio decreases when the rim angle increases. We note that the optical performance and the k ratio are proportional and depend essentially on the total error. The total error decreases when the total optical efficiency increases. As the intercept factor, SPC optical performance corresponded to the rim angle of our experimental device ($\psi = 0.44.\pi$) maintain a constant value equal to 85% for k = 5 which corresponds to a total optical error equal to 0.026 rad. Therefore for a lower error, intercept factor is equivalent to 85%. This optical performance provides a great thermal performance of the SPC.

SECTION V: FUTURE RESEARCH DIRECTIONS

The Photpgrammetry measurement technique has been carried out on the SPC in order to determine the surface deformations of the SPC reflector and its geometric

Figure 8. The optical efficiency variation for different rim angles and k ratio

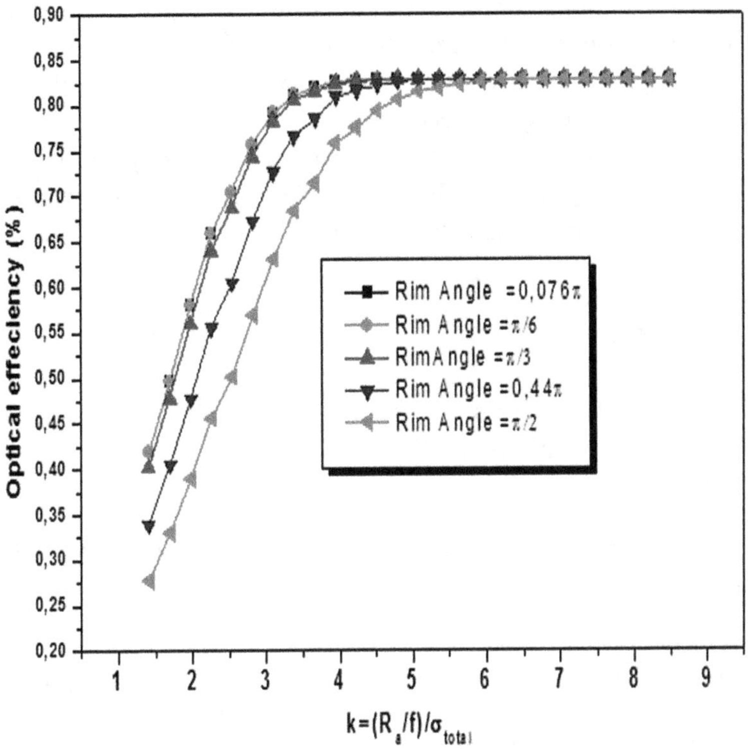

error compared to a required one, as well as the actual geometrical position of the receiver.

In future research works it can be applied for the concentrated solar technologies qualification and enhancement during three stages; the construction, and the mounting of the CSP facility also during operation time.

SECTION VII: CONCLUSION

This Chapter is an optical characterization study of a solar parabolic concentrator equipped with a programmed sun tracking system and a spherical receiver. The main findings of the present study are mentioned as follow:

- Developing an analytical function to calculate the intercept factor of point parabolic reflector with a spherical receiver under Gaussian distribution.

- Enhancing the geometrical and optical quality of a solar parabolic concentrator SPC: The Photpgrammetry measurement technique has been carried out on the SPC in order to determine the surface deformations of the SPC reflector and its geometric error compared to a required one, as well as the actual geometrical position of the receiver.
- In this paper photogrammetry have been used to determine the geometrical and optical quality of a SPC. It can be applied for the concentrated solar technologies qualification and enhancement.

REFERENCES

Alonso-Montesinos, J., & Batlles, F. J. (2015). Solar radiation forecasting in the short- and medium-term under all sky conditions. *Energy*, *1*(83), 387–393. doi:10.1016/j.energy.2015.02.036

Bendt, P., & Rabl, A. (1981). Optical analysis of point parabolic radiation. *Applied Optics*, (20): 674–683. doi:10.1364/AO.20.000674 PMID:20309177

Blanco, J., Malato, S., Fernandez, P., Vidal, A., Morales, A., Trincado, P., & Rangel, C. M. et al. (1999). Compound parabolic concentrator technology development to commercial solar detoxification applications. *Solar Energy*, *67*(4-6), 317–330. doi:10.1016/S0038-092X(00)00078-5

Bouvier, J. L., Michaux, G., Salagnac, P., Nepveu, F., Rochier, D., & Kientz, T. (2015). Experimental characterisation of a solar parabolic trough collector used in a micro-CHP (micro-cogeneration) system with direct steam generation. *Energy*, *83*, 474–485. doi:10.1016/j.energy.2015.02.050

Dascomb, J. (2009). *Low-cost concentrating solar collector for steam generation.* Thesis in department of mechanical engineering.

El Fadar, A. (2015). Thermal behavior and performance assessment of a solar adsorption cooling system with finned adsorber. *Energy*, *1*(83), 674–684. doi:10.1016/j.energy.2015.02.074

Fernández-Reche, J., & Valenzuela, L. (2012). Geometrical assessment of solar concentrators using close-range photogrammetry. *Energy Procedia*, *30*, 84–90. doi:10.1016/j.egypro.2012.11.011

García Cortés, S., Bello-García, A., & Ordóñez, C. (2012). Estimating intercept factor of a parabolic solar trough collector with new supporting structure using off-the-shelf photogrammetric equipment. *Applied Energy*, *92*, 815–821. doi:10.1016/j.apenergy.2011.08.032

Ho, C. K. (2008). *Software and codes for analysis of concentrating solar power technologies*. Albuquerque, NM: Sandia National Laboratories. doi:10.2172/946571

Jiang, S., Hu, P., Mo, S., & Chen, Z. (2010). Optical modeling for a two-stage parabolic trough concentrating photovoltaic/thermal system using spectral beam splitting technology. *Sol Energ Mater Sol Cells*, (94), 1686-96.

Khaled A. (2012). *Technical and economic performance of parabolic trough in Jordan*. Academic Press.

King, P., Comley, P., & Sansom, C. (2014). Parabolic trough surface form mapping using photogrammetry and its validation with a large Coordinate Measuring Machine. *Energy Procedia*, *49*, 118–125. doi:10.1016/j.egypro.2014.03.013

Kleih, J. (1991). Dish-Stirling test facility. *Solar Energy Materials*, *24*(1-4), 231–237. doi:10.1016/0165-1633(91)90063-Q

Lee, H. (2014). The geometric-optics relation between surface slope error and reflected ray error in solar concentrators. *Solar Energy*, *101*, 299–307. doi:10.1016/j.solener.2013.12.035

März, T., Prahl, C., Ulmer, S., & Weber, S. W. C. (2011). Validation of two Optical Measurement Methods for the Qualification of the Shape Accuracy of Mirror Panels for Concentrating Solar Systems. *Journal of Solar Energy Engineering*, 133.

Pottlera, K., Ulmera, S., Lüpferta, E., Landmanna, M., Rögerb, M., & Prahl, C. (2014). Ensuring performance by geometric quality control and specifications for parabolic trough solar fields. *Energy Procedia*, *49*, 2170–2179. doi:10.1016/j.egypro.2014.03.230

Prahl, C., Stanicki, B., Hilgert, C., Ulmer, S., & Roger, M. (2013). Airborne shape measurement of parabolic trough collector fields. *Solar Energy*, *91*, 68–78. doi:10.1016/j.solener.2013.01.012

Rodríguez, J. M., Sánchez, D., Martínez, G. S., Bennouna, E. G., & Ikken, B. (2016). Techno-economic assessment of thermal energy storage solutions for a 1 MWe CSP-ORC power plant. *Solar Energy*, *140*, 206–218. doi:10.1016/j.solener.2016.11.007

Saleh Ali, I. M., ODonovan, T. S., Reddy, K. S., & Mallick, T. K. (2013). An optical analysis of a static 3-D solar concentrator. *Solar Energy*, *88*, 57–70. doi:10.1016/j.solener.2012.11.004

Sellami, N., Mallick, T. K., & McNeil, D. A. (2012). Optical characterisation of 3-D static solar concentrator. *Energy Conversion and Management*, *64*, 579–586. doi:10.1016/j.enconman.2012.05.028

Shuai, Y., Xia, X. L., & Tan, H. P. (2008). Radiation performance of dish solar concentrator/cavity receiver systems. *Solar Energy*, *82*(1), 13–21. doi:10.1016/j.solener.2007.06.005

Shuai, Y., Xia, X. L., & Tan, H. P. (2008). Radiation performance of dish solar concentrator/ cavity receiver systems. *Solar Energy*, *82*(1), 13–21. doi:10.1016/j.solener.2007.06.005

Soria, R., Lucena, A. F. P., Tomaschek, J., Fichter, T., Haasz, T., Szklo, A., & Kern, J. et al. (2016). Modelling concentrated solar power (CSP) in the Brazilian energy system: A soft-linked model coupling approach. *Energy*, *1*(116), 265–280. doi:10.1016/j.energy.2016.09.080

Ulmer, S., Marz, T., Prahl, C., Reinalter, W., & Belhomme, B. (2011). Automated high resolution measurement of heliostat slope errors. *Solar Energy*, *85*(4), 681–687. doi:10.1016/j.solener.2010.01.010

Ulmer S., Pottler K., Lüpfert E., Röger M. (2007). Measurement techniques for the optical quality assessment of parabolic trough collector fields in commercial solar power plants. *Energy Sustainability*, 27-30.

Weber, C., Ulmer, S., & Koch, H. (2014). Enhancements in high-resolution dlope deviation measurement of solar concentrator mirrors. *Energy Procedia*, *49*, 2231–2240. doi:10.1016/j.egypro.2014.03.236

Wu, S. Y., Xiao, L., Cao, Y., & Li, Y. R. (2010). A parabolic dish/AMTEC solar thermal power system and its performance evaluation. *Applied Energy*, *87*(2), 452–462. doi:10.1016/j.apenergy.2009.08.041

Xiao, J., Wei, X., Lu, Z., Yu, W., & Wu, H. (2012). A review of available methods for surface shape measurement of solar concentrator in solar thermal power applications. *Renewable & Sustainable Energy Reviews*, *16*(5), 2539–2544. doi:10.1016/j.rser.2012.01.063

Zanganeh, G., Bader, R., Pedretti, A., Pedretti, M., & Steinfeld, A. (2012). A solar dish concentrator based on ellipsoidal polyester membrane facets. *Solar Energy*, *86*(1), 40–47. doi:10.1016/j.solener.2011.09.001

Zhiqiang, L., Lapp, J., & Lipinski, W. (2012). Optical design of a flat-facet solar concentrator. *Solar Engineering*, (86), 1962–6.

KEYS TERMS AND DEFINITIONS

Displacement Error: The receiver displacement error is defined as the angular deviation of the location of the receiver centerline from the design focus as seen from local points (*Pi*) on the reflector.

Intercept Factor: The ratio of the energy intercepted by the receiver of a solar concentrator to the energy reflected by the focusing device.

Photogrammetry: The Photogrammetry measurement technique determines the geometric properties of objects from photographic images.

Slope Error: The slope error is the deviation of the slope of the parabolic reflector surface from the required parabolic shape.

Chapter 8
New Advances in Multidimensional Processing for Thermal Image Quality Enhancement

Andrés David Restrepo-Girón
Universidad del Valle, Colombia

Humberto Loaiza-Correa
Universidad del Valle, Colombia

ABSTRACT

This chapter presents three recent methodologies about multidimensional processing for signal to noise ratio (SNR) and thermal contrast enhancement on sequences of thermographic images, acquired from active pulsed thermography experiments over composite slabs, mainly, carbon fiber reinforced plastic (CFRP). The first technique corresponds to noise pre-processing by means of iterative 3D filtering to take advantage of the high space-time correlation in thermal sequences; the other two techniques correspond to thermal contrast enhancement processing: one of them using an atypical median filtering scheme, and the other based on heat propagation discrete models. Beginning with their heuristic and mathematical foundations, and following with the algorithmic procedures development, advantages and limitations will be shown through suitable indexes for evaluation, and some comparisons with other similar techniques.

DOI: 10.4018/978-1-5225-2423-6.ch008

INTRODUCTION

Although thermography is not an area with recent application in engineering, the advances in informatics and computing systems, which have increased the performance of digital image processing algorithms, have done thermography to be one of the more promising techniques for Non-Destructive Testing (NDT) of materials because of non-contact and non-intrusion qualities (Bagavathiappan, Lahiri, Saravanan, Philip, & Jayakumar, 2013); because of these, thermography is gaining relevance in predictive and preventive maintenance activities. One of the more interesting and extensive NDT application for thermography focuses on the evaluation of composite materials, like *carbon fiber reinforced plastic* (CFRP), in automotive and aeronautical industries. This type of materials exhibits a better mechanical resistance/weight ratio than other very important materials, such as aluminum and steel (Snell Jr. & Spring, 2007), which is why NASA consider composites as the materials of choice for achieving lower weights and costs of aerospace vehicles (Tenney, Davis, Byron, & Johnston, 2009). However, excessive impacts or repetitive mechanical stress applied to them may create several types of failures: the most common being delamination between layers (Pohl, 1998), which may grow until breaking the material. The huge important role of a reliable NDT strategy for composites is demonstrated in the use of CFRP slabs, for example, as the main structural material for flight control surfaces, and fuselages of recent military and commercial planes, resulting in savings up to 20% of fuel consumption and decreasing the same quantity in CO_2 emissions (International Air Transport Association, 2009; Segui, 2014).

There are many challenges involved in application of thermography as an industrial NDT technique, but one of the most common and fundamental is an adequate *thermal contrast* of images for detection success: the more thermal contrast between defective regions and their healthy surroundings, the more probability of accurate detection of those defects. Subsequent characterization tasks depend on performance of the selected estimation methods, but only they will be run on the previous detected flaws. In that sense, it can be observed two fundamental trends about thermographic image processing for thermal contrast enhancement and internal flaws detection in materials:

- Processing focused on enhancing and finding the differential temperature between defective and healthy regions; in such a way that the expected resulting images consist of a certain intensity level of reference for sound substrate of the inspected material, and appreciable smaller or greater intensity levels (depending on thermal properties) for sought defects. An important feature of this group of techniques is that the information they give correspond to the temporal evolution of contrast temperature, either absolute, or relative, for every recorded pixel; for this reason they are called *time-resolved techniques* (Balageas, 2011; Larsen, 2011), many of them based on the heat propagation models (Grinzato, 2000; Rodríguez, Nicolau, Ibarra-Castanedo, & Maldague, 2014a), but taking timing information as the core of analysis.

- Processing using a mathematical transform (Ibarra-Castanedo, 2005; Marinetti et al., 2004) (Rodríguez, Ibarra-Castanedo, Nicolau, & Maldague, 2014b) or a statistical/probabilistic model (Balageas, Roche, Leroy, Liu, & Gorbach, 2015; Benítez, Loaiza, & Caicedo, 2011) in order to convert temperature information in a new type of data, sometimes compressed data without direct correspondence with the temperature, from which an intrinsic thermal contrast enhancement and subsequent detection of defects can be achieved. Many times these methods are combined with learning machines to carry detection and characterization tasks, but their performance decisively depends on the quality of preprocessing of thermal image sequences.

Since a few years, according to one or the other processing scheme, multidimensional processing has begun to be extensively applied in active thermography, in noise reduction (Valdés, Inamura, Valera, & Yao, 2006), thermal data compression (Alsaadawi & Netzelmann, 2010), simulation of temperature distribution and contrast enhancement (Grinzato, 2000; López, Nicolau, Ibarra-Castanedo, & Maldague, 2014), and solution of inverse problems (Cheng-Hung & Meng-Ting, 2008). However, in simulation and inversion tasks, working with multidimensional models usually leads to very heavy processing algorithms; decrease of these computational requirements constitutes one of the more active field of research in thermography.

Through next sections, different recent multidimensional processing strategies, from pre-processing to thermal contrast, are shown and discussed. They bring an alternative point of view in taking advantage of strong space-time correlation found in thermal image sequences, without using especial mathematical transformations, or learning machines, but making intensive use of median statistical operation and classical heat propagation model, in order to enhance thermograms quality.

3D ITERATIVE SPACE-TIME FILTERING FOR SNR ENHANCEMENT

About image noise filtering, the great diversity of filters that can be found for this purpose is justified by the key problem of how to vanish the random noise while preserving information corresponding to discontinuities in images (edges), which possibly delimit the objects of interest. This problem gets worse when the noise presents itself as a discontinuity that can be taken as a border.

Although there are several statistical and morphological techniques that can be extended easily from original 2D scheme to 3D implementation (Ponomaryov, 2007; Mertzios & Tsirikolias, 1998), in general, the multidimensional noise filtering develops under the following trends: 3D mathematical transformations of images (Dabov, Foi, Katkovnik, & Egiazarian, 2007; Hasan, 2014), video filtering, filtering of 3D surfaces or volumes (Mao, Ma, Zhao, & Li, 2006), and the combination of the last two methods for 4D and 5D processing in medical applications (Skala, Lipic, Sovic, Gjenero, & Grubisic, 2011; Skala, Kavanah, Lipic, Sovic, & Grubisic, 2013). In video filtering, motivation for multidimensional processing is based on the high correlation between space and time through consecutive frames (Meguro, Taguchi, & Hamada, 2009), even more, when the basic objective is to reduce the blur effect introduced by moving objects in the scene (Jetto, 1999). In typical active thermography experiments there is no relative movement among objects, but lateral diffusion of heat and optical aberrations causes blurring edges of possible defects, making more difficult detection and characterization when it combines with noise. For its hand, six noise sources can be found in thermal images (Valdés, Inamura, Valera, & Yao, 2006), which at the same time, can be grouped in two great classes: *fixed pattern noise*, and *random noise*. Fixed pattern noise is due to optical or electronic artifacts of the infrared sensor and may be almost cancelled when the use of *cold images* (Maldague, 2001). However, few works exist about random noise attenuation on thermographic images, different from classical 2D filters like Gaussian. This is partly because many thermograms processing techniques involving mathematical transformations, like *Thermal Signal Reconstruction* (TSR) or *Principal Component Thermography* (PCT), implicitly leads to a better SNR of images at the expense of changing the type of information contained in resultant sequences. About this issue, the work developed in (Valdés; Inamura, Valera, & Yao, 2006) establishes several algorithmic procedures that perform filtering on 2D signals, but able to be extrapolated to 3D arrays. Likewise, the research in (Alsaadawi & Netzelmann, 2010) highlights in proposing a spatio-temporal filtering based on volumes of dynamic sizes (instead of fixed size masks), which dimensions increasing according to a heat propagation model, while filter moves to thermograms corresponding to greater time instants. For its part, in (García, 2010) a multidimensional filter is presented based on *Discrete Cosine Transform* (DCT) to reduce considerably the noise level in thermal sequences, as long as the variable smoothing factor being adjusted manually.

In this chapter, a multidimensional processing approach is presented to attenuate noise in thermal imaging sequences, based on an extension of classical *convolution filters* to a 3D filtering that involves spatial and time information distributed through consecutive thermograms, but without compressing or transforming the thermal data into another type (Restrepo & Loaiza, 2012). Specifically, average, median, and directional average filters were implemented on MATLAB in a 3D fashion, and applied to a simulated thermal sequence with Gaussian noise added. Results obtained from tests show a consistent reduction of *root mean square error* (RMSE) on final images regard to 2D filtering performance; this reduction may become greater when 3D algorithms are executed iteratively.

3D Space-Time Filtering for Noise Attenuation

Starting from a thermal sequence S_T with N_k thermograms, each of them with N_i rows and N_j columns, the typical 2D filtering operation with an $m \times m$ mask (let us assume m as odd number) can be mathematically expressed as:

$$S_F\left(k\right) = F_{2D}^{m\times m}\left\{S_T\left(k\right)\right\}, \forall k \; with \; 1 \leq k \leq N_k \tag{1}$$

The intuitive implementation idea behind the extrapolation of a spatial filter to 3D space-time operation is to convert the 2D mask of pixels into a $m \times m \times l$ *parallelepiped*, as in Figure 1a (if $m = l = 3$, would be a *cube*, as in Figure 1b). The l planes conforming the parallelepiped correspond to regions of $m \times m$ pixels with identical positions, and belonging to thermograms: [$S_T(k-(l-1)/2),\dots, S_T(k),\dots, S_T(k+(l-1)/2)$]; and the pixel to be modified is the central one of the middle plane (thermogram k). However, in order to apply MATLAB instructions oriented to 2D arrays, a rearrangement of rows can be made to obtain a $\left(l \times m\right) \times m$ matrix kernel in which rows of the same index in each plane of the parallelepiped result consecu-

Figure 1. Space-time 3D kernel a) General 3D kernel b) Cube for 3D filtering with $m = l = 3$
Restrepo & Loaiza, 2012.

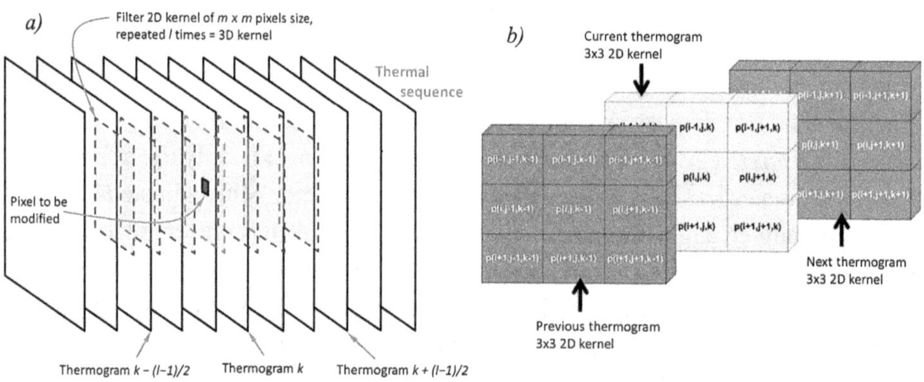

tively organized (one under the other) in this enlarged mask (see example in Figure 2b for a cube). After this reorganization the pixel to be modified will be located at the central point of this mask.

According to this strategy, the next steps summarize in general the 3D space-time noise filtering method.

Step 1: Each thermogram $S_T(k)$ of the sequence is expanded in $(m - 1)/2$ rows of pixels at each border (for $m = 3$, the expansion would aggregate only one row of pixels at each border), using a suitable criteria to reduce distortion at image periphery. The resulting sequence is an expanded version $S_{TE}(k)$.

Figure 2. Example of rearrangement of the 3D kernel of Figure 1b for 3D filtering implementation
Restrepo & Loaiza, 2012.

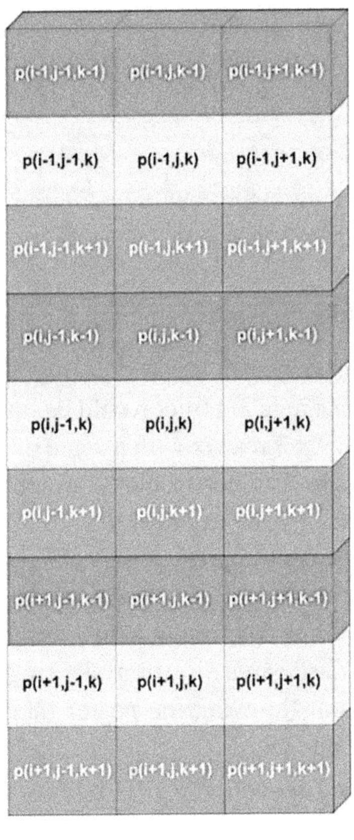

Step 2: The first and the last $(l-1)/2$ thermograms of $S_{TE}(k)$ are 2D filtered with the filter type selected. This step is needed because the 3D filter kernel needs $(l-1)/2$ temporal terms before and after the k term, so the first and the last $(l-1)/2$ thermograms could not complete de 3D kernel.

Step 3: To apply the 3D filtering to every thermogram $S_{TE}(k)$ (with $k = 2,..., N-1$), a new image $S_{TE}^{(l)}(k)$ must be created from it, with $(lx(N_i + 2)) \ x \ (N_j + 2)$ pixels size, by organizing consecutively groups of l rows with identical index in the thermograms $[S_T(k-(l-1)/2),..., S_T(k),..., S_T(k+(l-1)/2)]$ until all of their rows are swept (equation 2). Figure 2 shows an example of this rearrangement for $m = l = 3$.

$$\begin{gathered}
row\ l(i-1)+1\ \ of\ S_{TE}^{(l)}(k) = row\ i\ of\ S_{TE}\left(k-\frac{(l-1)}{2}\right) \\
\vdots \\
row\ l(i-1)+\frac{(l+1)}{2}\ \ of\ S_{TE}^{(l)}(k) = row\ i\ of\ S_{TE}(k) \\
\vdots \\
row\ l(i-1)+l\ \ of\ S_{TE}^{(l)}(k) = row\ i\ of\ S_{TE}\left(k+\frac{(l-1)}{2}\right)
\end{gathered} \tag{2}$$

Step 4: The selected 3D filter is applied through the sequence formed by thermograms $S_{TE}^{(l)}(k)$, with $k = 2,..., N-1$. With this methodology and taking $m = l = 3$, for example, an average 3D filter would be implemented filling all of the pixels of the enlarged 9 x 3 matrix with a value of 1/27 to create the convolution mask. The median 3D filter would be executed calculating the median over the values covered by the enlarged 9 x 3 matrix and taking this value to replace intensity of the central pixel of the enlarged matrix. For the directional-average filter, since it searches to preserve intensity transitions on edges better than simple average filtering, it is necessary to use four enlarged 9 x 3 masks depending on the desired direction of averaging (see Figure 3), to select later the filtering result throwing the nearer intensity value to the original intensity of the central pixel.

Step 5: A filtered expanded sequence $S_{TE}^{F_{3D}}(k)$ is organized, as equation (3) expresses and, later, the filtered sequence of size $N_i \times N_j \times N_k$ is recovered.

Figure 3. Organization scheme of the four planes forming the masks used in directional-average 3D filter
Restrepo & Loaiza, 2012.

1

1/9	0	0
1/9	0	0
1/9	0	0
0	1/9	0
0	1/9	0
0	1/9	0
0	0	1/9
0	0	1/9
0	0	1/9

2

0	1/9	0
0	1/9	0
0	1/9	0
0	1/9	0
0	1/9	0
0	1/9	0
0	1/9	0
0	1/9	0
0	1/9	0

3

0	0	1/9
0	0	1/9
0	0	1/9
0	1/9	0
0	1/9	0
0	1/9	0
1/9	0	0
1/9	0	0
1/9	0	0

4

0	0	0
0	0	0
0	0	0
1/9	1/9	1/9
1/9	1/9	1/9
1/9	1/9	1/9
0	0	0
0	0	0
0	0	0

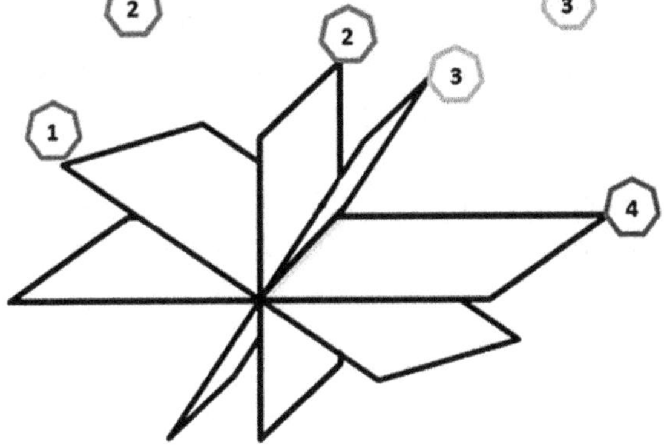

$$S_{TE}^{F_{3D}} = \left[F_{2D}^{m\times m} \left\{ S_{TE} \left(k\Big|_{k=1\ldots\frac{(l-1)}{2}} \right) \right\} \right] \left[F_{3D}^{m\times m\times l} \left\{ S_{TE} \left(k\Big|_{k=2\ldots N-1} \right) \right\} \Big| F_{2D}^{m\times m} \left\{ S_{TE} \left(k\Big|_{k=N-\frac{(l-1)}{2}+1\ldots N} \right) \right\} \right] \tag{3}$$

Selecting $m = l = 3$, the simplest 3D noise filter is obtained, and equation (3) would be transform to (4). This configuration was implemented to test performance of 3D filtering for thermograms quality enhancement.

$$S_{TE}^{F_{3D}} = \left[F_{2D}^{3\times3} \left\{ S_{TE} \left(1 \right) \right\} \Big| F_{3D}^{3\times3\times3} \left\{ S_{TE} \left(k\Big|_{k=2\ldots N-1} \right) \right\} \Big| F_{2D}^{3\times3} \left\{ S_{TE} \left(N \right) \right\} \right] \tag{4}$$

Regarding to implementation, average-based 3D filters does not cause too much problem about the runtime spent under previous scheme; but median 3D filters, even when small masks are used, can take too large execution times, overall if they are implemented in MATLAB. Although the best option is to codify these algorithms in languages like C++, there are fortunately, alternative ways to implement these types of filter in MATLAB, making use of *parallel computing toolbox* from three perspectives:

1. Taking advantage of multicore processors by means of parallel *for* cycles;
2. Dividing code and big size data into smaller parts to be executed by each core of a multicore processor;
3. Using *Graphics Processing Units* (GPU) to exploit their heavy multicore processing feature, in combination with a standard CPU.

3D Iterative Space-Time Filtering

In most of the cases, it is possible to introduce an iterative execution of classical 2D or 3D space-time filters to increase the smoothing, but preserving well enough the intensity transitions of image details; this is the key difference between iterative small- kernel filters and large-kernel filters. If we have an image *I(i,j)*, which can be expressed as a sum of an ideal image $I_e(i,j)$ and noise $\lambda(i,j)$, theory establishes that applying several times the same filter *F(i,j)* over *I(i,j),* the result would converge to (Maldague, 2001):

$$I_e\left(i,j\right) = I_{e-1}\left(i,j\right) + \left[I\left(i,j\right) - I_{e-1}\left(i,j\right)\right] * F\left(i,j\right) \tag{5}$$

The ideal image $I_e(i,j)$ corresponds to the image after the last iteration *e* of filtering, and the second term of previous summation is equivalent to a residual noise; If enough iterations are applied, this residual noise will tend to zero and the inner structure of the image is recovered, provided that filter *F(i,j)* be able to preserve relevant details of I_e.

Assuming that $F_{3D}^{m\times m\times l}$ is such a type of 3D filter with a $m \times m \times l$ kernel, and S_T is a sequence that can be seen as the sum of an ideal sequence S_{Te} and random noise, the root mean square of the discrepancies between S_T and S_{Te} becomes a good filtering assessment index according to equation (6):

$$RMSE = \sqrt{\sum_k \left[\sum_{i,j} \left(F_{3D}^{m\times m\times l}\left\{S_{Te}(i,j,k)\right\} - F_{3D}^{m\times m\times l}\left\{S_T(i,j,k)\right\}\right)^2\right]}, \forall i,j,k \tag{6}$$

Performance of $F_{3D}^{m\times m\times l}$ can be evaluated with respect to the known ideal sequence, what is useful when artificial clean and noisy images are utilized; but in practice, when S_{Te} is unknown and it can be just estimated, it is necessary to find a stop criterion for iterations, especially when the number of iterations may become large if a zero residual noise (equation (5)) is desired. In that sense, the RMSE of equation (6) was taken and modified to adapt it to a convenient stop criterion, as expressed by equation (7): $RMSE_{(\delta)}$ refers to the whole discrepancy between the filtered S_T in iteration δ and the filtered S_T one iteration before ($\delta-1$), taking the first sequence as that closer to the ideal S_{Te}.

$$RMSE_{(\delta)} = \sqrt{\sum_k \left[\sum_{i,j} \left(F^{m \times m \times l}_{3D(\delta)} \left\{ S_T(i,j,k) \right\} - F^{m \times m \times l}_{3D(\delta-1)} \left\{ S_T(i,j,k) \right\} \right)^2 \right]}, \forall i,j,k \tag{7}$$

The idea is that the inner subtraction in equation (7) will tend to zero (and so RMSE) as $F^{m \times m \times l}_{3D}\left(S_T\right)$ approximates to S_{Te}, and so, the change in RMSE will be smaller each iteration as it get close to zero. Then, the following stop criterion is established:

$$\frac{RMSE_{(\delta)} - RMSE_{(\delta-1)}}{RMSE_{(\delta-1)}} \leq \varphi, with\ 0 \leq \varphi < 1 \tag{8}$$

Space-Time Filtering Performance

In order to show performance of space-time noise filtering with average, directional-average and median 3D kernels, an artificial thermal sequence was synthesized with ThermoCalc6L software, simulating a CFRP slab with 9 internal air-filled squared flaws; later it was distorted by adding Gaussian noise, with zero mean and variance equal to 0.001. In Figure 4, two thermal profiles (in time) are taken from one pixel located on a sound area, and the other, on a defective area; each graph shows them through the sequence without noise, with added noise, filtered by a 2D median kernel, and filtered by a 3D median kernel, as examples. From this figure is evident that 3D filtering smooths the spatial temperature distribution at the same time that it is able to attenuate considerably fluctuations of pixels over the time, during one only filtering action. The result is a better separation among healthy and defective profiles, which favors discrimination and later detection.

Figure 5 shows graphically RMSE values (according to equation (6)) regard to the original artificial sequence without noise, S_{Te}, for the initial unfiltered noisy sequence S_T and filtered sequences using 2D and 3D average, directional-average and median filters, executed in iterative as in non-iterative fashion; for the case of iterative application of 3D filters, a value of $\varphi = 0.1$ was chosen as standard.

Figure 4. Thermal profiles of a healthy and a defective pixel from: a) original sequence; b) noisy sequence; c) filtered sequence with a 2D median filter; d) filtered sequence with a 3D median filter

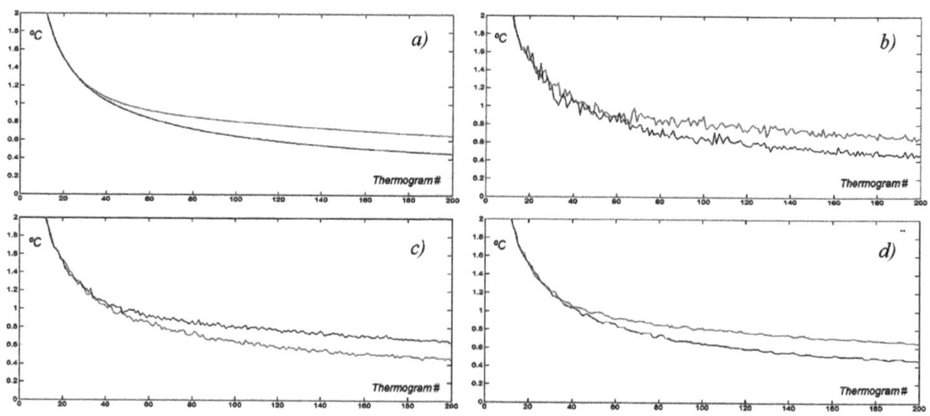

From this figure is clear that 3D filtering throws better results than their 2D versions: 26%, 43% and 34% in RMSE reduction for average, directional-average and median filters, respectively. This affirmation is true, even when 2D filters are iteratively applied. It is interesting how directional-average filter implemented in a space-time (3D) scheme brings results comparable to that obtained with 2D median filter, having in consideration that the first one is more conservative. Finally, the iterative 3D strategy offers the more decisive attenuation, especially for median filtering: additional RMSE reductions of 9% and 31% with directional-average and median

Figure 5. RMSE values for noisy sequence processed by different 2D and 3D filters

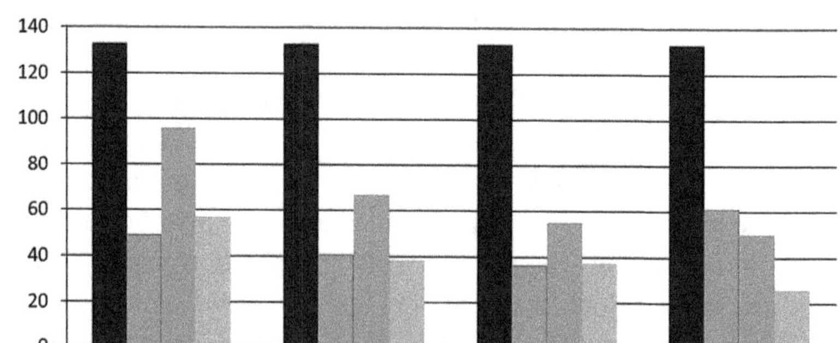

	2D	Iterative 2D	3D	Iterative 3D
■ Noisy sequence	133	133	133	133
■ Average filter	49	41	37	61
■ Directional-average filter	96	67	55	50
■ Median filter	57	39	38	26

filter, respectively, are obtained with respect to 3D non-iterative implementation. However, despite of RMSE reduction, it is possible that 3D and iterative filters above all, may introduce some artifacts to filtered images, depending on the type of filter. For example, iterative 3D median filters usually produce a *"stair"* effect on thermal profiles, so an extra 2D smoothing filtering may be advisable in such a case. Anyway, these results apply in similar situations whenever the main objective is to enhance the contrast between anomalies and sound areas of material. It is very possible that in images with natural scenes, some filters like directional-average filter, bring a better performance than the median one (Faúndez, 2001).

For comparison purpose, and observing the excellent results obtained from it, *multidimensional DCT-based filter* was probed with the same artificial sequence. It was found that for a small range of values of the called *smoothing factor*, the DCT-based filter reaches an RMSE better than those obtained for 3D average and median filters; only the iterative 3D median filter is able to produce an even smaller RMSE. Nevertheless, the basic problem with the DCT-based filter is that the optimal range for the smoothing factor does not correspond to that found automatically according to (García, 2010), which in fact, is greater; and the greater the smoothing factor, the greater the resulting RMSE.

Finally, from an active thermography inspection of a CFRP slab with 25 squared insertions of Teflon to simulate internal defects, a real thermal sequence was captured and used to assess those space-time filters. This time, a global signal to noise ratio (SNR) is calculated (as the mean of SNR in each thermogram) around two particular defects, to compare performances among filters with respect to the raw sequence. These two defects are of 15 mm side in size, one located at 0.2 mm and the other, at 0.8 mm from surface. Figure 6 details SNR values for the same filters analyzed in Figure 5. The way of calculation consists of defining two areas: one covering the defective zone, and the other corresponding to a squared ring that surrounds the first one, as in (Madruga, Albendea, Ibarra-Castanedo, & López-Higuera, 2010; Restrepo & Loaiza, 2014a).

The results about SNR evaluation are noticeable consistent with those obtained from artificial sequence, respecting the quality enhancement of thermograms as the filter considers spatial and temporal information in an iterative way of execution. This conclusion is more relevant regarding those defects whose depth and size hinder their detection and later characterization. Another example of the quality enhancement reached with space-time filtering is that shown in Figure 7, where it is seen an automotive part thermally excited by means of a modulated ultrasound wave to provoke peaks of heat which are acquired by an IR camera. One healthy (point 1) and two defective (point 2 and 3) points were selected to extract their profiles. It is clear that the defect with poorest SNR is made detectable after 3D filtering, though the drawback is the global attenuation in intensity of registered profiles.

Figure 6. SNR values from a noisy sequence processed by different 2D and 3D filters, for: a) the biggest and shallowest defect; b) the biggest and deepest defect

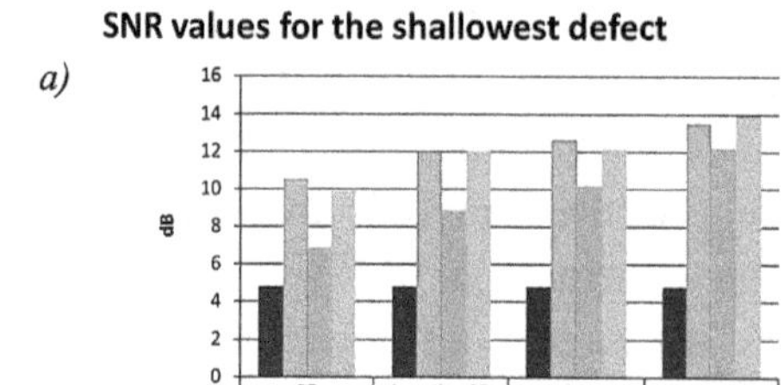

	2D	Iterative 2D	3D	Iterative 3D
Original sequence	4,8	4,8	4,8	4,8
Average filter	10,5	12,0	12,6	13,5
Directional-average filter	6,9	8,8	10,2	12,2
Median filter	9,9	12,0	12,1	13,9

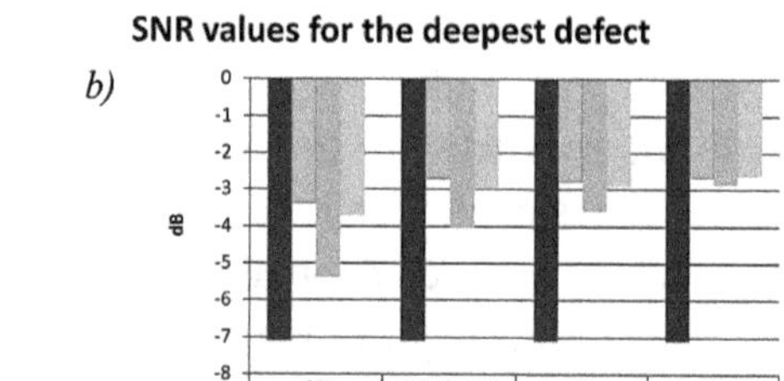

	2D	Iterative 2D	3D	Iterative 3D
Original sequence	-7,1	-7,1	-7,1	-7,1
Average filter	-3,4	-2,7	-2,8	-2,7
Directional-average filter	-5,4	-4,0	-3,6	-2,9
Median filter	-3,7	-3,0	-2,9	-2,6

Figure 7. Application of 3D space-time filtering to an automotive part: a) image of a mechanism similar to the used one; b) original thermogram; c) enhanced thermogram after 3D median filtering; d) enhanced thermogram after iterative 3D median filtering; e) thermal profiles from original sequence; f) the same thermal profiles after 3D median filtering; g) the same thermal profiles after iterative 3D median filtering

Courtesy of Visiooimage inc., Quebec City, Canada.

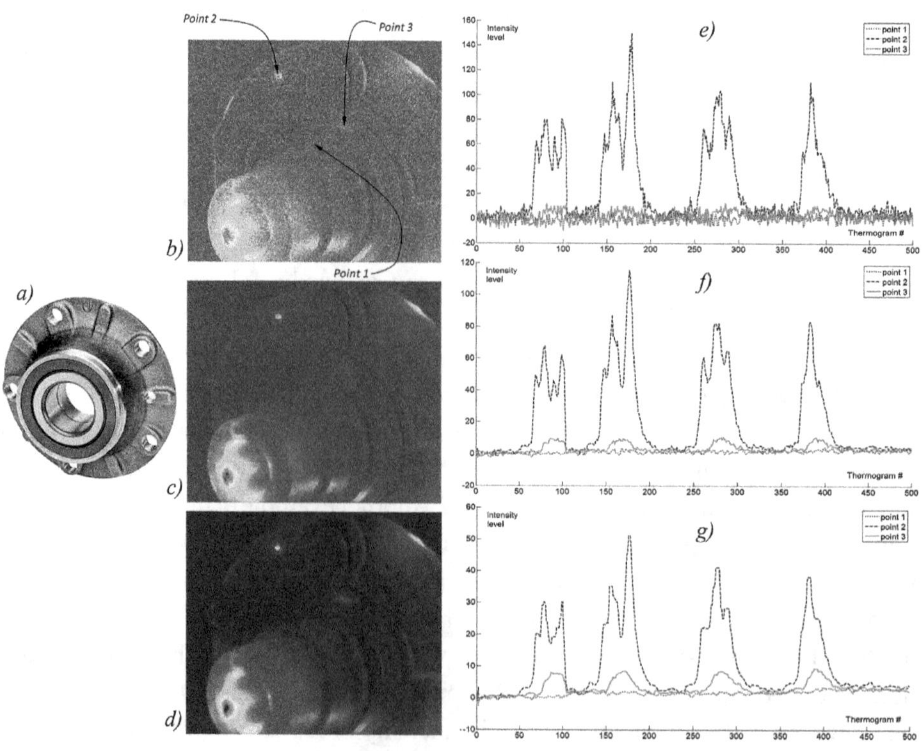

BACKGROUND THERMAL COMPENSATION BY MEDIAN FILTERING (BTCF)

Inside industrial world, two active thermographic techniques have reached a level of consolidation as reliable and practical methods for Nondestructive Testing (NDT) of materials: *Lock-In Thermography* (LT) and *Pulsed Thermography* (PT); however, the most common is the second one (Snell Jr. & Spring, 2007). In a PT experiment, the thermal excitation is generated by a heat pulse (using lamps or high intensity flashes) which width can be established from several milliseconds to a few seconds, according to the thermal conductivity of specimen under study; this fact makes it a relatively fast method, and hence, an attractive technique for industrial applications. The relevant information is found on the cooling curves from every point on the irradiated surface, as a defective point will tend to show a different behavior of its cooling profile respecting to a healthy point (Figure 8) from a certain time instant. This difference is called *thermal contrast* and represents a crucial requirement for detectability of possible defects (Maldague, 2001). Although many researchers consider this contrast decay with time as a weakness, a suitable pre-processing of pulsed thermograms can give advantages to this method over phase or lock-in techniques (Balageas, 2011).

Figure 8. Example of thermal profiles from a PT experiment, corresponding to a healthy and a defective point, and thermal contrast profile as difference between both

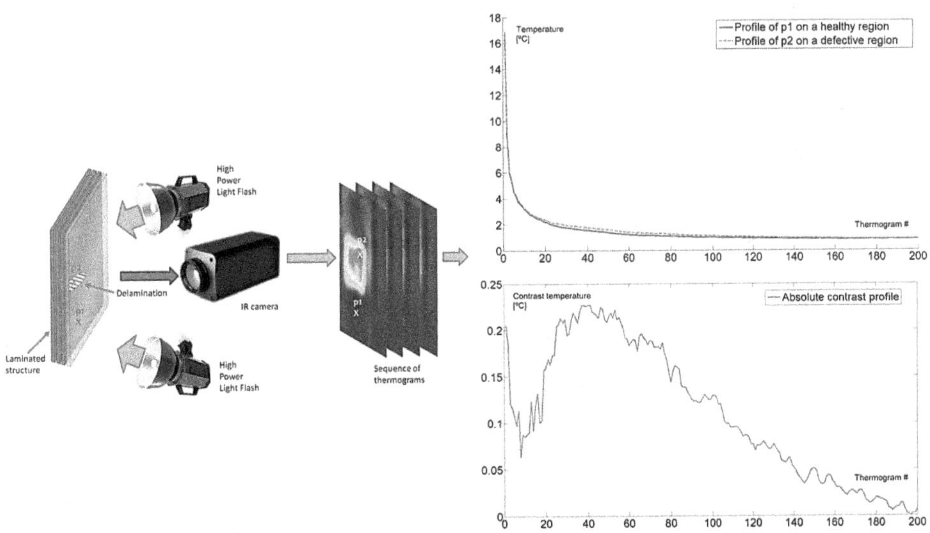

A quite remarkable effect of heat pulse excitation is that the background distribution of temperature (spatial profile) fits a concave surface. This behavior is basically due to the *non-uniform heating* caused by the natural dispersion of light coming from the lamps and radiating surface, which makes the focal lighting point brighter than its surroundings (Benítez, Ibarra-Castanedo, Bendada, Maldague, Loaiza, & Caicedo, 2009). Another phenomenon known as *vignetting*, contributes with this non-uniform background appearance by making the captured image to look darker as it get closer to borders. Vignetting is caused by distortion of camera lens, own camera thermal radiation, and conditions of the scene; but fortunately, modern cameras can automatically compensate this effect. Finally, variations in emissivity of the surface affect the apparent background heating, as well. The direct consequence of the presence of uneven heating in the thermograms acquired from PT experiments is the reduction in intensity contrast, which is a negative consequence for detection and subsequently characterization of any defective areas. So, the compensation of the uneven heating on thermal images improves the thermal contrast and the next image analysis. This section presents a thermal contrast enhancement technique able to compensate the non-uniform background heating resulting from heat pulse excitation and other types of apparent uneven heating. This technique presents the next key features (Restrepo & Loaiza, 2013):

- The method decouples the background thermal information from the information related to possible defects.
- The method does not depend on any mathematical model of heat diffusion.
- The method does not depend on determination of any intrinsic parameter of the inspected material.

Deterministic Mathematical Basis of Median Operations for Filtering

Median filtering is a very common strategy to remove impulse noise (*"salt & pepper"* noise) or little spurious details from images because it causes very low blurring effect over edges. This fact is due to two reasons: median operation does not attenuate the difference of intensity across edges, and also, it does not shift the borders of objects unlike linear mask filtering (Russ, 1998). For its part, the less common feature of median filtering consists of eliminating an unwanted region of certain area in pixels, whenever a kernel of enough size is used (González & Woods, 2002). This last fact can be easily seen thinking in a region R_A with area A in pixels, delimited by a border of intensity g, contained in a gray image I, like in Figure 9. Let us assume now that intensities of pixels in R_A are greater than g, and intensities of its surroundings (I-R_A) are smaller than g; if a squared mask K_M of M pixels side (M being odd) is used to perform a median filter over I, it can be demonstrated that result will be the suppression of the region R_A if the equation (9) is satisfied:

$$A \leq \frac{M^2 - 1}{2} \tag{9}$$

Pixels belonging to the region R_A will be replaced by intensities smaller than g, i.e. by pixels belonging to the region $I-R_A$ and included in K_M each time the median kernel operates. Obviously, in real situations where R_A has borders with smooth slopes, g must be little enough to contain the most of the region. R_A could be a cluster of multiple regions with total area A in pixels, as well.

A more analytical way to examine the capability of median operation in eliminating certain outstanding portions from a signal is sustained on deterministic properties demonstrated in (Barner & Arce, 1998). According to this analysis, restricted to 1D signals for clarity, it can be established four shape structures (shown in Figure 10) to which any discrete time series matches, in function of the length M of the median window:

- *Constant neighborhood* or sequence of at least $(M + 1)/2$ consecutive samples at a constant value.
- *Edge*, defined as a monotonic sequence between two constant neighborhoods at different value and adjacent to it. The monotonic segment must not have any constant neighborhood.
- *Impulse*, defined as a constant sequence of at most $(M - 1)/2$ samples, between two constant neighborhoods presenting the same value and adjacent to the first one. The three sections must have M samples at least.

Figure 9. Relation between sizes of median kernel and the area to be suppressed: a) example of R_A and gray level g, containing a little more than 400 pixels; b) R_A replaced by gray levels outside, using a median kernel of 29 pixels side

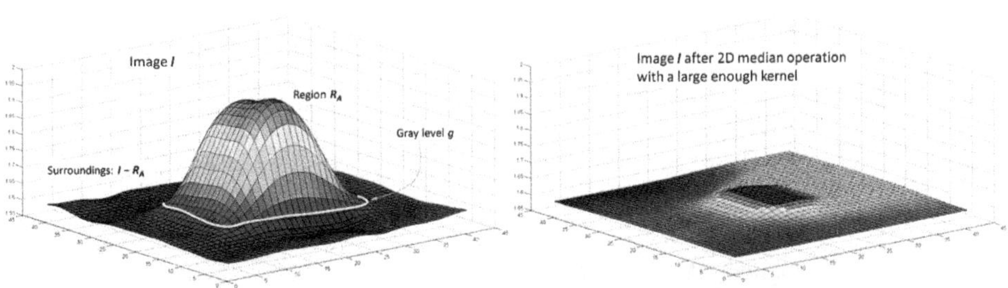

- *Oscillation* or any sequence which cannot be classified as one of the previous structures.

Consider a 1D median filter composed by an M element vector (M being odd) acting as a sliding window that shifts one sample at a time, and performing the median operation over M samples of a signal $s(n)$. The resulting signal $r(n)$ after application of this filter can be:

- *The same signal* $s(n)$, if $s(n)$ consists of a monotonic sequence of samples, an edge, a constant neighborhood, or a combination of the last two structures;
- Or *different to* $s(n)$, if $s(n)$ consists of any impulse or any oscillation (all of the rest complex signals). Specifically, any impulse of at most $(M - 1)/2$ samples will be eliminated by the median filtering of size M.

Observe that $r(n)$ will be equal or not to $s(n)$ depending on the length of $s(n)$ as much as its shape. In general, the signal structures which will not be modified by a median filtering of size M (the first case) are called *median filter roots*, and the most interesting feature is that any complex signal being not a median filter root will tend to become a root if a median filter of suitable length is *executed repeatedly*. This conclusion is too much important because it is saying that median operation is able to recover the basic structure of a signal (low frequency information). Actually, this

Figure 10. Example of 1D sequence presenting the four defined structures

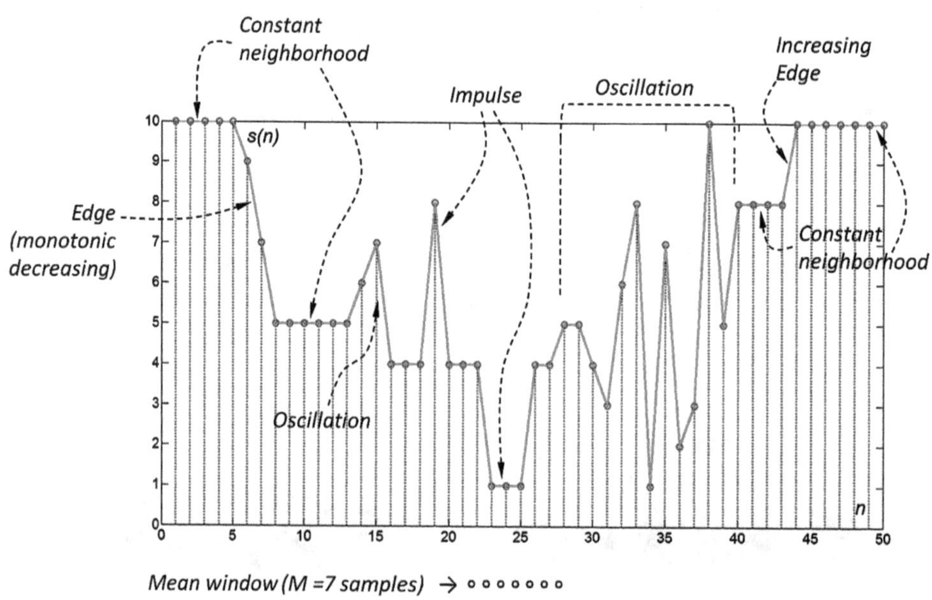

property supports theoretically the use of median filtering for thermal background compensation because, in active thermography experiments, this dynamic background profile coincides typically with a Gaussian function shape, as seen in the Figure 11. This usual shape could be segmented into one increasing monotonic sequence, one constant neighborhood containing the maximum intensity of the profile, and one decreasing monotonic sequence. A model of two only monotonic sequences for this case will suffer a smoothing effect at the point of union between them, producing the previous three-structure model, at the end. If this top segment is at least $M/2+1$ samples length, the whole structural shape for this exemplary background profile should be a root for a median filter of M samples length.

Going further in this reasoning, the mentioned 1D impulse structure of at most $(M-1)/2$ samples can be extrapolated to a 2D impulse structure of at most $(M^2-1)/2$ of area in pixels (according to heuristic expression in equation (9)), so that a squared median kernel of area M^2 pixels (with M odd) will remove it from original image. This type of filtering is like a *"masking"* operation where the region to be suppressed is replaced directly by values found in surrounding pixels, and quite different to a *smoothing* or low-pass filtering operation, where these values are calculated through any criterion involving both: surrounding pixels and pixels belonging the region to be eliminated.

Nevertheless, it is clear from Figure 11c that a discrepancy exists between the top of ideal profile and the flat shape which median filtering will introduce. If size of mask is made larger each time, the top segment of ideal profile is also larger from the filter point of view, and the median filter output will be an evident constant neighborhood. So, there should be a compromise between a mask large enough to deal with expected defects and not too large to distort the background profile. In previous works about BTCF (Restrepo & Loaiza, 2013), an automatic method was proposed to define a suitable mask size assuring very low distortion at top of the

Figure 11. Background thermal profile approximation to a median filter root structure: a) example of real background heating; b) Gaussian model curve; c) segmentation of Gaussian model in three basic segments
Restrepo & Loaiza, 2014b, 2016.

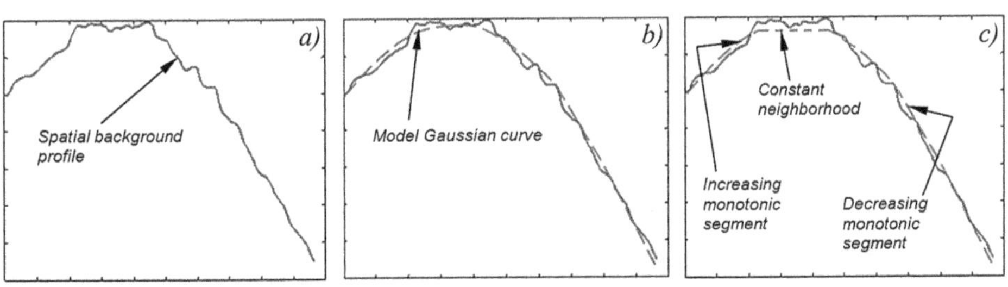

background profile. It was found that ¼ of the smaller side of captured thermograms is a good upper limit for this purpose in a typical experiment of slabs evaluation.

Thermal Background Estimation by Large Median Filtering

The work presented here starts assuming that detectable material heterogeneities, or defects, are the main objects to be removed from thermal images in order to estimate and preserve the background heating profile; for accomplishment of this task, the properties of median operation explained before are used. The next steps summarize the method of thermal background estimation by median filtering.

Step 1: The application of equation (9) brings a serious problem in the implementation of the median filter: the mask K_M could be considerably large depending on the expected size of defects; and the greater the median filter kernel size, the larger runtime. To decrease the filtering runtime while retaining the size of the filtered region (the same for K_M), a strategy was proposed, in which only some of the M^2 pixels are considered for the median operation: selected pixels are separated each other by a distance of ζ pixels along the four cardinal directions, in such a way that the relative position of the pixel to be modified does not change, and selected pixels along periphery of the processed region enclose the same area than the original K_M, as shown in Figure 12. With this strategy, expressed in equation (10), the median filtering value $F_{med}^{M,\zeta}$ for the pixel $p_{i,j}$ of the image I, using a mask of M pixels side and ζ distance between pixels, would demand approximately the same runtime as a median filter with mask of M/ζ pixels side; therefore, ζ plays the role of a *decimation factor*.

$$F_{med}^{M,\zeta}\left(I\left(p_{i,j}\right)\right) = median\left(I\left(p_{i\pm n\zeta,j\pm n\zeta}\right)\right) \; with \; n = 0,1,2,\ldots,\frac{M-1}{2\zeta} \; and \; M \; odd \tag{10}$$

In practice, user should choose a desired effective kernel side size η to evaluate the closer and smaller integer value for ζ. The value of η is related to M and ζ according to equation (11), so that the *effective kernel* of the final median filter is $\eta \times \eta$ pixels; and the final median mask is called *decimated maskK*$_{M\zeta}$.

$$\eta \leq \frac{M-1}{\zeta} + 1 \tag{11}$$

Figure 12. Example of a decimated kernel for median calculation with side M = 13 pixels and decimated factor ζ = 3, to form an effective mask of 5 × 5 pixels
Restrepo & Loaiza, 2014b, 2016.

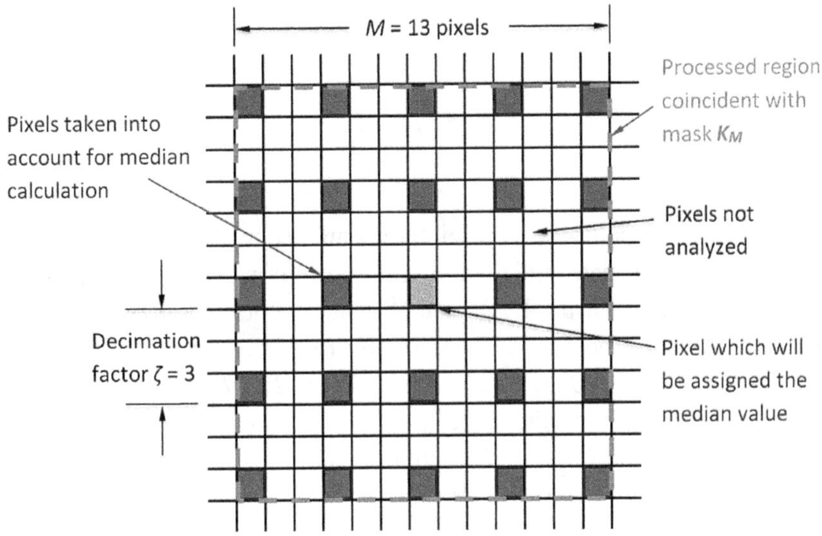

Step 2: Since the median mask will cover a considerably large area, image borders can become adversely affected depending on the extension or *padding* method used. Although, in absence of noise, the profile at each border of a typical thermogram is an almost perfect monotonic segment, there will be always the probability that residuary noise, emissivity fluctuations, or other artifacts introduce the risk of some lateral distortion if the values of the outmost pixels are simply sustained far away the image borders. Thereby, a *linear border expansion* method is proposed to extend image borders trying to maintain the tendency of peripheral temperature distribution.

Assuming a $N_i \times N_j$ pixels image *I*, and a decimated mask covering a M^2 area, this padding method consists of taking a rectangular region B_E formed by $(M - 1)/2$ adjacent pixel rows width from each border in order to calculate the intensity mean $\overline{x}_q$ (with $q = 1,..., (M\text{-}1)/2$ going into the image) for every row of B_E (see figure 13a). From all of the $(M - 1)/2$ mean values, a fitting straight line is found to define the tendency of temperature as it gets close to the image border. From this regression line, it is able to calculate the adjusted mean values $\hat{\overline{x}}_q$ and estimate the mean values of the rows beyond the image border $\hat{\overline{x}}_r$ (with $r = 1,..., (M - 1)/2$ going outside the image) which will create finally the *expanded border*B_{EX} of $(M - 1)/2$

pixels wide, as well. The intensity of any pixel p_{rv} contained in B_{EX} can be calculated by means of proportion between the mean of the artificial row r and the mean of the row just corresponding to border of the original image ($q = 1$), as defined by equation 12.

$$p_{r,v} = \frac{\hat{\bar{x}}_r}{\hat{\bar{x}}_{q=1}} p_{q=1,v} \tag{12}$$

This procedure remains the same either if r and q correspond to a matrix row or a matrix column coordinate, and then, v will be a column or row coordinate, respectively. The more efficient implementation of the method starts with two sides of image I, for example, the first and last matrix rows, and so it will build two expanded regions: B_{EX1} and B_{EX3}, each one of $(M - 1)/2 \times N_j$ pixels size. Later, B_{EX2} and B_{EX4} are made applying the same technique but including previously the expanded borders B_{EX1} and B_{EX3} in the regions B_{E2} and B_{E4}; it is to say that B_{E2} and B_{E4} result to be $(N_i + M - 1) \times (M - 1)/2$ pixels size, and then, B_{EX2} and B_{EX4} will be the same size.

Step 3: The decimated median filter, with mask of M^2 area and $\eta \times \eta$ pixels, can be executed taking care of returning the resulting image to the original size to obtain the estimation of the background temperature distribution, i.e. the thermal behavior of the material if it does not have defects. Calling S to the original thermographic sequence, composed by N_k thermograms, and $S(t_k)$ to the thermogram at instant t_k, equation (13) represents this operation to obtain the *background heating sequence*S^H.

$$S^H = F_{med}^{M,\varsigma}\left(S\left(t_k\right)\right), \forall t_k \ with\ k = 1, \ldots, N_k \tag{13}$$

Figure 14 shows two examples of thermal background estimation after application of the decimated median filter to real thermograms taken from a CFRP slab inspection. The global result is a decoupling of two types of information: one of low-frequency (healthy behavior) and the other of high frequency (defective behavior). This decoupling feature is exploited in (Florez-Ospina & Benítez, 2014) to propose a global signal to noise ratio (GNSR) in order to evaluate defect detectability in processed infrared images. In that technique, the decimated median filter is applied firstly to recover the background temperature distribution and develop a noise analysis of estimated thermograms without the presence of possible defects.

Figure 13. Example of linear border expansion: a) estimation of behavior tendency of borders according to side region B_E; b) estimation of intensities for new pixels belonging expanded region B_{EX}

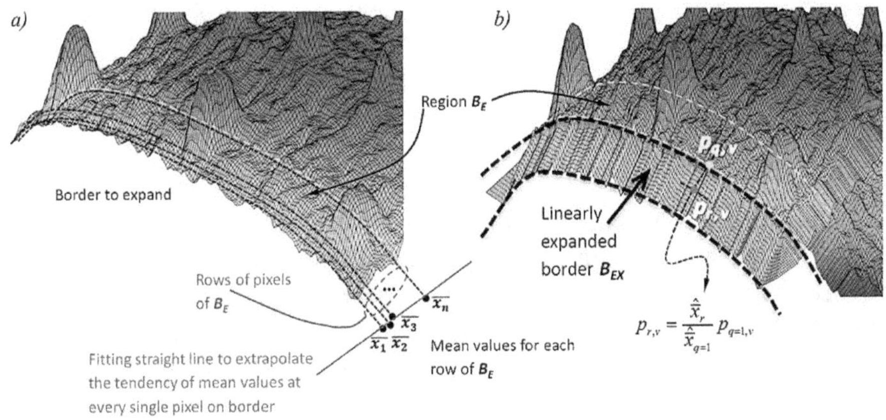

Figure 14. Example of decoupling of background thermal information by means of decimated median filtering on thermograms from CFRP slab inspection

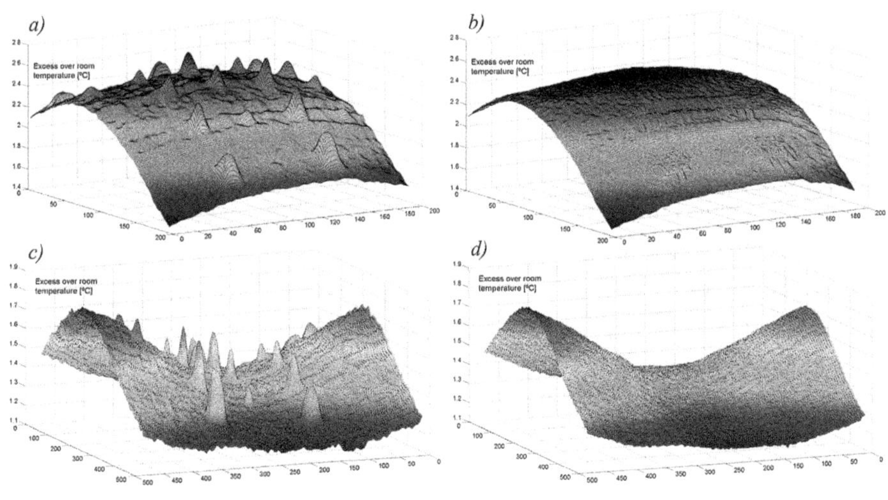

Background Thermal Compensation for Contrast Enhancement

The *background thermal compensation by filtering* (BTCF), for contrast enhancement of thermograms, bases on the *unsharp filtering* scheme, where details on an image I are emphasized by subtracting a smoothed version of that image $F_{smooth}(I)$ from itself: $I_{emphasized} = I - F_{smooth}(I)$ (González & Woods, 2002). In this sense, BTCF coincides with the strategy named *filtered contrast*, proposed by Grys and Minkina (2010); but in the case of BTCF method, the median filter is used to "*mask*" the heterogeneities which are looked for, instead of blurring them as classical low pass filtering does. Thereby, the final contrast sequence ΔT_{BTCF} is obtained by carrying out the difference in equation (14).

$$\Delta T_{BTCF} = S\left(t_k\right) - S^H\left(t_k\right), \forall t_k \ with \ k = 1,\dots, N_k \tag{14}$$

Additionally, it is possible to define a normalized version for BTCF technique, taking a reference thermogram at instant t_r, and doing the operation in (15); this reference image should be taken from the sequence at an instant when there are still no detectable defects at surface. The goodness of normalization locates in the ability to compensate certain noise and fluctuations patterns which may be not quite evident and are not adequately filtered by the decimated median filter, leading to even better results than ΔT_{BTCF}, though at the expense of requiring a reference thermogram selection.

$$\Delta T_{BTCF}^{Norm} = \frac{S\left(t_k\right)}{S\left(t_r\right)} - \frac{S^H\left(t_k\right)}{S^H\left(t_r\right)}, \forall t_k \ with \ k = 1,\dots, N_k \tag{15}$$

In order to show performance of BTCF in thermal contrast enhancement, a real thermal sequence was captured from an active thermography inspection of a CFRP slab with 25 squared insertions of Teflon to simulate internal defects, of different sizes and at several depths. Table 1 presents the acquisition conditions for this experiment, and Figure 15 shows a scheme of the CFRP slab used.

Figure 16 presents compensation results obtained with BTCF through different schemes of application, considering the thermogram taken as example in Figure 14b. For its part, in Figure 17 several thermal profiles appear from BTCF contrast sequence.

Table 1. Acquisition conditions in the experiment

Camera	Image Resolution [Pixels]	Intensity Resolution	Acquisition Frequency	Excitation Source	Time Window	Frames Captured
PHOENIX InSb CCD Matrix	512x640 max, 470x475 used	14 bits	87.8 Hz	2 Photograph flash	3.42 s	300

Figure 15. Scheme of the CFRP slab used to generate the thermal sequence for contrast enhancement evaluation
Restrepo & Loaiza, 2014b, 2016.

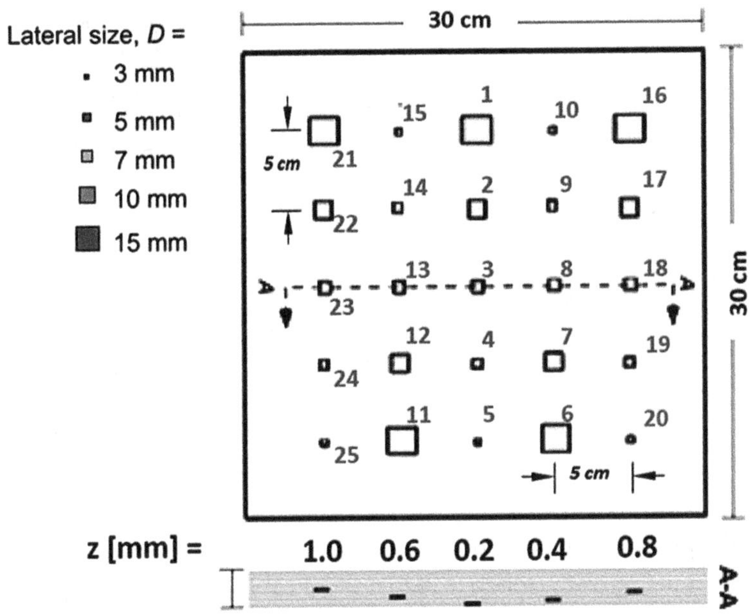

Figure 16. Example of contrast thermograms (3D surfaces) after BTCF application: a) normal BTCF (equation 14); b) BTCF + truncation; c) normalized BTCF (equation 15) + truncation

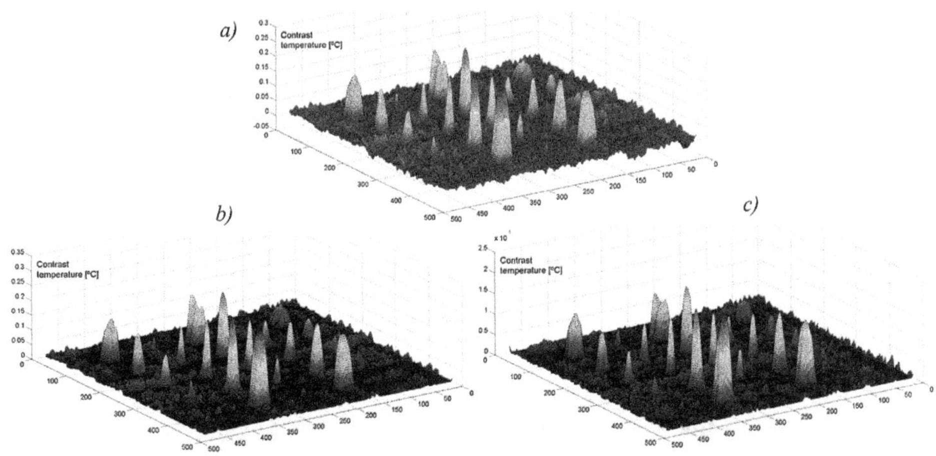

Figure 17. Example of BTCF thermal contrast profiles from the bigger defects
Restrepo & Loaiza, 2014b, 2016.

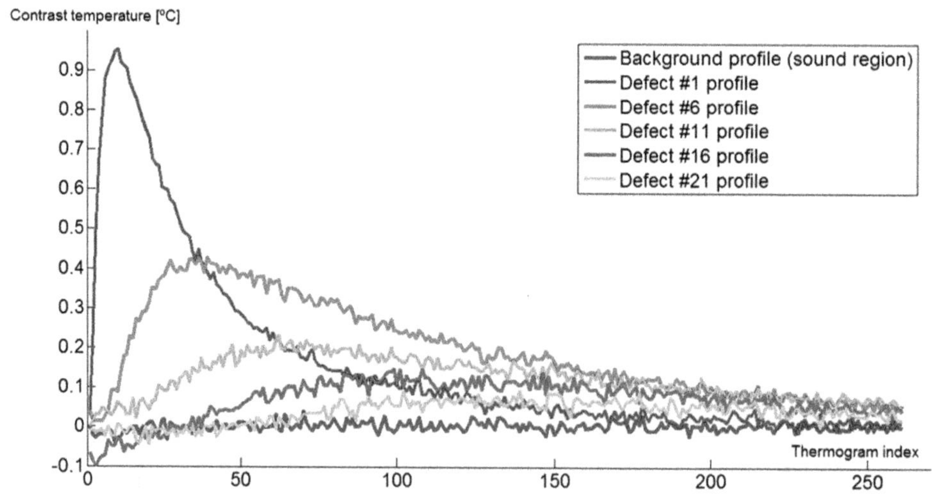

From the previous figure is noticeable that the contrast of sound regions, as in spatial as in temporal way, tends to stay around zero, from a global point of view; this fact allows negative values of contrast temperature can be cut or *truncated*, leading to a quite better contrast. Evolution of thermal profiles with time, also evidence that the background is almost canceled. In general, it can be said that all of the thermal profiles tend to be flat and close to zero when they do not lie over defects, and when they evolve to thermal equilibrium. Moreover, the normalization also brings certain attenuation of some patterns originated in structural fibers of the slab.

Although the main objective of any technique for thermal contrast enhancement is to attenuate the non-uniformity heating effect in a sequence *S*, definition of a parameter or index for assessment of thermal image quality is a recent research trend. An interesting referenced quality index, based on standard deviations of the original (reference *A*) and the resulting processed (objective *B*) images, was defined by Wang and Bovik (2002) and modified by Benítez (2012) to adapt it as a thermal contrast enhancement index *CE*:

$$CE = 1 - \frac{2\sigma_A \sigma_B}{\sigma_A^2 + \sigma_B^2} \qquad (16)$$

CE will give a value closer to one when $\sigma_B << \sigma_A$, what would indicate that the background heating was reduced; however, this result could be caused by an undesirable attenuation of defects intensity, and not really by a better contrast. Moreover, this index cannot be utilized when, in processing the reference image, temperature values are transformed to normalized values in the resulting image, for example. From this point, an index $\vartheta(k)$ was proposed for assessment of this purpose, in such a way that, for each thermogram $S(k)$, it relates the intensity span of all of the healthy regions of the material with the intensity span of the entire image, as expressed by equation (17); the healthy regions of material are represented in a subset H of predefined pixels p. This index, called *background relative non-uniformity* (BNU), will give 1 when there is no difference between healthy and defective zones (null contrast), and zero when sound regions of material exhibit an ideal constant intensity (maximum contrast). Since for a sequence of N_k thermograms we will have a series of N_k BNU values, a global BNU (GBNU) is defined as the mean of these values to condense in one only number the contrast feature of an entire sequence.

$$BNU\left(k\right)=\frac{\max\left(S\left(k,p\in H\right)\right)-\min\left(S\left(k,p\in H\right)\right)}{\max\left(S\left(k\right)\right)-\min\left(S\left(k\right)\right)} \tag{17}$$

$$GBNU=\frac{1}{N_k}\sum_{k=1}^{N_k}BNU\left(k\right) \tag{18}$$

Table 2 presents the performance of BTCF in different schemes, based on values for GBNU and global SNR, the last one, calculated for the shallowest (#1) and the deepest (#21) defects of 15 mm side. Also, NC and CAD contrast performances are included for comparison purposes. BTCF In all cases, M = 57 pixels (around 1/8 of the shorter side of images) and η = 5 pixels (resulting ζ = 14), because they represent the best compromise between performance and execution time for the given resolution. The sequence was noise filtered with a three-iteration Gaussian filter, taking $\sigma = 0.7$. For more knowledge about BTCF performance sensitivity with respect to median kernel parameters, read (Restrepo & Loaiza, 2016).

From this table it is clear that truncated versions of BTCF sequence (as in absolute as in normalized scheme) give appreciable better results about BNU and SNR than the filtered and CAD sequences; NC sequence offers comparable values than BTCF, but with the only drawback of requiring the *a priori* selection of a healthy area. For its part, the normalization procedure influences positively the flatness of background profiles according to smaller BNU values, though SNR over defects tends to decrease a little; on the contrary, normalization followed by truncation has really a very positive impact on SNR for deeper defects, increasing detectability for defects of smaller diameter/depth ratio.

Table 2. GBNU and SNR values for considered defects from the contrast sequence obtained after application of different BTCF schemes

	GBNU	SNR – Defect #1	SNR – Defect #21
Original thermal sequence	0.64	5.2	-6.4
Sequence after iterative Gaussian noise filtering	0.73	12.8	-2.8
NC Sequence	0.23	12.5	4.1
CAD Sequence	0.42	12.9	2.4
BTCF sequence	0.23	15.6	5.2
BTCF sequence + truncation	0.18	20.7	8.3
Normalized BTCF sequence	0.19	14.7	5.1
Normalized BTCF sequence + truncation	0.16	19.7	9.4

FINITE DIFFERENCE ESTIMATION OF SURFACE THERMAL BACKGROUND DISTRIBUTION FOR CONTRAST ENHANCEMENT

As said before, the main problem in obtaining contrast enhanced thermal images from a PT experiment is the fact that excitation heating does not distribute uniformly over the radiated surface, and in real cases, material emissivity variations and optical distortions of the IR sensor also contribute to this non-uniformity effect. The simplest contrast method, called *absolute contrast* (AC), is in general not suitable for this objective because it performs a subtraction between profiles from every spatial point on a sequence and the profile of an arbitrary healthy point (or the average profile of an arbitrary healthy region). Although there are refinements of the AC method, all of them have the crucial drawback (at least for automating purposes) of requiring a sound area selection before knowing where the internal flaws are, if they exist (Larsen, 2011; Benítez, Loaiza, & Caicedo, 2011). Among those improved techniques from AC, *normalized contrast* (NC) is the most important.

In order to overcome this challenge, the *differentiated absolute contrast* (DAC) method was proposed (Pilla, Klein, Maldague, & Salerno, 2002) to answer the problem of a priori selection of a sound area, using the 1D model of adiabatic heat transfer over the surface of a solid semi-infinite medium without internal flaws, heated by a thermal Dirac pulse. The technique pretends to predict the temperature evolution in every healthy point of the heated surface, to subtract it from the real temperature behavior, obtaining so greater discrepancies from defective regions than from sound areas. The good results of DAC execution depend mainly on the degree of approximation of the real specimen to a semi-infinite medium, and of the heating pulse to an impulse excitation, and so, DAC may be not an adequate choice for slabs evaluation. This difficulty was resolved by the definition of the *modified-DAC* (m-DAC) technique (Benítez, Ibarra-Castanedo, Bendada, Maldague, Loaiza, & Caicedo, 2008) that uses a corrected version of the previous 1D heat transfer model, by including the concept of *thermal quadrupoles* (Maillet, Houlbert, Didierjean, Lamine, & Degiovanni, 1993) working in Laplace space for simplicity. The new model needs values for thickness of the layer, time width of the heating pulse, and diffusivity of the material to attenuate distortion of thermal profiles introduced by the real shape of the heating pulse and reflection of the heat wave in the opposite side of the layer.

Nevertheless, the remaining problem was about using a 1D model to resolve a 3D problem. Though the first represents a more practical and light solution, it may evidence appreciable discrepancies with respect to real time evolution and spatial distribution of temperature. On the other hand, 3D methods have a drawback in the computational effort required to reconstruct the thermal distribution inside the evaluated specimen. The work in (Grinzato, 2000) is one of the first intentions to involve a 3D heat conduction model in thermal contrast, resulting in the *3D filtering* technique that estimates the 3D temperature evolution of healthy material to subtract it from the real one. In that scenario, the technique described in this section does not perform temperature estimation in three spatial axes throughout the material in every sample time (like *finite elements* reconstruction models): for each pixel, it estimates the value of temperature in the next time, only on the frontal surface of the specimen, from actual real temperature data of the same pixel and its neighbors (Restrepo & Loaiza, 2014a). This feature not only makes this technique to be able to run faster, but to be more reliable because the processing involved is quite different than estimating a detailed thermal distribution from ideal initial conditions and theoretical data.

3D and 1D Differential Thermal Estimation Model for Slabs and Semi-Infinite Mediums

The 3D model of heat propagation by conduction (convection and radiation not considered) through a specific medium is given by:

$$\frac{\partial}{\partial x}\left(\kappa_x \frac{\partial T(x,y,z,t)}{\partial x}\right)+\frac{\partial}{\partial y}\left(\kappa_y \frac{\partial T(x,y,z,t)}{\partial y}\right)+\frac{\partial}{\partial z}\left(\kappa_z \frac{\partial T(x,y,z,t)}{\partial z}\right)+w(x,y,z,t) = C\rho\frac{\partial T(x,y,z,t)}{\partial t} \tag{19}$$

where:

x,y,z,t: Spatial coordinates (z along depth axis) and time;
$T = T(x,y,z,t)$: Excess of temperature with respect to room temperature;
C, ρ: Specific heat and density;
κ_q: Thermal conductivity along q axis;
$w(x,y,z)$: Heat generation rate by volume unity [W/m^3] inside the propagation medium.

Assuming an homogeneous medium (even in an average way) with no heat generation inside, no loses by convection across its surfaces (after thermal excitation heat spreads throughout slab until thermal equilibrium is reached) and a thermal excitation very similar to an impulse, the previous model can be discretized by means of finite differences in explicit way, as follows:

$$\kappa_x\left(\frac{T^m_{(i-1)jk}-2T^m_{ijk}+T^m_{(i+1)jk}}{\Delta x^2}\right)+\kappa_y\left(\frac{T^m_{i(j-1)k}-2T^m_{ijk}+T^m_{i(j+1)k}}{\Delta y^2}\right)+\cdots+\kappa_z\left(\frac{T^m_{ij(k-1)}-2T^m_{ijk}+T^m_{ij(k+1)}}{\Delta z^2}\right)$$
$$=C\rho\left(\frac{T^{m+1}_{ijk}-T^m_{ijk}}{\Delta t}\right) \tag{20}$$

where:

$t_m = m.\Delta t$, con $m = 0, 1, 2,..., M - 1$, being M the total number of thermograms, and Δt, the acquisition period of frames;
$x_i = i.\Delta x$, con $i = 0,1,2,...,N_x - 1$, being N_x the total number of rows in each image; and $\Delta x = L_x/N_x$, being L_x the vertical length of image field;
$y_j = j.\Delta y$, con $j = 0,1,2,...,N_y - 1$, being N_y the total number of columns in each image; and $\Delta y = L_y/N_y$, being L_y the horizontal length of image field;
$z_k = k.\Delta z$, con $k = 0,1,2,...,N_k - 1$, being N_k the total number of steps dividing the thickness L_z of the slab; and $\Delta z = L_z/N_z$. Value of N_k is unknown in a real PT experiment;
T^m_{ijk} = temperature on pixel i,j,k for thermogram m;

Considering pixels as squared elements such that $\Delta x = \Delta y = \Delta p$, and rearranging terms, we have:

$$\frac{\kappa_x}{C\rho}\frac{\Delta t}{\Delta p^2}\left(T^m_{(i-1)jk}-2T^m_{ijk}+T^m_{(i+1)jk}\right)+\frac{\kappa_y}{C\rho}\frac{\Delta t}{\Delta p^2}\left(T^m_{i(j-1)k}-2T^m_{ijk}+T^m_{i(j+1)k}\right)$$
$$+\cdots+\frac{\kappa_z}{C\rho}\frac{\Delta t}{\Delta z^2}\left(T^m_{ij(k-1)}-2T^m_{ijk}+T^m_{ij(k+1)}\right)=T^{m+1}_{ijk}-T^m_{ijk}$$

Knowing that $\alpha_q = \kappa_q / C\rho$ is the *thermal diffusivity* of the specimen along q axis, if we define new variables β and Φ, and associate similar terms from previous equation, the desired basic model is obtained:

$$T^{m+1}_{ijk}=\left[1-2\left(\beta_x+\beta_y+\beta_z\frac{\Delta p^2}{\Delta z^2}\right)\right]T^m_{ijk}+\beta_x\Phi^m_{x,k}+\beta_y\Phi^m_{y,k}+\beta_z\frac{\Delta p^2}{\Delta z^2}\left(T^m_{ij(k-1)}+T^m_{ij(k+1)}\right) \tag{21}$$

where:

$$\beta_x=\alpha_x\frac{\Delta t}{\Delta p^2};$$
$$\beta_y=\alpha_y\frac{\Delta t}{\Delta p^2};$$
$$\beta_z=\alpha_z\frac{\Delta t}{\Delta z^2};$$
$$\Phi^m_{x,k}=T^m_{(i-1)jk}+T^m_{(i+1)jk};$$
$$\Phi^m_{y,k}=T^m_{i(j-1)k}+T^m_{i(j+1)k}$$

Considering that thermograms are referred to surface, then $k = 0$ and so:

$$T^{m+1}_{ij0}=\left[1-2\left(\beta_x+\beta_y+\beta_z\frac{\Delta p^2}{\Delta z^2}\right)\right]T^m_{ij0}+\beta_x\Phi^m_{x,0}+\beta_y\Phi^m_{y,0}+\beta_z\frac{\Delta p^2}{\Delta z^2}\left(T^m_{ij(-1)}+T^m_{ij(+1)}\right) \tag{22}$$

Here, $T^m_{ij(k-1)}$ and $T^m_{ij(k+1)}$ are unknown terms because the information captured by the IR camera only involves the temperature distribution along x and y axes, not along z (depth) axis. Since both terms are Δz away from T^m_{ijk}, it can be assumed that lateral diffusion of heat is negligible for such propagation. From this point, two approximated solutions of equation (19) proposed by Tadeu & Simões (2005) are taken to complete the model in equation (22).

Approximation Case for a Half-Space Propagation Medium

Tadeu and Simões used Green functions to propose approximated solutions for the general model of heat propagation by conduction (Tadeu & Simões, 2005, 2006) and convection (Tadeu & Simões, 2007), for *infinite* (or *unbounded*), *half-space* (or *semi-infinite*), *one layer* (or *slab*) and *multilayer* mediums. These models can be separated into two types of terms: the first one representing the influence of heat source; and the second one, the influence of border temperature and flux conditions (if convection effects are neglected). For the case of a half-space medium, assuming zero normal heat flux at the surface, the following equation arises:

$$T\left(x,y,z,t\right) = w\left(x,y,z,t\right)\frac{e^{-r_0/4\alpha\tau} + e^{-r_1/4\alpha\tau}}{\rho C\left(4\pi\alpha\tau\right)^{3/2}} \tag{23}$$

where:

$$r_0 = \left(x - x_0\right)^2 + \left(y - y_0\right)^2 + \left(z - z_0\right)^2;$$
$$r_1 = \left(x - x_0\right)^2 + \left(y - y_0\right)^2 + \left(z + z_0\right)^2;$$
$$\tau = t - t_0$$

Taking only the temperature evolution along z axis, and making $z_0 = 0$, $t_0 = 0$:

$$T\left(z,t\right) = w\left(z,t\right)\frac{2e^{-z^2/4\alpha_z t}}{\rho C\left(4\pi\alpha_z t\right)^{1/2}} \tag{24}$$

If equation (24) is evaluated for $z = z \pm \Delta z$, constant terms $\frac{2w\left(z,t\right)}{\rho C\left(4\pi\alpha_z t\right)^{1/2}}$ are grouped, and this result is used in equation (23), we obtain:

$$T\left(z \pm \Delta z,t\right) = T\left(z,t\right)e^{z^2/4\alpha_z t}e^{-\left(z\pm\Delta z\right)^2/4\alpha_z t}$$

Again, at surface $z = 0$, so:

$$T\left(\pm\Delta z,t\right) = T\left(0,t\right)e^{-\left(\pm\Delta z\right)^2/4\alpha_z t} = T\left(0,t\right)e^{-\Delta z^2/4\alpha_z t}$$

Returning to previous discrete notation:

$$T^{m}_{ij(\pm 1)} = T^{m}_{ij0} e^{-\Delta z^2 / 4\beta_z m \Delta p^2} \tag{25}$$

Replacing equation (25) in equation (22), the *discrete 3D differentiated thermal estimation model* for a half-space medium is finally obtained:

$$T^{m+1}_{ij0} = \left\{ 1 - 2\left[\beta_x + \beta_y + \beta_z \frac{\Delta p^2}{\Delta z^2}\left(1 - e^{-\Delta z^2 / 4\beta_z m \Delta p^2} \right) \right] \right\} T^{m}_{ij0} + \beta_x \Phi^{m}_{x,0} + \beta_y \Phi^{m}_{y,0} \tag{26}$$

This model is useful when the thickness of the propagation medium is great enough with respect to its diffusivity and the acquisition window of experiment; in this case, the effect of opposite side of the medium will not be seen in the thermal sequence. However, this model is also useful in depth estimation of defective zones, as explained in (Restrepo & Loaiza, 2015).

Approximation Case for a Slab Propagation Medium

According to usual PT experiments, it can be assumed that heat fluxes are null on both sides of the layer. For this situation, the temperature at point x,y,z inside the slab of thickness h, due to a heat point source placed at x_0, y_0, z_0, is determined by:

$$T\left(x,y,z,t\right) = \frac{w(x,y,z,t) e^{-r_0/4\alpha\tau}}{\rho C\left(4\pi\alpha\tau\right)^{3/2}} + \sum_{n=1}^{NS} w(x,y,z,t) \frac{e^{-r_1/4\alpha\tau} + e^{-r_2/4\alpha\tau} + e^{-r_3/4\alpha\tau} + e^{-r_4/4\alpha\tau}}{\rho C\left(4\pi\alpha\tau\right)^{3/2}} \tag{27}$$

where:

$$\begin{aligned}
r_0 &= \left(x - x_0\right)^2 + \left(y - y_0\right)^2 + \left(z - z_0\right)^2; \\
r_1 &= \left(x - x_0\right)^2 + \left(y - y_0\right)^2 + \left(z + z_0 + 2h\left(n-1\right)\right)^2; \\
r_2 &= \left(x - x_0\right)^2 + \left(y - y_0\right)^2 + \left(z + z_0 - 2hn\right)^2; \\
r_3 &= \left(x - x_0\right)^2 + \left(y - y_0\right)^2 + \left(z - z_0 + 2hn\right)^2; \\
r_4 &= \left(x - x_0\right)^2 + \left(y - y_0\right)^2 + \left(z - z_0 - 2hn\right)^2; \\
\tau &= t - t_0
\end{aligned}$$

NS is an arbitrary number of thermal wave reflections to be considered, depending on diffusivity of material: the smaller diffusivity, the higher the *NS* value. Taking the same assumptions than before, regarding to Δz propagation, equation (27) can be rewritten as:

$$T\left(z,t\right)=\frac{w(z,t)}{\rho C\left(4\pi \alpha t\right)^{1/2}}\left[e^{-r_0'/4\alpha_z t}+\sum_{n=1}^{NS}\left(e^{-r_1'/4\alpha_z t}+e^{-r_2'/4\alpha_z t}+e^{-r_3'/4\alpha_z t}+e^{-r_4'/4\alpha_z t}\right)\right] \tag{28}$$

with: $r_\xi' = r_\xi\Big|_{x=y=x_0=y_0=0}\ for\ \xi = 1,2,3,4$

If we assume again that heat pulse is applied at $t = t_0 = 0$ over the frontal surface ($z_0 = 0$) of the inspected slab, and define $\sigma_m = 4\alpha_z t_m = 4\beta_z m\Delta p^2$, equation 28 can be rewritten into discrete notation as:

$$T_{ijk}^{m} = \frac{w_{ijk}^{m}}{\rho C\left(4\pi\alpha_z t_m\right)^{1/2}}\left[A_k^{(NS)m}\right] \tag{29a}$$

where:

$$A_k^{(NS)m} = e^{-z_k^{\,2}/\sigma_m} + \sum_{n=1}^{NS}\left(e^{-\left(z_k+2h(n-1)\right)^2/\sigma_m} + 2e^{-\left(z_k-2hn\right)^2/\sigma_m} + e^{-\left(z_k+2hn\right)^2/\sigma_m}\right) \tag{29b}$$

As before, the incidence of the power density in the model can be eliminated by relating the actual temperature value, for any pixel, with its next temperature value; furthermore, evaluating in $k = 0$ to restrict the problem to the estimation of next temperature on the surface, follows:

$$T_{ij(\pm 1)}^{m} = T_{ijk}^{m}\frac{A_{(\pm 1)}^{(NS)m}}{A_0^{(NS)m}} \tag{30}$$

Hence, replacing unknown terms in equation (22) by their definitions according to equation (30), the *discrete 3D differentiated thermal estimation model* for a slab medium is obtained:

$$T_{ij0}^{m+1} = \left\{1 - 2\left(\beta_x + \beta_y\right) - \beta_z \frac{\Delta p^2}{\Delta z^2}\left[2 - \left(\frac{A_{(-1)}^{(NS)m} + A_{(1)}^{(NS)m}}{A_{0}^{(NS)m}}\right)\right]\right\} T_{ij0}^{m} + \beta_x \Phi_{x,0}^{m} + \beta_y \Phi_{y,0}^{m} \tag{31}$$

Models expressed in equations (26) and (31) only require knowledge about diffusivity of material and spatial resolution of the IR sensor (for estimation of Δp), since actual temperature values proceed from the pixel *i,j* of interest and its 4-neighbors. Finally, in order to ensure model stability and according to (Hensen & Nakhi, 1994), the values of Δp and Δt should satisfy the equation (32); the greatest value between α_x and α_y will have more incidence on this inequality. Parameter Δz does not affect stability of the model because the estimation of thermal evolution along a Δz displacement depends on a unidimensional function evaluated at z axis only.

$$\alpha_{x,y} \frac{\Delta t}{\Delta p^2} < \frac{1}{4} \tag{32}$$

Variants of Previous Models

Equations (26) and (31) are the more generic versions of the proposed differentiated thermal estimation models. Nevertheless, it is possible to work with more simplified versions when suitable conditions are satisfied. Specifically, when it is viable to affirm that material is an isotropic medium, diffusivity is the same in all directions: $\alpha = \alpha_x = \alpha_y = \alpha_z$. So, half-space and slab differentiated models become respectively (Restrepo & Loaiza, 2014c):

$$T_{ij0}^{m+1} = \left\{1 - 2\beta\left[2 + \frac{\Delta p^2}{\Delta z^2}\left(1 - e^{-\Delta z^2/4\beta m \Delta p^2}\right)\right]\right\} T_{ij0}^{m} + \beta \Phi_{0}^{m} \tag{33}$$

$$T_{ij0}^{m+1} = \left\{1 - \beta\left[4 + \frac{\Delta p^2}{\Delta z^2}\left(2 - \frac{A_{(-1)}^{(NS)m} + A_{(1)}^{(NS)m}}{A_{0}^{(NS)m}}\right)\right]\right\} T_{ij0}^{m} + \beta \Phi_{0}^{m} \tag{34}$$

where: $\beta = \beta_x = \beta_y = \beta_z$ and $\Phi_k^m = \Phi_{x,k}^m + \Phi_{y,k}^m$.

Furthermore, lateral heat diffusion is due to non-uniformity thermal excitation; but in inspection of thin layers the thickness of specimen is usually little enough to consider that lateral propagation is negligible. A similar case would happen when the inspection window is superficial regarding to thickness of a specimen, in such a way that it can be considered as a semi-infinite medium. So, it is possible to convert 3D discretization into a 1D operation as follows:

$$\kappa_z \left(\frac{T^m_{ij(k-1)} - 2T^m_{ijk} + T^m_{ij(k+1)}}{\Delta z^2} \right) = C\rho \left(\frac{T^{m+1}_{ijk} - T^m_{ijk}}{\Delta t} \right) \tag{35}$$

Finally, making the same applicable mathematical steps and approximations than before, the *discrete 1D differentiated thermal estimation models* for half-space and slab mediums are respectively:

$$T^{m+1}_{ij0} = \left\{ 1 - 2\beta_z \left(1 - e^{-\Delta z^2 / 4\alpha_z m \Delta t^2} \right) \right\} T^m_{ij0} \tag{36}$$

$$T^{m+1}_{ij0} = \left\{ 1 - \beta_z \left[2 - \left(\frac{A^{(NS)m}_{(-1)} + A^{(NS)m}_{(1)}}{A^{(NS)m}_{0}} \right) \right] \right\} T^m_{ij0} \tag{37}$$

Finite Difference Thermal Contrast (FDTC)

The objective in using the previous discrete models is to try to estimate the next temperature value over the inspected surface $\tilde{T}^m_{ij0}$, for every pixel at each time instant, as if the considered specimen was completely sound. For that purpose, their constitutive terms can be arranged in a convolution mask or kernel, as shown in Figure 18. In this manner models can be applied to every thermogram like a 2D convolution filtering, to obtain new thermograms containing estimated temperature values. The idea behind is to assume that any healthy point over surface would present low estimation error with respect to real temperature values; while any other point for which estimation error increase consistently over time, could be catalogued as a defective point. Then, the output of the FDTC convolution filter (Figure 18) is:

$$F^m_{ij0} = \frac{T^{m+1}_{ij0} - \tilde{T}^{m+1}_{ij0}}{T^{m+1}_{ij0}} \tag{38}$$

Figure 18. Mask for FDTC method implemented as a convolution filter

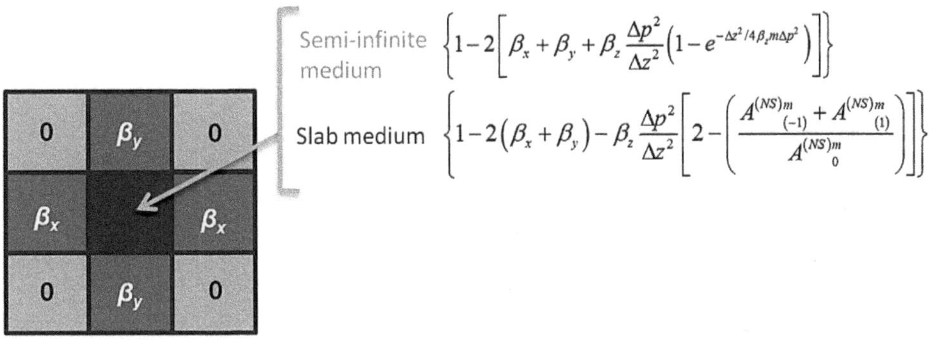

Clearly, this is a dynamic and differentiated filter in the sense that the kernel changes for one pixel to another, and its output corresponds to relative excess of temperature between the real and the estimated value (Figure 19a). The typical time evolution of this excess temperature on a defective point creates the *differentiated profile*$P_D$shown in Figure 19b. This behavior corresponds to an initial slower cooling phase and a later faster cooling phase, as it occurs in reality (this is true for flaw conductivities smaller than layer conductivities). However, in practice it is better to work with *accumulated profiles*P_A (equation 39), because subtraction in equation (38) is very susceptible to noise, even when noise has been filtered from raw sequences.

$$P_A\left(i,j,k\right)=\sum_{v=1}^{k}P_D\left(i,j,v\right) \tag{39}$$

Using the same real CFRP slab of Figure 15, the acquisition conditions of Table 1, and FDTC parameters in Table 3, profiles of figure 20 were obtained. As seen in figure 20a, and contrary to artificial test sequences, it is very likely at working with real IR images that profiles belonging to healthy points diverge in some degree from an ideal behavior around zero value. This probability is due to saturation of IR sensors at flashing procedure, intrinsic phenomena, material non-homogeneities and optical effects that FDTC model cannot deal with. To overcome this fact, an adjustment to FDTC method was proposed (named FDTCa) consisting of applying BTCF contrast to profiles, like those in Figure 20a, to obtain adjusted profiles like those in Figure 20b. For its hand, Figure 21 shows an example of thermograms before and after FDTC execution.

Figure 19. Differentiated filtering: a) excess between real and simulated temperature throughout time; b) example of differential profile in FDTC
Restrepo & Loaiza, 2014a.

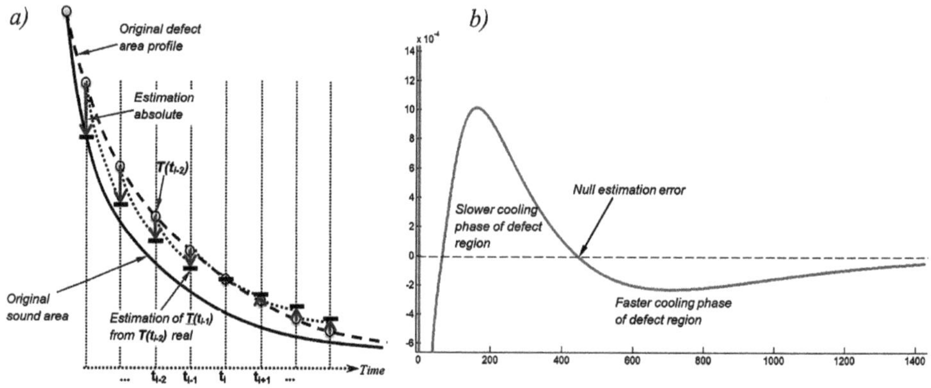

Table 3. FDTC execution parameters

Model Version	Depth Step (Δz) [m]	Spatial Resolution (Δp) [mm/pixel]	Number of Thermal Reflections (*NS*)	Diffusivity (α) [m²/s]
Isotropic slab medium	6.2×10^{-6}	0.62	10	4.6×10^{-7}

Figure 20. Thermal profiles for defects #1, 6, 11, 16 and 21, and a sound point, after application of: a) FDTC; b) FDTC with adjustment (FDTCa)
Restrepo & Loaiza, 2014a.

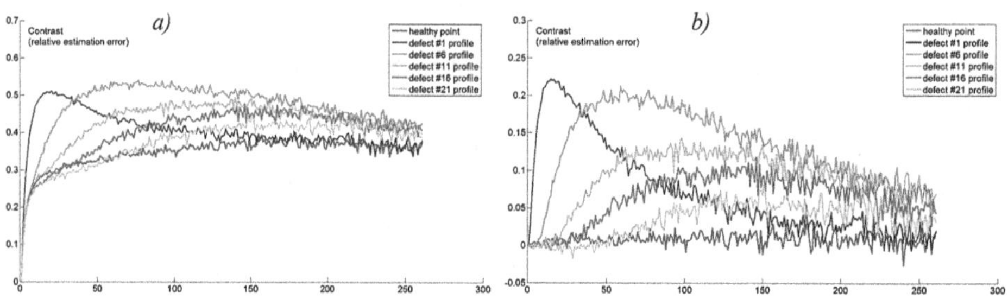

Like in BTCF profiles, the magnitude peak of any FDTCa accumulated profile decays as defects are deeper. However, the most relevant fact from comparison between previous graphs and those in Figure 17, is that this magnitude decay with defects depth is smaller than the same decay evidenced in BTCF and other techniques (see Figure 22), like *normalized contrast* (NC) or *differentiated absolute contrast*

Figure 21. Example of contrast thermograms (3D surfaces) after FDTC application: a) original thermogram; b) FDTC thermogram; c) FDTCa thermogram

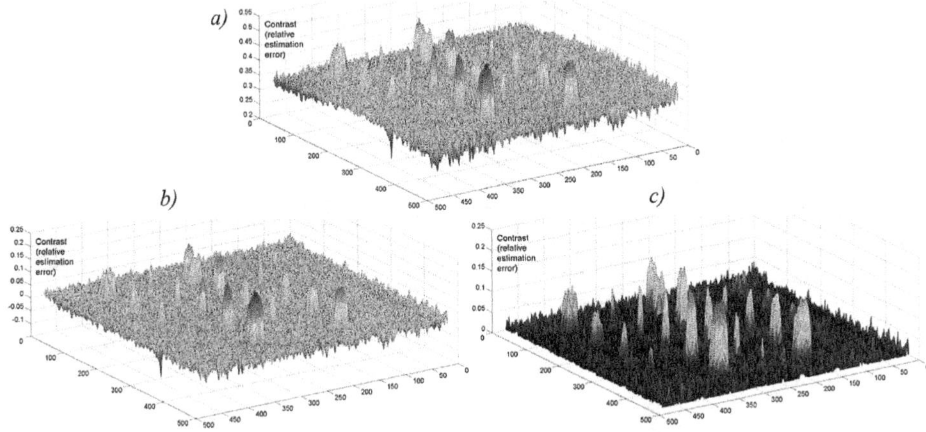

Figure 22. Comparison between contrast magnitude peak decay of NC and FDTC profiles for the five defects of greater area
Restrepo & Loaiza, 2014a.

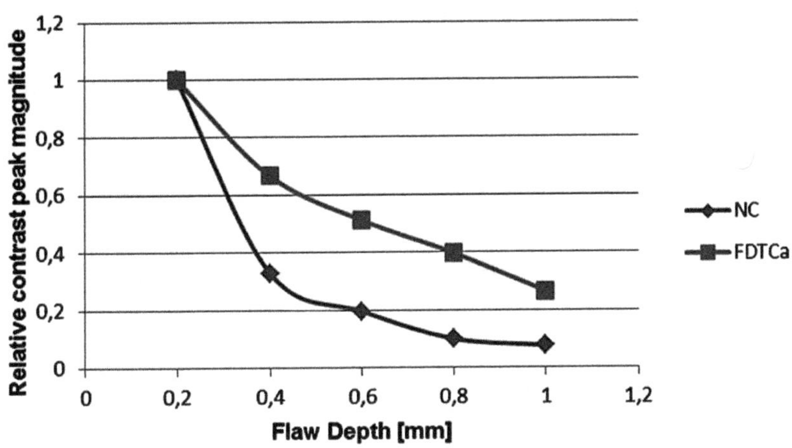

(DAC) (Restrepo & Loaiza, 2014a). This fact results in better global contrast and consequently greater probability of defects detection with low diameter/depth ratio.

Finally, Table 4 presents GBNU and global SNR values obtained for defects #1 and #21, from original and FDTC sequences. For this analysis, the same raw and noise filtered thermal sequences were used; in addition, the FDTC adjusted version used the same BTCF decimated kernel specifications than those used in previous section. From this table, it is clear from the better GBNU and SNR scores that BTCF

Table 4. GBNU and SNR values for considered defects from the contrast sequence obtained after application of different FDTC schemes

	GBNU	SNR – Defect #1	SNR – Defect #21
Original thermal sequence	0.64	5.2	-6.4
Sequence after iterative Gaussian noise filtering	0.73	12.8	-2.8
NC Sequence	0.23	12.5	4.1
CAD Sequence	0.42	12.9	2.4
CTDF sequence	0.23	12.0	3.8
Adjusted CTDF sequence	0.20	13.1	5.0
Adjusted CTDF sequence + truncation	0.18	18.8	8.4

median filtering contributes to flatten the intensities of healthy regions in the FDTCa sequence. Also, the values of SNR and GBNU from the FDTCa sequence with intensity truncation are comparable with those from the Normalized BTCF sequence with truncation, and much better than those obtained from NC and CAD sequences; nevertheless, NC sequence presents greater values of SNR than FDTC without adjusting. These results confirm that normalization of temperature data improves considerably the thermal contrast, although normalization in BTCF technique depends only on one reference thermogram (more general), while in CTDF, normalization is based on estimated temperature values for every pixel at each time instant (more particular).

BTCF AND FDTC PERFORMANCE ABOUT DEFECTS DETECTION

Figure 23 compares the results in detecting defects on the test CFRP slab of Figure 15, using BTCF (absolute and normalized), FDTCa, NC, CAD and CAD-m contrast sequences. The original sequence was filtered by an iterative 3D median filter, followed by an iterative 2D average filter ($\varphi = 0.5$ for both) in order to smooth the "stairs" effect of median operation. After applying each of the previous contrast enhancement techniques, two images of synthesis were generated: the *energy image*, with the energy of contrast of every pixel, and the image containing the *sum of contrast values* through all of the frames after maximize the intensity dynamic range of each one. So, it was performed a preliminary detection of defective regions from the energy image, and later, a final detection of closed borders from the *sum image* and only around the previously segmented regions; in both cases, a combination of

Laplacian and Canny edge detectors was used. This procedure was implemented because the energy image has an outstanding SNR that is suitable for detection, but reduces the segmented area of defective regions when classical detectors are used.

The more relevant aspect seen at this figure is that the result for CTDFa presents a greater number of detected defects coinciding to the real flaws of the test slab of Figure 15, what confirms that CTDFa thermal profiles have more relative energy than other time-resolved techniques; nevertheless, because of the same reason, the inner fibers structure of the slab has more probability to be detected as defects. On the other side, both versions of BTCF throw results very similar to NC application, but without requiring the *a priori* selection of a healthy area of material. At the same time, CAD and CAD-m produce results that coincide with defective regions detected by BTCF, but with greater areas and more false alarms; this is due to the less effective performance of CAD techniques in compensating the non-uniform background heating through all of the sequence.

CONCLUSION

The iterative 3D filtering is an interesting alternative for pre-processing thermal images organized in sequence, because it uses the high correlation between spatial and temporal information. The fundamental idea is to convert a 2D classical kernel into a volumetric kernel; but, in order to utilize the built functions in MATLAB this

Figure 23. Comparative results in defects detection with: a) absolute BTCF; b) normalized BTCF; c) CTDFa; d) NC; e) CAD; f) CAD-m

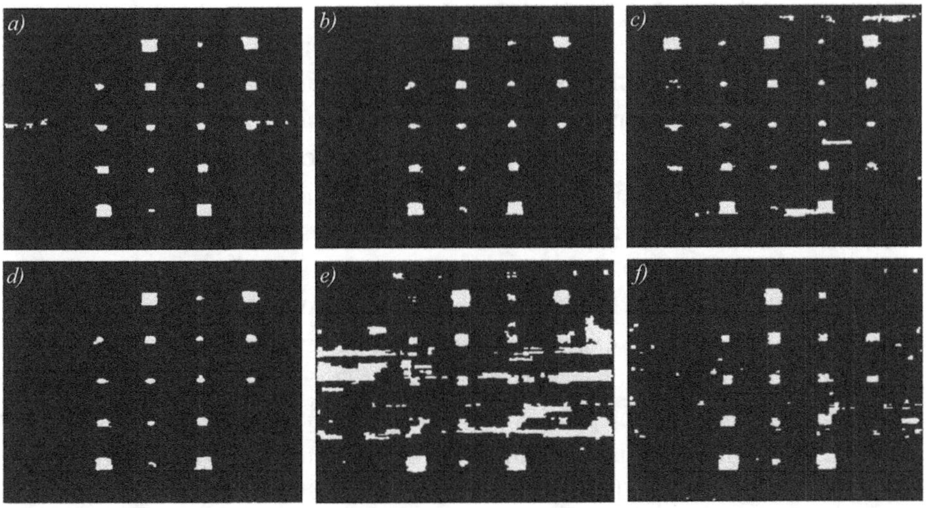

3D kernel can be transformed into a matrix. The results obtained from application of 3D filters are better than its 2D counterparts, and even better than the 2D classic filters iteratively executed. For that reason, addition of iterative execution to some 3D filters gives the better performance for thermograms pre-processing, regarding compromise between noise attenuation and edge preservation. Specially, iterative 3D median filter stands out among analyzed filters, though it presents the drawback of much larger runtime.

On the other side, BTCF contrast enhancement method is based on a decimated median filter covering an area big enough to "mask" the defective regions without prohibited execution times; it extends image borders with an adaptive procedure to diminish possible distortions over thermograms periphery; and finally, it subtracts the decoupled background surfaces from original thermograms to obtain the enhanced contrast sequence. The decoupling of background information makes possible BTCF to be applied for characterization of sound regions of the material under inspection, without the problem of knowing previously where defects are located. The enhanced contrast, even applied as a complement of another technique like FDTCa, is supported on the resulting flatness around zero intensity at healthy areas of each contrast image, as at thermal contrast profiles belonging healthy points. These facts help detectability of low diameter/depth ratio flaws (even more with normalization and suitable contrast truncation), and accuracy of depth characterization without using computational intelligence tools.

Finally, FDTC contrast enhancement technique is based on a simplified 3D discrete model of heat propagation to estimate the next temperature on each pixel according to its own current temperature and those from its 4-neighbors, in such a way that the constitutive terms can be arranged in a spatial dynamic filtering mask. The algorithm does not need to estimate temperature along three spatial axes throughout the material in every sample time: it only estimates temperature over the surface of the inspected specimen, calculating a very short propagation of heat along z (depth) axis by means of suitable approximation models. This feature makes possible to achieve a substantial reduction in execution time compared to techniques using 3D models for simulation and inversion tasks. More than this, FDTC strategy does not perform estimation from ideal initial conditions and theoretical data as done by usual 3D thermal reconstruction, but it estimates temperatures from real values in each thermogram.

For each pixel and time, FDTC gives the relative error of the estimated next temperature with respect to the experimental next temperature over the frontal surface, so that this error increases over defective regions. This feature makes the obtained profiles to be differential and prone to noise; for that reason, it is better to work with accumulated profiles to enhance contrast and subsequent defects detection. Also, in real cases, noise and optical distortions may cause residual background be less flat and deviate from zero; but the background irregularity can be compensated with the BTCF method, preserving the greater relative energy of FDTC profiles with respect to other time-resolved techniques.

As future work, it should be investigated in more detail the use of BTCF and FDTC for detection and characterization of defects in CFRP and other materials, with different types of defects and other active thermography schemes, as well as the use of learning machines in the processing stages. Also, research on the development of optimized BTCF and FDTC algorithms running on faster software solutions, or implemented on concurrent hardware platforms, can lead to important advances in order to achieve efficiency and operability standards for industrial application.

REFERENCES

Alsaadawi, Z., & Netzelmann, U. (2010). Spatio-temporal filtering of active thermography data for noise reduction and data compression. In *Proceedings of 10th International Conference on Quantitative InfraRed Thermography (QIRT 2010)*. Québec, Canada: QIRT conference open archives. doi:10.21611/qirt.2010.006

Bagavathiappan, S., Lahiri, B., Saravanan, T., Philip, J., & Jayakumar, T. (2013). Infrared thermography for condition monitoring – A review. *Infrared Physics & Technology*, *60*, 35–55. doi:10.1016/j.infrared.2013.03.006

Balageas, D. (2011). Defense and Illustration of Time Resolved Pulsed Thermography for NDE. In *Proceedings of Thermosense: Thermal Infrared Applications XXXIII*. SPIE. doi:10.1117/12.882967

Balageas, D., Roche, J., Leroy, F., Liu, W., & Gorbach, A. (2015). The thermographic signal reconstruction method: A powerful tool for the enhancement of transient thermographic images. *Biocybernetics and Biomedical Engineering*, *35*(1), 1–9. doi:10.1016/j.bbe.2014.07.002 PMID:25678731

Barner, K., & Arce, G. (1998). Order-Statistic Filtering and Smoothing of Time Series: Part II. In N. Balakrishnan & C. R. Rao (Eds.), *Handbook of Statistics* (Vol. 17, pp. 525–554). Elsevier. doi:10.1016/S0169-7161(98)17023-2

Benítez, H. (2012). *Image quality indexes for infrared non-destructive testing*. Paper presented at Infrared Non-Destructive Evaluation Seminar, Cali, Colombia.

Benítez, H., Ibarra-Castanedo, C., Bendada, A., Maldague, X., Loaiza, H., & Caicedo, E. (2008). Definition of a new thermal contrast and pulse correction for defect quantification in pulsed thermography. *Infrared Physics and Technology Journal, 51*(3), 160-167.

Benítez, H., Ibarra-Castanedo, C., Bendada, A., Maldague, X., Loaiza, H., & Caicedo, E. (2009). Defect Characterization in Infrared Nondestructive Testing with Learning Machines. *NDT&E, 42*(7), 630–643. doi:10.1016/j.ndteint.2009.05.004

Benítez, H., Loaiza, H., & Caicedo, E. (2011). *Termografía activa pulsada en inspección de materiales. Técnicas avanzadas de procesado*. Cali, Colombia: Universidad del Valle—Pontificia Universidad Javeriana.

Cheng-Hung, H., & Meng-Ting, C. (2008). A transient three-dimensional inverse geometry problem in estimating the space and time dependent irregular boundary shapes. *International Journal of Heat and Mass Transfer, 51*(21-22), 5238–5246. doi:10.1016/j.ijheatmasstransfer.2008.03.019

Dabov, K., Foi, A., Katkovnik, V., & Egiazarian, K. (2007). Image denoising by sparse 3D ransform-domain collaborative filtering. *IEEE Transactions on Image Processing, 16*(8), 2080–2095. doi:10.1109/TIP.2007.901238 PMID:17688213

Faúndez, M. (2001). *Tratamiento Digital de Voz e Imagen, y Aplicación a la Multimedia*. Alfaomega Grupo Editor.

Florez-Ospina, J., & Benítez, H. (2014). From local to global analysis of defect detectability in infrared non-destructive testing. *Infrared Physics & Technology, 63*, 211–221. doi:10.1016/j.infrared.2013.12.017

García, D. (2010). Robust smoothing of gridded data in one and higher dimensions with missing values. *Computational Statistics & Data Analysis, 54*(4), 1167–1178. doi:10.1016/j.csda.2009.09.020 PMID:24795488

González, R., & Woods, R. (2002). *Digital Image Processing*. Prentice Hall.

Grinzato, E., Bison, P., Marinetti, S., & Vavilov, V. (2000). *Thermal NDE enhanced by 3D numerical modeling applied to works of art.* Paper presented at 15th World Conference on Non-destructive Testing, Rome, Italy. Retrieved from http://www.ndt.net/article/wcndt00/papers/idn909/idn909.htm

Grys, S., & Minkina, W. (2010). Filtered thermal contrast – error analysis. In *Proceedings of 10th International Conference on Quantitative InfraRed Thermography (QIRT 2010)*. Québec, Canada: QIRT conference open archives. doi:10.21611/qirt.2010.051

Hasan, M. (2014). *Adaptive Edge-guided Block-matching and 3D filtering (BM3D) Image Denoising Algorithm* (Master of Science Thesis). University of Western Ontario, Canada.

Hensen, J., & Nakhi, A. (1994). Fourier and Biot numbers and the accuracy of conduction modeling. In *Proceedings of Building Environmental Performance Conference (BEP'94) "Facing the Future"*. University of York.

Ibarra-Castanedo, C. (2005). *Quantitative Subsurface Defect Evaluation by Pulsed Phase Thermography: Depth Retrieval with the Phase* (Doctoral dissertation). Available from Collection Mémoires et thèses électroniques, Université Laval.

International Air Transport Association. (2009). A global approach to reducing aviation emissions. First stop: carbon-neutral growth from 2020. Author.

Jetto, L. (1999). Stochastic modelling and 3D minimum variance recursive estimation of image sequences. *Multidimensional Systems and Signal Processing*, *10*(3), 225–254. doi:10.1023/A:1008406707503

Larsen, C. (2011). *Document flash thermography* (Master of Science Thesis). Available from Utah State University DigitalCommons@USU.

López, F., Nicolau, V., Ibarra-Castanedo, C., & Maldague, X. (2014). Thermal-numerical model and computational simulation of pulsed thermography inspection of carbon fiber reinforced composites. *International Journal of Thermal Sciences*, *86*, 325–340. doi:10.1016/j.ijthermalsci.2014.07.015

Madruga, F., Albendea, P., Ibarra-Castanedo, C., & López-Higuera, J. (2010). Signal to noise ratio (SNR) comparison for lock-in thermographic data processing methods in CFRP specimen. In *Proceedings of 10th International Conference on Quantitative InfraRed Thermography (QIRT 2010)*. Québec, Canada: QIRT conference open archives. doi:10.21611/qirt.2010.079

Maillet, D., Houlbert, A., Didierjean, S., Lamine, A., & Degiovanni, A. (1993). Non-destructive thermal evaluation of delaminations in a laminate: Part I – Identification by measurement of thermal contrast. *Composites Science and Technology*, *47*(2), 137–153. doi:10.1016/0266-3538(93)90043-G

Maldague, X. (2001). *Theory and Practice of Infrared Technology for Non Destructive Testing*. New York, NY: John Wiley & Sons.

Mao, Z., Ma, L., Zhao, M., & Li, Z. (2006). Feature-preserving mesh denoising based on contextual discontinuities. *Journal of Zhejiang University SCIENCE A*, *7*(9), 1603–1608. doi:10.1631/jzus.2006.A1603

Marinetti, S., Grinzato, E., Bison, P. G., Bozzi, E., Chimenti, M., Pieri, G., & Salvetti, O. (2004). Statistical analysis of IR thermographic sequences by PCA. *Infrared Physics & Technology*, *46*(1-2), 85–91. doi:10.1016/j.infrared.2004.03.012

Meguro, M., Taguchi, A., & Hamada, N. (2009). Data-Dependent Weighted Average Filtering For Image Sequence Enhancement. In *Proceedings of 1999 International Conference on Image Processing (ICIP'99)*. Kobe, Japan: IEEE Computer Society.

Mertzios, B. G., & Tsirikolias, K. (1998). Coordinate Logic Filters and Their Applications in Image Processing and Pattern Recognition. *Circuits Systems Signal Processing*, *17*(4), 517–538. doi:10.1007/BF01201506

Pilla, M., Klein, M., Maldague, X., & Salerno, A. (2002). New Absolute Contrast for Pulsed Thermography. In *Proceedings of 6th International Conference on Quantitative InfraRed Thermography (QIRT 2002)*, Dubrovnik, Croacia: QIRT conference open archives. doi:10.21611/qirt.2002.004

Pohl, J. (1998). Ultrasonic Inspection of Adaptive CFRP-Structures. *NDT Database & Journal*, *3*(9). Retrieved from: http://www.ndt.net/article/ecndt98/aero/015/015.htm

Ponomaryov, V. (2007). Real-time 2D–3D filtering using order statistics based algorithms. *Journal of Real-Time Image Processing*, *1*(3), 173–194. doi:10.1007/s11554-007-0021-5

Restrepo, A., & Loaiza, H. (2012). Filtrado 3D Espacio-Temporal Iterativo para la Atenuación de Ruido en Secuencias de Imágenes Térmicas para END. *Ingenium*, *6*(13), 27–34. doi:10.21774/ing.v6i13.85

Restrepo, A., & Loaiza, H. (2013). Non-uniform heating compensation on thermal images using median filtering. *DYNA*, *80*(182), 74–82.

Restrepo, A., & Loaiza, H. (2014a). New 3D finite difference method for thermal contrast enhancement in slabs pulsed thermography inspection. *Journal of Nondestructive Evaluation*, *33*(1), 62–73.

Restrepo, A., & Loaiza, H. (2014b). Background thermal compensation by filtering (BTCF) for infrared thermography evaluation. In *Proceedings of the XIX International Simposium of Signal and Images Processing and Artificial Vision (STSIVA, 2014)*. IEEEXplore. doi:10.1109/STSIVA.2014.7010162

Restrepo, A., & Loaiza, H. (2014c). 3D discrete model for thermal contrast enhancement and defects depth estimation in CFRP slabs. *Ingeniería y Competitividad*, *16*(2), 143–153.

Restrepo, A., & Loaiza, H. (2015). Enhanced method for flaws depth estimation in CFRP slabs from FDTC thermal contrast sequences. *Ingeniería e Investigación*, *35*(3), 61–68. doi:10.15446/ing.investig.v35n3.50552

Restrepo, A., & Loaiza, H. (2016). Background thermal compensation by filtering for contrast enhancement in active thermography. *Journal of Non-destructive Evaluation, 35*(1), 20. doi:10.1007/s10921-016-0336-x

Rodríguez, F. L., Ibarra-Castanedo, C., Nicolau, V., & Maldague, X. (2014b). Optimization of pulsed thermography inspection by partial least-squares regression. *NDT & E International, 66*, 128–138. doi:10.1016/j.ndteint.2014.06.003

Rodríguez, F. L., Nicolau, V., Ibarra-Castanedo, C., & Maldague, X. (2014a). Pulsed Thermography Signal Processing Techniques Based on the 1D Solution of the Heat Equation Applied to the Inspection of Laminated Composites. *Materials Evaluation, 72*(1), 91–102.

Russ, J. (1998). *The Image Processing Handbook* (3rd ed.). CRC Press, Springer, IEEE Press.

Segui, J. (2014). Boeing Simulates Thermal Expansion in Composites with Expanded Metal Foil for Lightning Protection of Aircraft Structures. *COMSOL News - The Multiphysics Simulation Magazine, 2014*, 4-7.

Skala, K., Kavanah, H., Lipic, T., Sovic, I., & Grubisic, I. (2013). *5D thermal imaging system for body dissipation monitoring in motion*. Paper presented at 18th International Conference on Thermal Engineering and Thermogrammetry (THERMO), University of Technology and Economics (BME), Budapest, Hungary.

Skala, K., Lipic, T., Sovic, I., Gjenero, L., & Grubisic, I. (2011). 4D thermal imaging system for medical applications. *Periodicum Biologorum, 113*(4), 407–416.

Snell, J., Jr., & Spring, R. (2007). Infrared Thermography Advances. *Featured Articles – NDT Magazine*. Retrieved November, 2010, from http://www.marinecompositesnde.com/images/Snell,_2007.pdf

Tadeu, A., & Simões, N. (2005). Fundamental solutions for transient heat transfer by conduction and convection in an unbounded, half-space, slab and layered media in the frequency domain. *Engineering Analysis with Boundary Elements, 29*(12), 1130–1142. doi:10.1016/j.enganabound.2005.06.002

Tadeu, A., & Simões, N. (2006). Three-dimensional fundamental solutions for transient heat transfer by conduction in an unbounded medium, half-space, slab and layered media. *Engineering Analysis with Boundary Elements, 30*(5), 338–349. doi:10.1016/j.enganabound.2006.01.011

Tadeu, A., & Simões, N. (2007). Conduction and convection phenomena through a slab with thermal heterogeneities. *Applied Mathematical Modelling*, *31*(7), 1444–1459. doi:10.1016/j.apm.2006.04.010

Tenney, D. R., Davis, J. G., Byron, R., & Johnston, N. (2009). NASA Composite Materials Development: Lessons Learned and Future Challenges. *NATO RTO AVT-164 Workshop on Support of Composite Systems*. Bonn, Germany: NASA Technical Reports Server (NTRS).

Valdés, M., Inamura, M., Valera, J., & Yao, L. (2006). Multidimensional filtering approaches for pre-processing thermal images. *Multidimensional Systems and Signal Processing*, *17*(4), 299–325. doi:10.1007/s11045-006-0001-0

Wang, Z., & Bovik, A. (2002). A universal image quality index. *IEEE Signal Processing Letters*, *9*(3), 81–84. doi:10.1109/97.995823

Chapter 9
Personal Authentication through Finger Knuckle Geometric and Texture Feature Measurements:
Finger Knuckle Biometrics

Usha Kazhagamani
Pondicherry Engineering College, India

M. Ezhilarasan
Pondicherry Engineering College, India

ABSTRACT

Finger Knuckle biometric is an emerging automated human identification approach that has received extensive significance in the area of research and real time applications in the recent past. Generally, a typical finger knuckle biometric system investigates the finger knuckle patterns present in the outer bend surface of the finger back region i.e., proximal phalanx region. In contrast, this paper focuses on the entire finger back region which includes proximal and distal phalanx of the finger knuckle surface for recognition. Further, this paper investigates a novel approach to achieve improved performance by simultaneous extraction and integration of finger knuckle geometric and texture features from a captured finger knuckle region. The geometric measures are derived by means of angular geometric analysis method which extracts angular-based feature information for unique identification. Similarly, texture measures are derived through statistical-based texture analysis methods.

DOI: 10.4018/978-1-5225-2423-6.ch009

INTRODUCTION

Automated method of personal identification which utilizes physiological features has been widely used in highly critical security applications and forensics. The high level security applications include access control in physical or logical systems that impose new challenge of handling large volume of biometric data for accurate personal identification with high speed (Wayman et al., 2005). In the recent past, hand based biometric systems has drawn considerable attention of researchers due to its merits like i) hand traits can be easily captured using low cost acquisition devices ii) hand traits have highly discriminative features which are potential high enough for personal recognition iii) hand traits has high user acceptance rates since they were captured in a user friendly manner and iv) biometric system using hand traits yields high performance in terms of speed and accuracy (Hand-based Biometrics, 2003). Researchers have proposed various personal recognition algorithms using hand-based biometric technologies such as finger print, palm print, hand geometry, hand vein structures and finger knuckle print (Ribaric, & Fratric, 2005, Jain et al., 1999, Kumar & Prathyusha, 2008)

Recently, finger knuckle print technology was incorporated in personal identification since it has unique potential biometric identifier (Qiu et al., 2004). Generally, finger knuckle print is defined as texture patterns present in the proximal phalangeal region of the finger dorsum surface. In contrast, this paper investigates on the finger knuckle surface containing dermal patterns of both proximal and distal phalanx regions (Lin Zhang et al., 2009). This paper focuses on deriving both geometrical and texture features from finger knuckle surface patterns in order to improve the performance with respect to accuracy. This paper analyses the captured finger knuckle regions by extracting ROI using binarization and threshold methodology. From the ROI of the finger knuckle print, the angular based geometrical feature information is extracted since the magnitude based feature information extracted by the traditional geometrical methods is not so discriminative to identify the individuals. Further, the texture pattern of the finger knuckle print need to be exploited to achieve improved performance in personal identification. The texture patterns of the captured finger knuckle features are studied based on statistical methods. Statistical methods are computationally simple methods which could yield better performance in terms of accuracy and speed. In this paper, co-occurrence matrix method and grey level difference vector method are implemented to extract statistical based texture feature information from the captured finger knuckle surface. This paper also focuses on integration angular geometric feature information and statistical texture feature information to attain improved performance in personal identification.

The organization of the paper is as follows. Section 2 presents the categorization of existing personal recognition algorithms and in-depth analysis of existing literatures under each category. Section 3 presents the design of the proposed personal authentication system. Section 4 illustrates the steps involved in preprocessing and region of interest segmentation process. Section 5 details the algorithmic steps involved in the extraction of geometric measurements from finger knuckle surface. Section 6 presents statistical-based textural analysis method which extracts local texture feature information from finger knuckle surface. Section 7 discusses various methods involved in fusion process for combining scores obtained through geometric measurements and texture analysis of finger knuckle surface. Extensive experimental analysis which has been conducted to assess the performance of proposed personal authentication system along with their detailed results discussions are presented in section 8. Section 9 concludes the paper.

RELATED WORK

In the literature, researchers have proposed various promising methods for finger knuckle biometrics. These promising methods are broadly classified into two categories viz., geometric based method and texture based method. Generally, geometrical based feature extraction methods utilizes a number of edge detecting approaches for extracting features like edge points, lines, creases, wrinkles etc., from the finger knuckle print. Then, the extracted edge information is converted into geometrical feature information to represent the feature vector for matching (Sanchez-Reillo et al., 2000). Similarly, the knuckle texture based feature extraction methods are classified into three categories viz., i) transform-based texture analysis ii) statistical-based texture analysis iii) model-based texture analysis methods (Kumar and Zhang,, 2006). In transform-based texture analysis, the mathematical models are used to characterize the spatial variation present in an image in terms of feature information. Likewise, the statistical-based texture analysis methods represent the image texture feature in terms of parameters which could able to quantify relationship exists among the grey level pixels of an image. In case of model-based texture analysis, the texture feature information is derived by quantifying the image texture using fractal and stochastic models (Wang, et al., 2007). Some of the finger knuckle print recognition methods are discussed below.

Finger Knuckle Print Recognition Using Geometric Analysis

Kumar and Ravikanth (2009), proposed a novel approach for personal authentication using finger knuckle surface. In this work, finger knuckle shape and texture features were extracted for reliable human identification. The shape oriented feature information like finger length, width, area and perimeter were extracted and integrated with texture features. Authors incorporated appearance based methods such as Principle Component Analysis (PCA), Linear Discriminant Analysis and Independent Component Analysis for extracting texture feature information. In this work, authors examined the presence of ring or any other metal like elements in the finger back region. The shape oriented feature extraction method of this paper is termed as Finger Geometric Feature Extraction Method (FGFEM) for the sake of comparative analysis. Kumar el al., (2010) in his further work contributed a novel approach for personal authentication using hand images. In this work, feature information is extracted by means of hand geometry, palm print texture features and finger knuckle shape features. The finger knuckle shape features are extracted by means of Knuckle Geometric Feature Extraction Method (KGFEM). Further, in this work authors quantified the quality features of the captured hand trait and integrated with the feature information based on quality dependent fusion.

Finger Knuckle Print Recognition Using Texture Analysis

Transform Based Texture Analysis Methods

Goh Kah Ong Michael et al, (2010) have proposed bimodal verification system using palm print and finger knuckle print. In this system, the entire hand image is captured by the image acquisition device. From the captured hand image, the normalized palm and knuckle regions are extracted by means of competitive hand valley detection algorithm. Feature information from the ROI of the palm print are extracted by means of directional coding techniques and represented in the form of bit strings. The feature information from the ROI of knuckle image is obtained by subjecting it to ridgelet transform. Support vector machine with radial basis function is used for classification. Michal Chora and Rafal Kozik (2010) uses Probabilistic Hough Transform (PHT) to code line features of finger knuckle print. In this, robust features are also deriving by means of SURF algorithm for performance improvement. Euclidean metric is used to find the similarity matching.

Further, Yang Wankougmet et al, (2011) characterizes finger knuckle print using Gabor filters, linear discriminant analysis and principle component analysis. These techniques are the prominent methods for face recognition. In this paper, authors have exploited the advantages of these methods to quantify the knuckle texture patterns. Xiaoyang xing et al, (2011) proposes a new model based texture analysis methods known as complex locality preserving projection approach for discriminating the finger knuckle surface images. This approach extracts the low dimensional features by preserving the manifold of input data set and also derives the orthogonal based functions to overcome the redundant information

Furthermore, Lin Zhang et al, (2010) recommended a novel coding method known as monogenic code which is generated by applying monogenic signal to each of the pixel of finger knuckle image and binarizing of signal output. Lin Zhang et al, (2011) proposed a new model for personal authentication using finger knuckle print based on texture analysis method. The Riesz transform method was used to extract the feature information from the captured FKP images which results in RieszCompCode. The support vector machines with radial basis function were used to classify the finger knuckle print input images according to the generated feature information of the reference image. Lin Zhang and Hailong Zhu, (2011) investigated a phase congruency model to extract local features of finger knuckle print. These local features include local orientation and phase information as feature information for reliable identification. In addition to this, authors in their further work (Lin Zhang, Hongyu Li, 2012) introduced two coding mechanisms named as Rcode1 and Rcode2 obtained by subjecting the finger knuckle image to the first and second order of Riesz transform.

Statistical Based Texture Analysis Methods

The statistical approach for feature extraction from finger knuckle print is explored in the following contributions. In (Pengfei et al., 2014), a statistical approach known as Average Absolute Deviation is used to quantify the feature information from the extracted ROI image of the finger knuckle print. Matching scores were generated by computing the Euclidean distance between reference and input image. Zhang et al, (2009) presented a band limited phase only correlation method (BPLOC) for matching finger knuckle prints. The matching process is done by means of analyzing peak values obtained from the phase only correlation method. Linlin Shen et al, (2010) contributed a bi-modal personal authentication system using palm print and finger knuckle print. In this work, authors incorporated two dimensional Gabor filtering methods in order to extract texture feature information from the captured palm print and finger knuckle print. Hamming distance metric is used to identify the matching between the reference and input images of the captured trait.

Further, Merounica et al., (2011) have shown the implementation of Fourier transform functions for deriving the feature information from the finger knuckle region and palm region. Matching process is done by identifying the linear phase shift in frequency using phase correlation algorithm. The identified phase shift are taken inverse Fourier transform to define the cross correlation between phase components. Shoichiro Aoyama et al, (2011) presented a novel method for finger knuckle print recognition using phase-based local block matching performed through Band Phase-Only Limited Correlation. This phase correlation method could able to extract local and global information for reliable personal authentication.

Furthermore, Guangwei Gao et al., (2013) addressed the issue of rotational variances by reconstructing the finger knuckle print image using dictionary learning method. In this authors incorporated a novel fusion method known as adaptive score level fusion. Authors in their further work (Guangwei Gao et al., 2014) contributed multiple orientation coding method which extracts texture information in terms of local binary pattern. Recently, Kumar (2014) presented a personal recognition system using minor finger knuckle regions. In this work authors incorporated four different methods such as, i) Local Binary Pattern (LBP) generation method ii) Improved LBP generation method iii) BPLOC method iv) 1-D log filtering method in order to prove the potentiality of minor finger knuckle region towards personal authentication.

Model Based Texture Analysis Methods

Madasu Hammandlu and Jyotsana Grover, (2012) contributed a model based texture analysis method for finger knuckle print recognition. In this work, authors incorporated Sobel and other divergence operators for representing the feature information of line and curved line features of finger knuckle print. The obtained feature information is represented in a binary pattern. Hamming distance exist between the binary patterns obtained from the registered and input image. Additionally, Chetaua Hedge et al, (2011) contributed a novel finger knuckle print recognition system which provides dual modes of security viz., normal and advanced mode. Sobel operator is incorporated to model the obtained finger knuckle print image for providing security in the normal mode. To provide security in advanced mode, the preprocessed FKP image is subjected to Random transform and Eigen values are computed by means of generating Radon graphs.

Ming Liu et al., (2014) presented a novel approach for the recognition of an individual using inner knuckle print patterns. In this paper, a new method is designed to capture inner knuckle patterns and ROI extraction of inner knuckle print is performed using hybridization of Gabor filtering and derivative line detection method. From this binary line images are extracted and matched by means of cross-correlation-based method. Score level fusion is implemented to classify the input data. AjayKumar and BichaiWang, (2015) introduced a novel methodology for matching finger knuckle print is based on minutiae patterns. Recovery of finger knuckle regions with sample regions can also be done based on this methodology. These types of analysis are found to be useful in forensic analysis. AdityaNigam and PhalguniGupta (2015) presented a multimodal biometric system which makes use of palm print and finger knuckle print patterns for identifying an individual. Sign of local gradient is used to transform the captured biometric images and corner features such as vcode and hcode are extracted for matching palm and knuckle images. Highly uncorrelated features measure is used to compare the registered and input image of the palm and finger knuckle print images.

Extracts of the Literature

The detailed analysis of existing finger knuckle print recognition methods result in the following pitfalls. They are:

1. Geometrical methods presented in the literature extracts knuckle shape oriented information such as, length, width, area and perimeter which has lower degree of discriminatory power when the system scales for a large number of users.
2. In model-based texture analysis methods, the variations in the appearance of the finger knuckle image have been quantified to generate unique feature information. This approach shows lower level of performance when the finger knuckle images were acquires using low resolution capturing devices.
3. Transform-based texture analysis methods presented in the literature lack in projecting scaling and orientation parameters which in turn fail to handle higher dimensional singularities.
4. The presented statistical-based texture analysis extracts and utilizes either local or global phase information for finger knuckle print recognition which could lead to performance degradation for the image set which has high value of intra-class and inter-class similarities respectively.

Hence, we are motivated to develop a finger knuckle recognition system by incorporating both geometrical and local texture measures. The geometrical method incorporated in this work derives angular-based feature information for personal identification. Similarly, statistical-based texture analysis method derives both local and global phase information for identification. Finally, integration of finger knuckle geometrical measures and texture features is performed to achieve higher accuracy rates.

PROPOSED SYSTEM DESIGN

The following Figure 1 illustrates the block diagram of the proposed finger knuckle biometric system. The first phase of finger knuckle biometric system is image capturing/acquiring phase. In this phase, the finger knuckle surface is captured by placing it on a white surface illuminated in front of the camera. Initially, the captured

Figure 1. Block diagram of the proposed finger knuckle biometric system

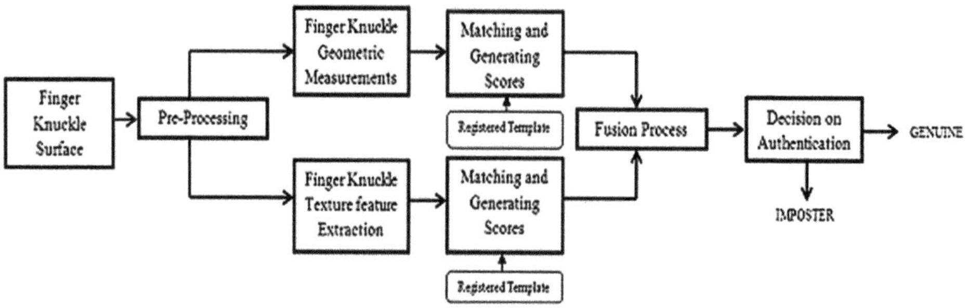

finger knuckle surface is subjected to pre-processing in order to extract proximal and distal finger knuckle regions. Secondly, geometric and texture feature information are simultaneously extracted. Thirdly, the matching between the input image and the registered template is performed and matching scores are derived for geometric and texture methods individually. Finally, the obtained matching scores are fused to conclude the decision on authentication.

Preprocessing and ROI Extraction

The physiological features of the finger back region are shown in the Figure 2 (b) and described as follows. Each finger back region of the hand has three phalangeal joints. The joint which links each finger with the hand surface is called as Metacapophalangeal joint. The joint which is formed in the middle surface of the finger is called as Proximal Inter Phalangeal (PIP) joint. The presence of PIP joint in the finger back knuckle surface forms the flexion shrinks on the inner surface of the skin which creates unique dermal patterns. These patterns are denoted as proximal knuckle patterns. The joint present in the tip of the FBKS is called as distal inter phalangeal joint. This joint also forms the unique dermal patterns which are denoted as distal patterns. In this paper, the inherent skin patterns of both proximal and distal knuckle are considered for processing.

The preprocessing and ROI extraction involves the process of mining the proximal and distal knuckle regions from the FBKS. Also it does the process of alignment of finger knuckle images captured in various scenarios in the acquisition process. As a first step in preprocessing, the acquired FBKS is subjected to a thresholding operation to obtain binarized images. The threshold value is obtained by means of empirical analysis on the training samples of the finger knuckle images. From the obtained binarized images, the boundary of the image is extracted by means of contour tracing (Lin Zhang et al., 2009). The exact boundary of the image is obtained by tracing the largest possible contour in the binarized finger knuckle image by eliminating the small contours. The Figures 3 (a), 3(b) 3 (c), 3(d) and 3 (e) shows the acquired finger back knuckle surface, the binarized image of the finger back knuckle surface and contour extracted images from the binarized image respectively. The boundary image obtained through contour tracing is called as a contour image of FBKS and named as $I_{contour}$.

Finger Knuckle Geometric Measures

The main objective of this study is to evaluate the improvement in performance induced by the integration of geometric and texture features of a finger knuckle biometric system. The geometric measurements are extracted from the ROI images of proximal and distal knuckle regions using angular geometric analysis method (AGAM) as discussed in (Usha and Ezhilarasan, 2013). As detailed in (Usha and Ezhilarasan, 2013), the angular geometric analysis method extracts six geometric features from proximal knuckle and six from distal knuckle region. Hence totally, 12 geometric measurements were derived from a finger knuckle surface which includes, two finger knuckle length, six finger knuckle widths and four finger knuckle angular information. Totally, 48 geometric features were extracted and registered in a feature vector. The distance between the input finger knuckle geometric measures $\left(fks_i\right)$ and the registered feature vector $\left(fks_r\right)$ is computed through Weighted Euclidean Distance $\left(WED\right)$ rule, which is given by (3)

$$D\left(fks_i, fks_r\right) = \sqrt{\sum_{i=1}^{k} w_i \left(fks_i - fks_r\right)^2} \tag{3}$$

where,

fks_i : Represents the geometric measurement vector of an input finger knuckle surface image.

fks_r : Represents the geometric measurement of registered finger knuckle surface image.

w_i : Corresponds to the weight which is assigned a lower value for lower variance between input and registered value and assigned higher value for higher variance between input and registered value.

The key significance of the proposed angular geometric analysis method is that it extracts angular-based feature information which is highly potential enough to distinguish the individuals.

Finger Knuckle Texture Feature Extraction

Co-Occurrence Matrix Method (CMM)

Co-occurrence Matrix Method (CMM) estimates finger knuckle image texture features based on second order statistics which manipulates the spatial relationship between the pixels and among the group of pixels. The proposed method incorporates a texture feature extraction mechanism contributed by Haralick (1973), which represents the texture pattern of an image by manipulating joint probability distributions based on the frequency of occurrence of each grey level pixel values located at fixed geometric positions relative to each neighborhood pixels.

In general, the co-occurrence matrix is represented by $C\left(i,j,d,\theta\right)$ for a knuckle image of size $I \times I$. Further, this matrix denotes the number of occurrence of a pair of pixels with grey level intensity (i, j) present at a distance'd' and direction 'θ'. Furthermore, the pixel values are quantified into I levels according to the size of the image. The derived CMM values are normalized based on the sum of all elements in the matrix. In addition, each element in the normalized CMM represents the probability of occurrence of pair of pixels in a specified spatial relationship with fixed spatial layout. From the normalized CMM, fourteen statistical based texture feature information such as energy, correlation, contrast, variance, inverse difference moment, entropy, sum average, sum variance, sum entropy, difference entropy, difference variance, information measures of correlation and maximal correlation variance are derived for each distance and directional values.

However, in our proposed method, the CMM values are derived from the finger knuckle sub-image of size $S \times S$, where $S < D$, which is cropped from the ROI image of finger knuckle surface of size $D \times D$. The ROI of finger knuckle print image of $D \times D$ is divided into the set of N sub-images or sub-bands by implementing sliding window approach. The size of the sliding window is chosen based on empherical analysis. In this approach, the distance d varies from 1 to n, $(d = 1, 2, \ldots, n)$ where 'n' represents the number of pixels that denotes the size of the image or the size of the area of the texture elements present in the image and the value of the directional element θ are considered as $0^{\circ}, 45^{\circ}, 90^{\circ}\ and\ 135^{\circ}$ respectively. Based on the directional elements, four CMMs are derived and the average values of the obtained four matrices are manipulated to obtain a new co-occurrence matrix. Finally, the statistical based texture feature information is derived from the newly generated CMM based on average mean values. This method results in a feature information, that is invariant to rotational and transformation characteristics with reduced dimensionalities.

For each finger knuckle sub-image, the five selected statistical based finger knuckle texture features viz., Energy $\left(E_n\right)$, Inverse differential moment $\left(I_{DM}\right)$, Correlation $\left(C_r\right)$, Contrast $\left(C_t\right)$ and Information Measure of correlation $\left(C_{im}\right)$ derived to emphasize the superiority of the CMM approach in finger knuckle print recognition. The above chosen statistical texture features are derived as follows.

1. Energy $\left(E_n\right)$ is obtained by manipulating the sum of squares of CMM entries which is given derived through (4),

$$E_n = \sum_{i=0}^{I-1}\sum_{j=0}^{I-1}\left(P(i,j)\right)^2 \tag{4}$$

where,

$P\left(i,j\right)$: Represents a matrix denotes the number of occurrence of a pair of pixels with grey level intensity (i, j).

2. Inverse differential moment indicates the homogeneity of a finger knuckle image that quantifies the closeness of distribution of pixel elements in CMM as given by (5)

$$I_{DM} = \frac{\sum_{i=1}^{I-1}\sum_{j=1}^{I-1}P_{ij}}{1+\left(i-j\right)^2} \tag{5}$$

3. Correlation $\left(C_r\right)$ quantifies the value of pixel that has the maximum likelihood probability with reference to its neighbor pixel derived through (6)

$$C_r = \frac{1}{\sigma_i\sigma_j}\sum_{i=0}^{I-1}\sum_{j=0}^{I-1}\left(ij\right)P\left(i,j\right) - \mu_i\mu_j \tag{6}$$

where,

$$\mu_i = \sum_{i=1}^{I-1}\sum_{j=1}^{I-1} i * P(i,j),$$

$$\mu_j = \sum_{i=1}^{I-1}\sum_{j=1}^{I-1} j * P(i,j),$$

$$\sigma_i = \sqrt{\sum_{i=0}^{I-1}\sum_{j=0}^{I-1} P(i,j)\left(\left(i-\mu_i\right)\right)^2},$$

$$\sigma_j = \sqrt{\sum_{i=0}^{I-1}\sum_{j=0}^{I-1} P(i,j)\left(\left(i-\mu_j\right)\right)^2}.$$

4. Contrast $\left(C_t\right)$ quantifies the difference in intensities between a pixel and its neighbor which is given by (7)

$$C_t = \sum_{i=0}^{I-1}\sum_{j=0}^{I-1} (i-j)^2 P(i,j) \tag{7}$$

5. Information Measure of correlation $\left(C_{im}\right)$ derives addiional properties on correlation values which is not defined by rectangular correlation values. The value of $\left(C_{im}\right)$ is manipulated through the entropy values obtained from $\left(P(i,j)\right)$ as given in (8) and (9)

$$C_{im}(p) = \frac{Exy - Exy1}{Max\left(Ex, Ey\right)} \tag{8}$$

$$C_{im}(s) = \left(1 - \exp\left[-2.0(Exy2 - Exy)\right]\right)^{1/2} \tag{9}$$

where,

$$Exy = -\sum_{i=0}^{I-1}\sum_{j=0}^{I-1} P(i,j), \log P(i,j)$$

$$Exy1 = -\sum_{i=0}^{I-1}\sum_{j=0}^{I-1} P\left(i,j\right), \log\left\{P\left(i\right), P\left(j\right)\right\}$$

$$Exy2 = -\sum_{i=0}^{I-1}\sum_{j=0}^{I-1} P\left(i\right), P\left(j\right) \log\left\{P\left(i\right), P\left(j\right)\right\}$$

$P\left(i\right)$: Probability mass function at the pixel position i.

$P\left(j\right)$: Probability mass function at the pixel position j.

The CMM approach results in a statistical based texture feature vector of dimension 128 x 6 x 5 = 3840, where 128 represents the number of sub-bands, 6 represents the number of directions and 5 represents the number of statistical features derived.

Grey Level Difference Vector Method (GDVM)

In this personal authentication system, the Grey level difference vector method is incorporated for texture feature measurements. In addition, this GDVM is identified as one of the successful statistical texture analysis method that yields high prominent features when compared to the other statistical features. Moreover, the GDVM approach extracts set of statistical parameters similar to that of the CMM approach. The proposed GDVM approach manipulates probability density function through the difference of values derived between image intensities. Whereas, the CMM approach only manipulates the joint grey level distributions between two pixels or among the group of pixels present at distance 'd' and direction 'θ'. Furthermore, in GDVM approach, the resulting statistical based parameters are derived in terms of first order statistics, whereas second order statistical approach is incorporated in CMM approach. However, both the methods derive statistical based texture features based on three basic parameters viz., Entropy, Energy and Correlation.

In this work, feature set extracted from finger knuckle surface using GDVM consists of same parameters as that of the CMM approach, viz., Energy, Inverse differential moment, Correlation, Contrast and Information Measure of correlation $\left(C_{im}\right)$. Each texture values are derived at difference convolution sizes which is given as $n = \left\{3, 5, 7, 9, 11, 32 \textit{ and } 64\right\}$ and normalized feature vector is obtained by dividing each element in feature vector by sum of all the elements. The directional angle $\left(\theta\right)$ is considered as $\theta = \left\{0°, 45°, 90° \textit{ and } 135°\right\}$. The GDVM approach results in feature vector with dimension 3840 similar to that of the CMM approach.

FUSION PROCESS

In this section, we discuss fusion process that combines matching scores obtained by means of geometric analysis method and two individual texture feature extraction methods (GLDV, CCM) to achieve better performance. Since, texture pattern exhibited by all the finger knuckle regions are found to be different, the matching scores obtained from these regions must be combined to yield performance improvement at a greater level. In this work, we employ three basic rules viz., SUM, SUM-WEIGHTED and PRODUCT rule for combining matching scores since they are computationally efficient. The combined matching score $\left(C_M\right)$ obtained through SUM $\left(C_S\right)$, SUM-WEIGHTED $\left(C_{SW}\right)$ and PRODUCT $\left(C_P\right)$ rule is given by (11), (12) and (13) respectively.

$$C_S = \frac{1}{C}\sum_{j=1}^{C} D_{ij} \tag{10}$$

$$C_{SW} = \prod_{j=1}^{n} D_{ij} \tag{11}$$

$$C_P = \sum_{j=1}^{n} W_i \times D_{ij} \tag{12}$$

where,

$$W_i = \frac{\dfrac{1}{\sum_{j=1}^{n}\left[\dfrac{1}{EER_j}\right]}}{EER_i}$$

In above equations, the D_{ij} represents matching score derived from the ith user using j^{th} classifier.

EXPERIMENTAL ANALYSIS AND RESULTS DISCUSSIONS

The performance of the proposed finger knuckle print recognition methods described in section 4 and 5 are analyzed by conducting three experiments using a newly created biometric data on finger knuckle surface. This database consists of finger knuckle surface images captured from 150 subjects using the acquisition setup as detailed in section 3. This finger knuckle database was collected from 80 males and 70 female subjects belonging to the age group of 18-40 years. The database was collected in three different sessions with a time interval of 5-6 weeks. In each session, three images of finger knuckle regions from four fingers viz., fore finger, middle finger, ring finger and little finger are captured for processing. Thus, 12 images are collected from each subject in one session. Hence, totally 36 images are collected from one subject in all the three sessions. Finally, the finger knuckle dataset comprises of 150 x 36 = 5400 images in which 600 (150 x 4) images were different finger knuckle regions.

In all the three experiments conducted, the images collected during the first session are taken as gallery set, whereas images collected during the second and third session are taken as probe set. The evaluation metrics taken for performance assessment of proposed finger knuckle recognition system were Genuine Acceptance Rate (GAR) and Equal Error Rate (EER). The GAR is derived by assessing the

Figure 2. Receiver operation characteristics

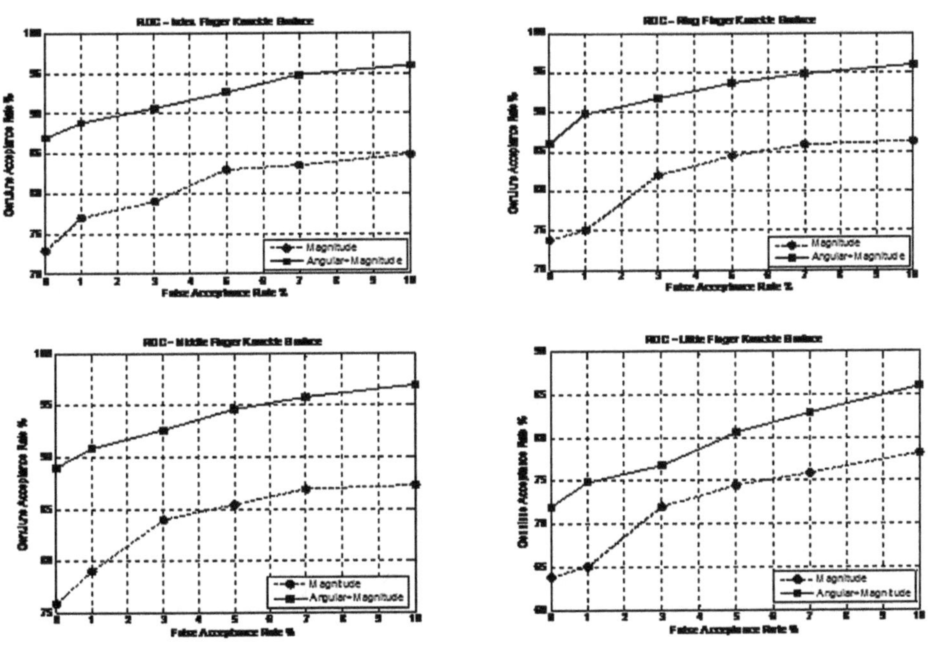

number of genuine matches towards the total number of matches performed by the system. Equal Error Rate is derived by assessing the number of false acceptances and false rejections. The point at which false acceptance rate and false rejection rate becomes equal derives the ERR of a system. In addition, the Decidability Threshold is defined as the normalized distance between genuine and imposter matching scores and given by (14)

$$\frac{\mu_g - \mu_i}{\sqrt{\left(\sigma_g^2 + \sigma_i^2\right)/2}} \tag{14}$$

where, μ_g *and* μ_i are the mean values of genuine and imposter matching scores respectively, σ_g *and* σ_i are the standard deviation values of genuine and imposter matching scores respectively.

Experiment 1

The main aim of this experiment is to evaluate the performance of finger knuckle recognition system based on geometrical measurement. The proposed angular geometric analysis method derives both magnitude and angular-based feature information in recognition is analyzed here. In this experiment, all the four finger knuckle surface images were used. Therefore, the number of gallery images is 150 X 4 = 600 images and number of images in the probe set is 150 X 6 X 4 = 3600 images.

For the entire gallery set images, the geometric measurements were derived and stored in a registered template database. Similarly, for the probe set images, geometric measurements are derived and matched with each and every gallery images. Thus, the number of genuine and imposter matches is found to be 8769 and 6947645 respectively. The Figure 3 (a), (b), (c) and (d), shows the ROC curves derived for all the four finger knuckle surface regions by incorporating geometric measurement in two ways viz., i) incorporating magnitude-only information ii) incorporating magnitude + angular information. Figure 3 (a), (b) and (c) show higher genuine acceptance rates when angular + magnitude information were incorporated in fore, middle and ring finger which is 16.12%, 14.82% and 19.12% higher than results obtained from magnitude-only information. The performance shown by the little finger is lower when compared to other finger knuckle surface regions due to its very unclear texture patterns.

Figure 4 shows the ROC curves derived by fusing the entire four fingers knuckle region scores using SUM, SUM-WEIGHTED and PRODUCT rule of matching

Figure 3. Receiver operational characteristics plot depicting combined performance using geometric measure

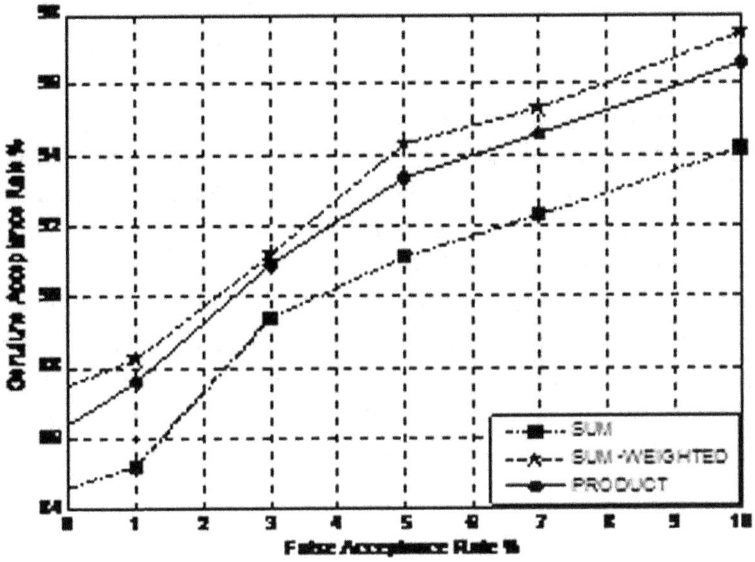

Figure 4. Receiver operational characteristics plot depicting comparative analysis rsults obtained using geometric measure

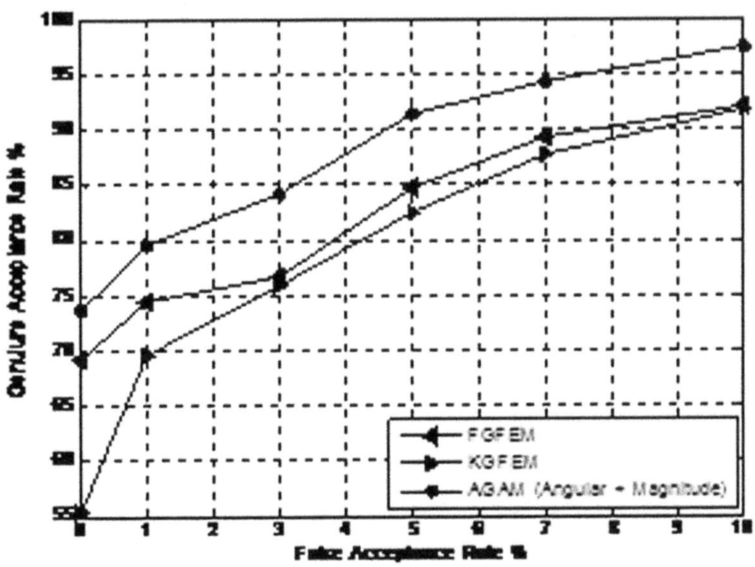

score level fusion. The performance shown by the SUM-WEIGHTED rule is found to be higher than the other two rules of fusing scores.

The proposed AGAM approach for finger knuckle geometric measures is compared with some of the existing shape oriented analysis methods such as a) Finger Geometric Feature Extraction Method (FGFEM) b) Knuckle Geometric Feature Extraction Method (KGFEM) (description of these methods were given in section 2). All the above stated methods are implemented in the same experimental set up proposed in this paper. The following figure 5 shows the ROC plots obtained through various recognition methods by fusing the matching scores obtained from all the four finger knuckle regions using SUM-WEIGHTED rule.

The obtained experimental results indicate that, the proposed AGAM approach decreases the error rate in an average value of 57% when compared with FGFEM, and 48% when compared with KGFEM. This significant improvement in performance is due to AGAM approach extracts shape oriented feature information of finger knuckle region in terms of both magnitude and angular information.

Figure 5. Receiver operational characteristics plot depicting performance in texture measure

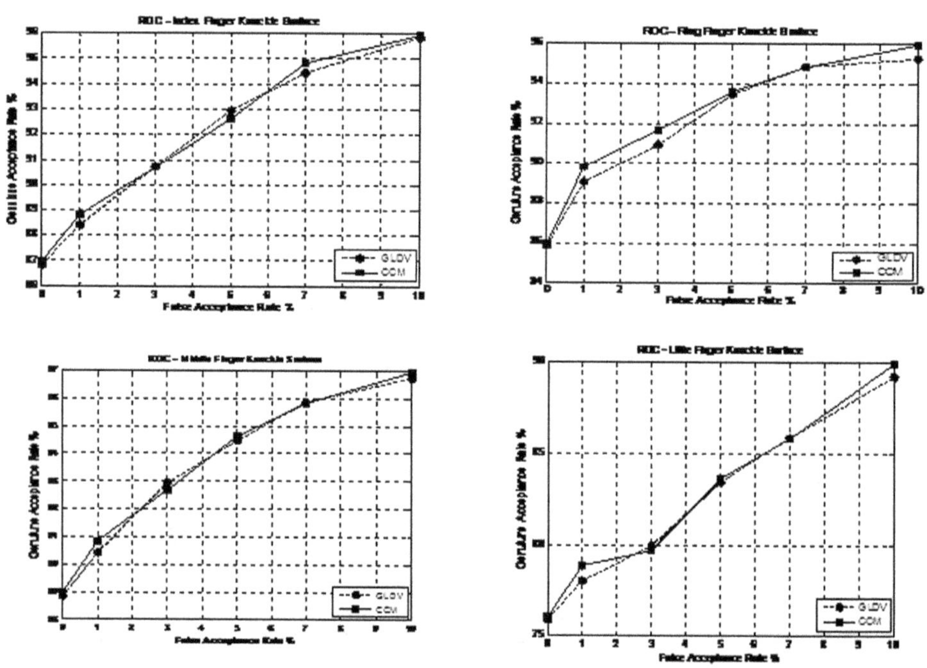

Experiment 2

The main aim of this experiment is to evaluate the performance of the finger knuckle biometric system based on texture analysis method. Since, the texture features are derived by means of two different statistical-based texture analysis methods viz., GLDV and CCM, the evaluation of these two methods were presented in this session. This experiment is conducted by considering all the four finger knuckle regions. Therefore, the number images in the gallery set are 600 (150 X 4) and number of images in the probe set is 3600 (150 X 6 X 4). The number of genuine matches is 8765and imposter matches is 6947564. The Figure 6 (a), (b), (c) and (d) show the ROC curves derived for all the four finger knuckle regions by different types of statistical feature extraction methods. The results show that, the both GLDV and CCM method derive almost equivalent performance on different finger knuckle regions. Among these, the finger knuckle patterns of little finger region shows considerably lower GAR values when compared with the other finger knuckle regions.

Even though, CCM extracts feature parameters based on second order statistical method whereas GDVM approach derives feature information based on first order statistics, the performance assessment of two statistical-based recognition method does not show any significant difference. Moreover, they results in similar perfor-

Figure 6. Receiver operational characteristics plot depicting combined performance using texture measure

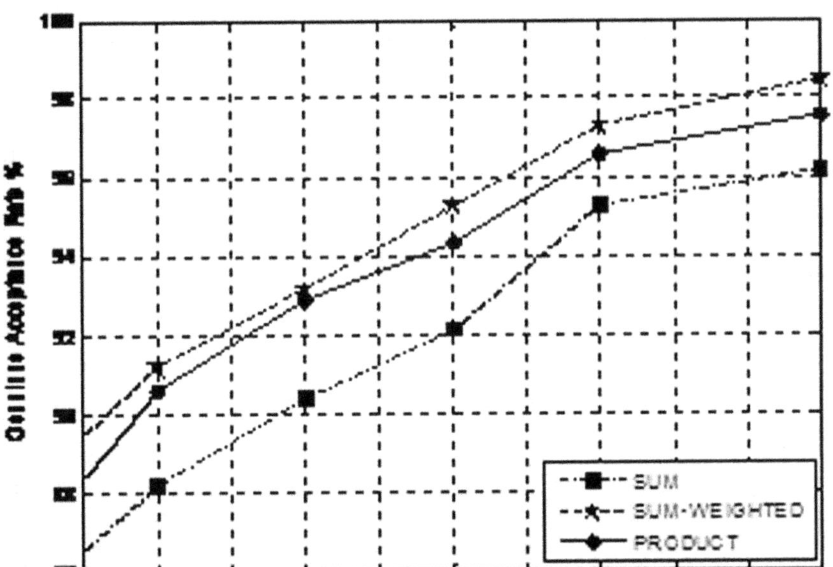

mance since both the methods derive finger knuckle texture features based on three basic statistical parameters viz., Entropy, Energy and Correlation.

Further, the ROC curves derived by fusing all the four finger knuckle regions' texture features using three basic rules of matching score level fusion are shown in Figure 7. The results show that ROC derived using the SUM-WEIGHTED rule of the matching score level fusion yields highest genuine acceptance rate of 98.14%, which is 2.034% higher than that of the performance obtained through geometric measurements proposed in this paper (AGAM).

Experiment 3

In the third experiment, the performance of the proposed finger knuckle biometric system is analyzed by integrating finger knuckle geometric measurements and texture features. The obtained individual matching scores from different finger knuckle regions are combined by means of the SUM-WEIGHTED rule, since this method could able to identify the inter-class variations of an image. The following Figure 7 shows the ROC curves derived by fusing all the four finger knuckle regions obtained by integrating geometric and texture features. The figure also depicts the comparative analysis of integrated approach with the geometric and texture feature extraction

Figure 7. Receiver OPERATIONAL CHARACTERISTICS plot depicting the integrated performance using geometric and texture measure

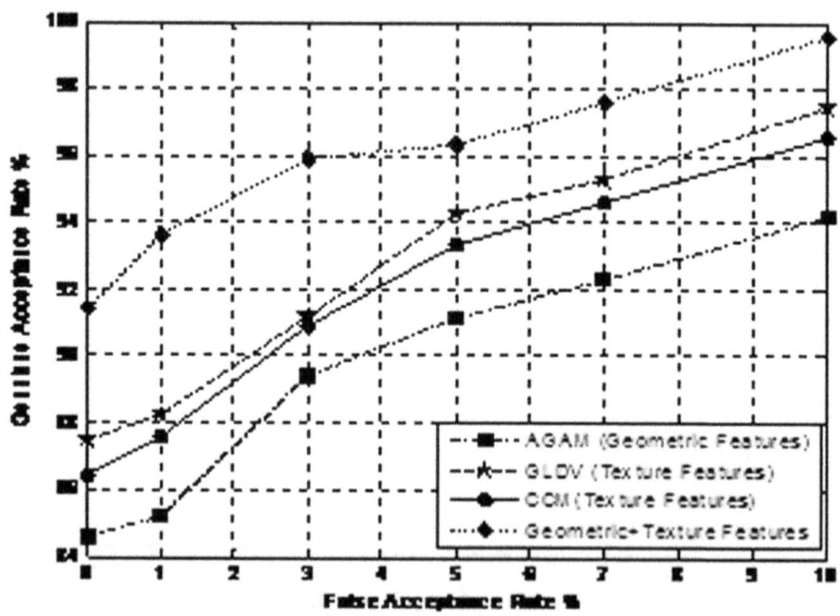

approaches. The results indicate that the integration of geometric and texture feature information for finger knuckle print recognition shows higher genuine acceptance rate of 99.12% which differs from incorporating only geometric measurements at an average rate of 4.23% and similarly, from incorporating only texture features it differs at an average rate of 2.29%.

This considerable increase in performance is obtained by combining the angular-based feature information with finger knuckle texture features of finger knuckle surface image. The results of recognition methods in terms of EER and its corresponding decidability threshold along with the fixed value of FAR are also presented in Table 1.

The results of Table 1 indicates that the integration of geometric and texture feature measures yields lowest error rate of 0.69% and 0.65% with lowest false acceptance value as 1.9587 and 1.9647 respectively for two different combination as shown in the table. This integrated approach reduces the error rate at the average of 56% when compared to the individual evaluation of geometric and texture analysis methods.

CONCLUSION

This paper has presented an intramodal biometric-based personal authentication system using finger knuckle surface. The presented biometric system in this paper acquires the finger dorsum region in a contact-free manner and simultaneously extracts the proximal and distal knuckle regions. The linear factor analysis based segmentation technology has been quite efficient and plays vital role in achieving high accuracy rate. In addition to this, the geometric measurements and texture feature extraction methods has been quite promising in deriving highly potential feature vector features for matching and improves the overall system performance.

Table 1. EER (%) values obtained for various feature recognition methods

Recognition Methods	EER (%)	DT	FRR (%) (when FAR =)
FGFEM	4.53	4.5684	8.5964
FKFEM	5.28	4.8694	7.6522
AGAM	1.67	4.7231	3.8974
CMM	1.38	4.5685	3.4567
GDVM	1.32	4.2354	3.4568
AGAM +CMM	0.69	4.3258	1.9587
AGAM+GDVM	0.65	4.3258	1.9647

Experimental results illustrated in this paper shows that integration of geometric and texture measurements obtained from both proximal and distal knuckle regions produces high accuracy rate of 99.12%, which may not be possible with any one of the measurements and by using either proximal or distal knuckle features alone.

REFERENCES

Aoyama, Ito, & Aoki. (2013) A Finger-Knuckle-Print Recognition Algorithm Using Phase-Based Local Block Matching, Elsevier. *Journal of Information Science*, *4*(5), 47–59.

Choras, M., & Kozik, R. (2010). Knuckle Biometrics Based on Texture Features. *IEEE International Workshop on Emerging Techniques and Challenges (ETCHB)*, *1*(1), 315-322.

Gao, , Yang, Qian, & Zhang. (2014). Integration of multiple orientation and texture information for finger-knuckle-print verification. *Neurocomputing*, *135*(1), 180–191.

Gao, G., Zhang, L., Yang, J., Zhang, L., Zhang, D., & Reconstruction, B. F.-K.-P. V. W. S. L. A. B. F. (2013, December). Reconstruction Based Finger-Knuckle-Print Verification With Score Level Adaptive Binary Fusion. *IEEE Transactions on Image Processing*, *22*(12), 5050–5062. doi:10.1109/TIP.2013.2281429 PMID:24043391

Hand-Based Biometrics. (2003). Hand-based biometrics. *Biometric Technology Today.*, *11*(7), 9–11. doi:10.1016/S0969-4765(03)07018-8

Hanmandlu, M., & Grover, J. (2012). Feature Selection for Finger Knuckle Print-based Multimodal Biometric System. *International Journal of Computers and Applications*, *38*(10), 27–33. doi:10.5120/4645-6905

Haralick, R. M., Shanmugam, K., & Dinstein, I. (1973). Texture features for image classification. *IEEE Transactions on Systems, Man, and Cybernetics*, *3*(6), 610–621. doi:10.1109/TSMC.1973.4309314

Jain, A. K., Ross, D., & Pankanti, S. (1999). A prototype hand geometry based verification system. *Proc. AVBPA*, *1*, 166-171.

Jing, X., Li, W., & Lan, C. (2011) Orthogonal Complex Locality Preserving Projections Based on Image Space Metric for Finger-Knuckle-Print Recognition. *Proc., International Conference on Hand-Based Biometrics (ICHB)*, *4*(1), 1-6.

Kumar, A. (2014). Importance of Being Unique From Finger Dorsal Patterns: Exploring Minor Finger Knuckle Patterns in Verifying Human Identities. *IEEE Transactions on Information Forensics and Security.*, *9*(8), 1288–1298. doi:10.1109/TIFS.2014.2328869

Kumar, A., & Prathyusha, K. V. (2008). Personal authentication using hand vein triangulation. *Proc., SPIE Biometric Technology for Human Identification*, *6*(1), 640-644.

Kumar, A., & Ravikanth, Ch. (2009). Personal Authentication Using Finger Knuckle Surface. *IEEE Transactions on Information Forensics and Security*, *4*(1), 98–110. doi:10.1109/TIFS.2008.2011089

Kumar, A., & Wang, B. (2015). Recovering and matching minutiae patterns from finger knuckle images. *Pattern Recognition*, *68*(1), 361–367. doi:10.1016/j.patrec.2015.08.013

Kumar, A., & Zhang, D. (2006). Personal recognition using hand shape and texture. *IEEE Transactions on Image Processing*, *15*(8), 2454–2461. doi:10.1109/TIP.2006.875214 PMID:16900698

Kumar, A., & Zhang, D. (2010). Improving Biometric Authentication Performance from the User Quality. *IEEE Transactions on Instrumentation and Measurement*, *59*(3), 730–735. doi:10.1109/TIM.2009.2028773

Li, Q., Qiu, Z., Sun, D., & Wu, J. (2004). Personal Identification using knuckle print. *Sinobiometrics, 1*(1), 680-689.

Liu, Tian, & Li. (2014). A new approach for inner-knuckle-print recognition. *Journal of Visual Languages and Computing, 25*(1), 33-42.

Meraoumia, Chitroub & Bouridane. (2011). Fusion of Finger-Knuckle-Print and Palm print for an Efficient Multi-biometric System of Person Recognition. In *Proc., IEEE International Conferences on Communications (ICC), 1*(1), 1-5.

Michael, Connie & Jin. (2010). Robust Palm Print and Knuckle Print Recognition System Using a Contactless Approach. *Proc., 5th IEEE Conference on Industrial Electronics & Application (ICIEA 2010), 1*(1), 15-17.

Nigam & Gupta. (2015). Designing an accurate hand biometric based authentication system fusing finger knuckle print and palmprint. *Neurocomputing, 15*(1), 1120-1132.

Yu, Zhou, & Li. (2014). Personal Identification Using Finger-Knuckle-Print Based On Local Binary Pattern, Applied. *Mechanics of Materials*, *44*(1), 703–706.

Ribaric, S., & Fratric, I. (2005). A biometric identification system based on Eigen palm and Eigen finger features. *IEEE Transactions on Pattern Analysis and Machine Intelligence*, *27*(11), 1698–1709. doi:10.1109/TPAMI.2005.209 PMID:16285370

Sanchez-Reillo, R., Sanchez-Avila, C., & Gonzales-Marcos, A. (2000). A. Biometric identification through hand geometry measurements. *IEEE Transactions on Pattern Analysis and Machine Intelligence*, *22*(10), 1168–1171. doi:10.1109/34.879796

Shen, Bai, & Zhen. (2010). Hand-based biometrics fusing palm print and finger-knuckle-print. *Proc., IEEE International Workshop on Emerging Techniques and Challenges (ETCHB)*, *1*(1), 123-129.

Usha, K., & Ezhilarasan, M. (2013). Personal Authentication based on Angular Geometric Analysis using Finger Back Knuckle Surface. *Proceedings of 3rd International Conference on Advances in Communication Network and Computing*, *1*(1), 51-58.

Wang, Y., Hu, J., & Phillips, D. (2007, April). A Fingerprint Orientation Model Based on 2D Fourier Expansion (FOMFE) and Its Application to Singular-Point Detection and Fingerprint Indexing. *IEEE Transactions on Pattern Analysis and Machine Intelligence*, *29*(4), 573–585. doi:10.1109/TPAMI.2007.1003 PMID:17299215

Wayman, J., Jain, J. K., Maltoni, D., & Maio, D. (2005). Biometric Systems, Technology, Design and Performance Evaluation. Springer-Verlag.

Yang, Sun, & Sun., (2011). Finger-Knuckle-Print Recognition using Gabor Feature and OLDA. *Proc., of the 30th Chinese Control Conference*, *4*(1), 78-84.

Zhang, Zhang, & Zhang. (2009). Finger-knuckle-print: a new biometric identifier. *Proc.IEEE International Conference on Image Processing*, *1*(1), 76-82. doi:10.1109/ICIP.2009.5413734

Zhang, Zhang, & Zhang. (2010). Monogenic Code: A Novel Fast Feature Coding Algorithm with Applications to Finger-Knuckle-Print Recognition. *IEEE International Workshop on Emerging Techniques and Challenges (ETCHB)*, *1*(1), 222-231. doi:10.1109/ETCHB.2010.5559286

Zhang, Li, & Shen. (2011). A Novel Riesz Transforms based Coding Scheme for Finger-Knuckle-Print Recognition. *Proc., IEEE International Conference on Hand-Based Biometrics (ICHB)*, *1*(1), 1-6.

Zhang, L., & Li, H. (2012). Encoding local image patterns using Riesz transforms With applications to palm print and finger-knuckle-print recognition. *Image and Vision Computing*, *30*(1), 1043–1105. doi:10.1016/j.imavis.2012.09.003

Zhang, L., Zhang, L., & Zhang, D. (2009). Finger-Knuckle-Print Verification Based on Band-Limited Phase-Only Correlation. *LNCS, 5702*, 141–148.

Zhang, L., Zhang, L., Zhang, D., & Zhu, H. (2011). Ensemble of local and global information for finger–knuckle-print recognition. *Pattern Recognition*, *44*(1), 1990–1998. doi:10.1016/j.patcog.2010.06.007

Compilation of References

AB, F. S. (2011a). Thermal Imaging Guide Book for Building and Renewable Applications. FLIR System AB.

AB, F. S. (2011b). *Thermal Imaging Guide Book for Industrial Applications*. Retrieved from: http://www.flirmedia.com/MMC/THG/Brochures/T820264/T820264_EN.pdf

Adler, F. (1965), Physiology of the Eye: Clinical Application (4th ed.). London: The C.V. Mosby Company.

Aguiar, Y. P., Maria de Fátima, Q. V., Galy, E., & Santoni, C. (2014). Accounting for Individual and Situation Characteristics to Understand the User Behaviour when Interacting with Systems during Critical Situations. *International Journal of Ambient Computing and Intelligence*, *6*(2), 29–55. doi:10.4018/IJACI.2014070103

Ahmed, S. S., Dey, N., Ashour, A. S., Sifaki-Pistolla, D., Bălas-Timar, D., Balas, V. E., & Tavares, J. M. R. (2016). Effect of fuzzy partitioning in Crohn's disease classification: A neuro-fuzzy-based approach. *Medical & Biological Engineering & Computing*, 1–15. PMID:27106754

Alonso-Montesinos, J., & Batlles, F. J. (2015). Solar radiation forecasting in the short- and medium-term under all sky conditions. *Energy*, *1*(83), 387–393. doi:10.1016/j.energy.2015.02.036

Alsaadawi, Z., & Netzelmann, U. (2010). Spatio-temporal filtering of active thermography data for noise reduction and data compression. In *Proceedings of 10th International Conference on Quantitative InfraRed Thermography (QIRT 2010)*. Québec, Canada: QIRT conference open archives. doi:10.21611/qirt.2010.006

Anbar, M., Milescu, L., Naumov, A., Brown, C., Button, T., Carly, C., & AlDulaimi, K. (2001). Detection of cancerous breasts by dynamic area telethermometry. *Engineering in Medicine and Biology YK Magazine, IEEE*, *20*(5), 80–91. doi:10.1109/51.956823 PMID:11668900

Anonymous, . (2002). *Infrared training: Level 1 course manual*. North Billerica, MA: Infrared Training Centre.

Aoyama, Ito, & Aoki. (2013) A Finger-Knuckle-Print Recognition Algorithm Using Phase-Based Local Block Matching, Elsevier. *Journal of Information Science*, *4*(5), 47–59.

Araki, T., Ikeda, N., Molinari, F., Dey, N., Acharjee, S. M., Saba, L., & Suri, J. S. et al. (2014). Effect of geometric-based coronary calcium volume as a feature along with its shape-based attributes for cardiological risk prediction from low contrast intravascular ultrasound. *Journal of Medical Imaging and Health Informatics*, *4*(2), 255–261. doi:10.1166/jmihi.2014.1251

Arora, N., Martins, D., Ruggerio, D., Tousimis, E., Swistel, A. J., Osborne, M. P., & Simmons, R. M. (2008). Effectiveness of a noninvasive digital infrared thermal imaging system in the detection of breast cancer. *American Journal of Surgery*, *196*(4), 523–526. doi:10.1016/j.amjsurg.2008.06.015 PMID:18809055

Ashour, A. S., Beagum, S., Dey, N., Ashour, A. S., Pistolla, D. S., Nguyen, G. N., & Shi, F. et al. (2017). Light microscopy image de-noising using optimized LPA-ICI filter. *Neural Computing & Applications*, 1–17.

Ashour, A. S., Samanta, S., Dey, N., Kausar, N., Abdessalemkaraa, W. B., & Hassanien, A. E. (2015). Computed Tomography Image Enhancement Using Cuckoo Search: A Log Transform Based Approach. *Journal of Signal and Information Processing*, *6*(03), 244–257. doi:10.4236/jsip.2015.63023

Aslam, Ramashri, & Ahsan. (2013, March). A New Approach to Image Segmentation for Brain Tumor detection using Pillar K-means Algorithm. *International Journal of Advanced Research in Computer and Communication Engineering*.

Bagavathiappan, S., Saravanan, T., Philip, J., Jayakumar, T., Raj, B., Karunanithi, R., … Jagadeesan, K. (2009). Infrared Thermal Imaging for Detection of Peripheral Vascular Disorders. *Journal of Medical Physics/Association of Medical Physicists of India*, *34*(1), 43.

Bagavathiappan, S., Lahiri, B., Saravanan, T., Philip, J., & Jayakumar, T. (2013). Infrared thermography for condition monitoring – A review. *Infrared Physics & Technology*, *60*, 35–55. doi:10.1016/j.infrared.2013.03.006

Balageas, D. (2011). Defense and Illustration of Time Resolved Pulsed Thermography for NDE. In *Proceedings of Thermosense: Thermal Infrared Applications XXXIII*. SPIE. doi:10.1117/12.882967

Balageas, D., Roche, J., Leroy, F., Liu, W., & Gorbach, A. (2015). The thermographic signal reconstruction method: A powerful tool for the enhancement of transient thermographic images. *Biocybernetics and Biomedical Engineering*, *35*(1), 1–9. doi:10.1016/j.bbe.2014.07.002 PMID:25678731

Banzi & Xue. (2015). An Automated Tool for Non-contact, Real Time Early Detection of Diabetes by Computer Vision. *International Journal of Machine Learning and Computing*, *5*(3), 225–229. doi:10.7763/IJMLC.2015.V5.511

Barner, K., & Arce, G. (1998). Order-Statistic Filtering and Smoothing of Time Series: Part II. In N. Balakrishnan & C. R. Rao (Eds.), *Handbook of Statistics* (Vol. 17, pp. 525–554). Elsevier. doi:10.1016/S0169-7161(98)17023-2

Beagum, S., Dey, N., Ashour, A. S., Sifaki-Pistolla, D., & Balas, V. E. (2016). Nonparametric de-noising filter optimization using structure-based microscopic image classification. *Microscopy Research and Technique*. doi:10.1002/jemt.22811 PMID:27901295

Belongie, S., Malik, J., & Puzicha, J. (2002). Shape Matching and Object Recognition Using Shape Contexts. *IEEE Transactions on Pattern Analysis and Machine Intelligence*, *24*(24), 509–521. doi:10.1109/34.993558

Bendt, P., & Rabl, A. (1981). Optical analysis of point parabolic radiation. *Applied Optics*, (20): 674–683. doi:10.1364/AO.20.000674 PMID:20309177

Benítez, H. (2012). *Image quality indexes for infrared non-destructive testing*. Paper presented at Infrared Non-Destructive Evaluation Seminar, Cali, Colombia.

Benítez, H., Ibarra-Castanedo, C., Bendada, A., Maldague, X., Loaiza, H., & Caicedo, E. (2008). Definition of a new thermal contrast and pulse correction for defect quantification in pulsed thermography. *Infrared Physics and Technology Journal, 51*(3), 160-167.

Benítez, H., Ibarra-Castanedo, C., Bendada, A., Maldague, X., Loaiza, H., & Caicedo, E. (2009). Defect Characterization in Infrared Nondestructive Testing with Learning Machines. *NDT&E*, *42*(7), 630–643. doi:10.1016/j.ndteint.2009.05.004

Benítez, H., Loaiza, H., & Caicedo, E. (2011). *Termografía activa pulsada en inspección de materiales. Técnicas avanzadas de procesado*. Cali, Colombia: Universidad del Valle—Pontificia Universidad Javeriana.

Bersch, S., Azzi, D., Khusainov, R., & Achumba, I. E. (2013). Artificial immune systems for anomaly detection in ambient assisted living applications. *International Journal of Ambient Computing and Intelligence*, *5*(3), 1–15. doi:10.4018/ijaci.2013070101

Bezdek, J. C. (1981). *Pattern Recognition with Fuzzy Objective Function Algorithms*. New York: Plenum Press. doi:10.1007/978-1-4757-0450-1

Bezerra, L. A., Oliveira, M. M., Araújo, M. C., Viana, M. J., Santos, L. C., Santos, F. G., ... & Resmini, R. (2013). Infrared Imaging for Breast Cancer Detection with Proper Selection of Properties: From Acquisition Protocol to Numerical Simulation. *Multimodality Breast Imaging: Diagnosis and Treatment, E. YK Ng; U. Rajendra Acharya*, 285-332.

Blanco, J., Malato, S., Fernandez, P., Vidal, A., Morales, A., Trincado, P., & Rangel, C. M. et al. (1999). Compound parabolic concentrator technology development to commercial solar detoxification applications. *Solar Energy*, *67*(4-6), 317–330. doi:10.1016/S0038-092X(00)00078-5

Boles, W., & Boashash, B. (1998). A Human Identification Technique Using Images of the Iris and Wavelet Transform. *IEEE Transactions on Signal Processing*, *46*(4), 1185–1188. doi:10.1109/78.668573

Bouvier, J. L., Michaux, G., Salagnac, P., Nepveu, F., Rochier, D., & Kientz, T. (2015). Experimental characterisation of a solar parabolic trough collector used in a micro-CHP (micro-cogeneration) system with direct steam generation. *Energy*, *83*, 474–485. doi:10.1016/j.energy.2015.02.050

Bouwmans, T., El Baf, F., & Vachon, B. (2008). Background Modeling using Mixture of Gaussians for Foreground Detection—A Survey. *Recent Patents Comput. Sci.*, *1*(3), 219–237. doi:10.2174/2213275910801030219

Bradley, D., & Roth, G. (2007). Adaptive thresholding using the integral image. *Journal of Graphics, GPU, and Game Tools*, *12*(2), 13–21. doi:10.1080/2151237X.2007.10129236

Bronzino, J. D. (2006). Infrared imaging of the breast–an overview. In Biomedical Engineering handbook–Medical Systems and Devices (3rd ed.). CRC.

Buddharaju, P., & Pavlidis, I. (2008). Physiology-Based Face Recognition in the Thermal Infrared Spectrum. In Medical Infrared Imaging Handbook (pp. 13-1 - 13-14).

Cao, M., & Cui, W. (2016). Study on the Measurement Technique of the Alignment Variation of Medical Infrared Thermal Imaging System. *International Journal of Smart Home*, *10*(1), 81–88. doi:10.14257/ijsh.2016.10.1.09

Carmeliet, P., & Jain, R. K. (2000). Angiogenesis in cancer and other diseases. *Nature*, *407*(6801), 249–257. doi:10.1038/35025220 PMID:11001068

Chandhok, C., Chaturvedi, S., & Khurshid, A. A. (2012). An approach to image segmentation using k-means clustering algorithm. *International Journal of Information Technology*, *1*(1), 11–17.

Chebrolu, S., Abraham, A., & Thomas, J. P. (2005). Feature detection and ensemble design of intrusion detection systems. *Computers & Security*, *24*(4), 295–307. doi:10.1016/j.cose.2004.09.008

Chen, T. W., Hsu, S. C., & Chien, S. Y. (2007, July). Robust Video Object Segmentation Based on K-Means Background Clustering and Watershed in Ill- Conditioned Surveillance Systems. *Proceedings of theIEEE Int. Conf. on Multimedia and Expo* (pp. 787 – 7900). doi:10.1109/ICME.2007.4284768

Chen, Z., Gao, L., & Kwiat, K. (2003). Modeling the Spread of Active Worms. *INFOCOM. Twenty-Second Annual Joint Conference of the IEEE Computer and Communications Societies*. IEEE. doi:10.1109/INFCOM.2003.1209211

Cheng-Hung, H., & Meng-Ting, C. (2008). A transient three-dimensional inverse geometry problem in estimating the space and time dependent irregular boundary shapes. *International Journal of Heat and Mass Transfer*, *51*(21-22), 5238–5246. doi:10.1016/j.ijheatmasstransfer.2008.03.019

Chen, Y., Abraham, A., & Yang, J. (2005). Feature deduction and intrusion detection using flexible neural trees. *Second IEEE International Symposium on Neural Networks*.

Cheriguene, S., Azizi, N., Zemmal, N., Dey, N., Djellali, H., & Farah, N. (2016). Optimized Tumor Breast Cancer Classification Using Combining Random Subspace and Static Classifiers Selection Paradigms. In *Applications of Intelligent Optimization in Biology and Medicine* (pp. 289–307). Springer International Publishing. doi:10.1007/978-3-319-21212-8_13

Choras, M., & Kozik, R. (2010). Knuckle Biometrics Based on Texture Features. *IEEE International Workshop on Emerging Techniques and Challenges (ETCHB)*, *1*(1), 315-322.

Chou, T.-S., & Yen, K. K. (2007). Fuzzy belief k-nearest neighbors anomaly detection of user to root and remote to local attacks, The2007IEEE workshop on information assurance, 207–213. doi:10.1109/IAW.2007.381934

Cinque, M., Coronato, A., & Testa, A. (2012). Dependable services for mobile health monitoring systems. *International Journal of Ambient Computing and Intelligence*, *4*(1), 1–15. doi:10.4018/jaci.2012010101

Clark, M. R., McCann, D. M., & Forde, M. C. (2003). Application of infrared thermography to the non-destructive testing of concrete and masonry bridges. *NDT & E International*, *36*(4), 265–275. doi:10.1016/S0963-8695(02)00060-9

Cohen, F. (1987). Computer viruses: Theory and experiments. *Computers & Security*, *6*(1), 22–35. doi:10.1016/0167-4048(87)90122-2

Colthup, N. (2012). *Introduction to infrared and Raman spectroscopy*. Elsevier.

Communications, A. (2009). *Some like it hot: Thermal cameras in surveillance* (White Paper). Retrieved from http://textlab.io/doc/538217/some-like-it-hot--thermal-cameras-in-surveillance

Corporation, F. (2009). *Introduction to Thermography Principles.* Author.

Costea, I., Dumitrescu, C., Minea, M., & Banica, A. (2016). Analysis of facial expressions using thermal imaging. In *Electronics Technology (ISSE), 2016 39th International Spring Seminar on* (pp. 306-309). doi:10.1109/ISSE.2016.7563210

Dabov, K., Foi, A., Katkovnik, V., & Egiazarian, K. (2007). Image denoising by sparse 3D ransform-domain collaborative filtering. *IEEE Transactions on Image Processing*, *16*(8), 2080–2095. doi:10.1109/TIP.2007.901238 PMID:17688213

Dascomb, J. (2009). *Low-cost concentrating solar collector for steam generation.* Thesis in department of mechanical engineering.

Daugman, J. (1994). *Biometric Personal Identification System Based on Iris Analysis.* United States Patent no. 5291560.

David, M., Friedrich, L., & Kurt, H. (2003). The Support Vector Machine Under Test. *Neurocomputing*, *55*(1-2), 169–186. doi:10.1016/S0925-2312(03)00431-4

Davis, A. P., & Lettington, A. H. (1988). Principles of thermal imaging. *Applications of Thermal Imaging*, 1-34.

Devi & Preethi. (2014). Disease Identification in Iris Using Gabor Filter. *International Journal of Engineering And Computer Science*, *3*(4), 5396–5399.

Dey, N., Roy, A. B., Pal, M., & Das, A. (2012). *FCM based blood vessel segmentation method for retinal images*. arXiv preprint arXiv: 1209.1181

Dey, N., & Ashour, A. S. (Eds.). (2016). *Classification and Clustering in Biomedical Signal Processing*. IGI Global. doi:10.4018/978-1-5225-0140-4

Dey, N., Ashour, A. S., Ashour, A. S., & Singh, A. (2015). Digital Analysis of Microscopic Images in Medicine. *Journal of Advanced Microscopy Research*, *10*(1), 1–13. doi:10.1166/jamr.2015.1229

Dey, N., Ashour, A. S., Beagum, S., Pistola, D. S., Gospodinov, M., Gospodinova, E. P., & Tavares, J. M. R. S. (2015). Parameter Optimization for Local Polynomial Approximation based Intersection Confidence Interval Filter Using Genetic Algorithm: An Application for Brain MRI Image De-Noising. *Journal of Imaging*, *1*(1), 60–84. doi:10.3390/jimaging1010060

Dey, N., Ashour, A. S., Chakraborty, S., Banerjee, S., Gospodinova, E., Gospodinov, M., & Hassanien, A. E. (2017). Watermarking in Biomedical Signal Processing. In *Intelligent Techniques in Signal Processing for Multimedia Security* (pp. 345–369). Springer International Publishing. doi:10.1007/978-3-319-44790-2_16

Dey, N., Dey, G., Chakraborty, S., & Chaudhuri, S. S. (2014). Feature analysis of blind watermarked electromyogram signal in wireless telemonitoring. In *Concepts and Trends in Healthcare Information Systems* (pp. 205–229). Springer International Publishing. doi:10.1007/978-3-319-06844-2_13

DG Topteches Electronic, Limited. (2015). *Walk Through Forehead Thermometer - Medical Thermometer Manufacturer*. curiousexpeditions.org Redirected from http://topteches1.sell.curiousexpeditions.org/pz5a00393-walk-through-forehead-thermometer-medical-thermometer-manufacturer.html

Diakides, N. A., & Bronzino, J. D. (2008). *Medical Infrared Imaging*. Taylor & Francis Group, LLC.

Dunn, J. C. (1973). A fuzzy relative of the ISODATA process and its use in detecting compact well-separated clusters. *Journal of Cybernetics*, *3*(3), 32–57. doi:10.1080/01969727308546046

Eastman, J. R. (2006). Introduction to Remote Sensing and Image Processing. In *IDRISI Guide to GIS and Image Processing* (Vol. 1, pp. 17 - 34). DRISI Production.

Egawa, S. (1997). *Clinical Radiation Thermometer*. WIPO patent: WO1997017887 A1.

El Fadar, A. (2015). Thermal behavior and performance assessment of a solar adsorption cooling system with finned adsorber. *Energy*, *1*(83), 674–684. doi:10.1016/j.energy.2015.02.074

Fadlallah, S. A., Ashour, A. S., & Dey, N. (2016). Advanced Titanium Surfaces and Its Alloys for Orthopedic and Dental Applications Based on Digital SEM Imaging Analysis. *Advanced Surface Engineering Materials*, 517-560.

Faúndez, M. (2001). *Tratamiento Digital de Voz e Imagen, y Aplicación a la Multimedia*. Alfaomega Grupo Editor.

Fernández-Reche, J., & Valenzuela, L. (2012). Geometrical assessment of solar concentrators using close-range photogrammetry. *Energy Procedia*, *30*, 84–90. doi:10.1016/j.egypro.2012.11.011

Florez-Ospina, J., & Benítez, H. (2014). From local to global analysis of defect detectability in infrared non-destructive testing. *Infrared Physics & Technology*, *63*, 211–221. doi:10.1016/j.infrared.2013.12.017

Fouad, K. M., Hassan, B. M., & Hassan, M. F. (2016). User Authentication based on Dynamic Keystroke Recognition. *International Journal of Ambient Computing and Intelligence*, *7*(2), 1–32. doi:10.4018/IJACI.2016070101

Fundamentals of Remote Sensing, Natural Resources Canada. (n.d.). Retrieved from http://www.nrcan.gc.ca/node/9309

Gao, , Yang, Qian, & Zhang. (2014). Integration of multiple orientation and texture information for finger-knuckle-print verification. *Neurocomputing*, *135*(1), 180–191.

Gao, G., Zhang, L., Yang, J., Zhang, L., Zhang, D., & Reconstruction, B. F.-K.-P. V. W. S. L. A. B. F. (2013, December). Reconstruction Based Finger-Knuckle-Print Verification With Score Level Adaptive Binary Fusion. *IEEE Transactions on Image Processing*, *22*(12), 5050–5062. doi:10.1109/TIP.2013.2281429 PMID:24043391

García Cortés, S., Bello-García, A., & Ordóñez, C. (2012). Estimating intercept factor of a parabolic solar trough collector with new supporting structure using off-the-shelf photogrammetric equipment. *Applied Energy*, *92*, 815–821. doi:10.1016/j.apenergy.2011.08.032

García, D. (2010). Robust smoothing of gridded data in one and higher dimensions with missing values. *Computational Statistics & Data Analysis*, *54*(4), 1167–1178. doi:10.1016/j.csda.2009.09.020 PMID:24795488

Gaussorgues, G., & Chomet, S. (2012). *Infrared thermography* (Vol. 5). Springer Science & Business Media.

Ghosh, A. K., Schwartzbard, A., & Schatz, M. (1999). *Learning program behavior profiles for intrusion detection*. Workshop on intrusion detection and network monitoring, Santa Clara, CA.

Ghosh, A.K., & Schwartzbard, A. (n.d.). *A study in using neural networks for anomaly and misuse detection*. In Usenix security symposium, Washington, DC.

Ghosh, A., Sarkar, A., Ashour, A. S., Bălas-Timar, D., Dey, N., & Balas, V. E. (2015). Grid Color Moment Features in Glaucoma Classification. *International Journal of Advanced Computer Science and Applications*, *6*(9). doi:10.14569/IJACSA.2015.060913

Gmb, H. O. (n.d.). *Basic Principles of Non-Contact Temperature Measurement* (IR Basics Brochure-E2014-01-A). Retrieved from www.optris.de

Golestani, N., EtehadTavakol, M., & Ng, E. Y. K. (2014). Level set method for segmentation of infrared breast thermograms. *Experimental and Clinical Sciences*, *13*, 241–251. PMID:26417258

Gonzalez, R. C., & Richard, E. (2002). *Woods Digital Image Processing* (2nd ed.). Upper Saddle River, NJ: Prentice hall.

González, R., & Woods, R. (2002). *Digital Image Processing*. Prentice Hall.

Goodhew, J., Boomsma, C., Pahl, S., & Goodhew, S. (2015). It's fffffffrrreeezsing,.... thought thermal imaging might shed some light as to where the heats going!'Can visualisation shed a light on heating. *Proceedings from ECEEE.*

Gordillo, N., Montseny, E., & Sobrevilla, P. (2013, October). State of the art survey on MRI brain tumor segmentation. *Magnetic Resonance Imaging, 31*(8), 1426–1438. doi:10.1016/j.mri.2013.05.002 PMID:23790354

Goswami, S., Dey, U., Roy, P., Ashour, A., & Dey, N. (2016). Medical Video Processing: Concept and Applications. *Feature Detectors and Motion Detection in Video Processing*, 1.

Green, M. W. (1999). The appropriate and effective use of security technologies in U.S. schools, A guide for schools and law enforcement agencies (research report). Sandia National Laboratories, Albuquerque, NM, USA.

Grinzato, E., Bison, P., Marinetti, S., & Vavilov, V. (2000). *Thermal NDE enhanced by 3D numerical modeling applied to works of art.* Paper presented at 15th World Conference on Non-destructive Testing, Rome, Italy. Retrieved from http://www.ndt.net/article/wcndt00/papers/idn909/idn909.htm

Grys, S., & Minkina, W. (2010). Filtered thermal contrast – error analysis. In *Proceedings of 10th International Conference on Quantitative InfraRed Thermography (QIRT 2010).* Québec, Canada: QIRT conference open archives. doi:10.21611/qirt.2010.051

Hampapur, A., Brown, L., Connell, J., Ekin, A., Haas, N., Lu, M., ... & Pankanti, S. (2005). Smart Video Surveillance. *IEEE Signal Processing Magazine*, 22(2), 38–51.

Hampapur, A., Brown, L., Connell, J., Pankanti, S., Senior, A., & Tian, Y. (2003). Smart Surveillance: Applications, Technologies and Implications. *Information, Communications and Signal Processing*, *2*, 1133–1138.

Hand-Based Biometrics. (2003). Hand-based biometrics. *Biometric Technology Today., 11*(7), 9–11. doi:10.1016/S0969-4765(03)07018-8

Hanmandlu, M., & Grover, J. (2012). Feature Selection for Finger Knuckle Print-based Multimodal Biometric System. *International Journal of Computers and Applications*, *38*(10), 27–33. doi:10.5120/4645-6905

Haralick, , & Shanmugam, & Dinstein. (1973). Textural features for image classification. *IEEE Transactions on Systems, Man and Cybernetics, 3*(6), 610-621. doi:10.1109/TSMC.1973.4309314

Hasan, M. (2014). *Adaptive Edge-guided Block-matching and 3D filtering (BM3D) Image Denoising Algorithm* (Master of Science Thesis). University of Western Ontario, Canada.

Heikkilä, M., & Pietikäinen, M. (2006). A Texture-Based. Method for Modelling the Background and Detecting Moving Objects. *IEEE TPAMI*, *28*(4), 657–662. doi:10.1109/TPAMI.2006.68

Heitz, D. (2015). *Infrared Thermometer to Fight Ebola Premiers at CES 2015*. Healthline Media. Retrieved from http://www.healthline.com/health-news/infrared-thermometer-to-fight-ebola-premiers-at-ces-010515#1

Hellier, C. J. (2003). *Thermal Infrared Testing Handbook of Nondestructive Evaluation*. The McGraw-Hill Companies, Inc.

Hemalatha, S., & Anouncia, S. M. (2016). A Computational Model for Texture Analysis in Images with Fractional Differential Filter for Texture Detection. In *Biometrics: Concepts, Methodologies, Tools, and Applications: Concepts, Methodologies, Tools, and Applications*. Academic Press.

Hensen, J., & Nakhi, A. (1994). Fourier and Biot numbers and the accuracy of conduction modeling. In *Proceedings of Building Environmental Performance Conference (BEP'94) "Facing the Future"*. University of York.

Heriansyah, R., & Abu-Bakar, S. A. R. (2009). Defect detection in thermal image for nondestructive evaluation of petrochemical equipments. *NDT & E International*, *42*(8), 729–740. doi:10.1016/j.ndteint.2009.06.008

Hildebrandt, C., Ammer, K., & Raschner, C. (2010). An Overview of Recent Application of Medical Infrared Thermography in Sports Medicine in Austria. *Sensors (Basel, Switzerland)*, *10*(5), 4700–4715. doi:10.3390/s100504700 PMID:22399901

Ho, C. K. (2008). *Software and codes for analysis of concentrating solar power technologies*. Albuquerque, NM: Sandia National Laboratories. doi:10.2172/946571

Hore, S., Chakroborty, S., Ashour, A. S., Dey, N., Ashour, A. S., Sifaki-Pistolla, D., & Chaudhuri, S. R. et al. (2015). Finding Contours of Hippocampus Brain Cell Using Microscopic Image Analysis. *Journal of Advanced Microscopy Research*, *10*(2), 93–103. doi:10.1166/jamr.2015.1245

Huang, C. L., Wu, Y. W., Hwang, C. L., Jong, Y. S., Chao, C. L., Chen, W. J., & Yang, W. S. et al. (2011). The application of infrared thermography in evaluation of patients at high risk for lower extremity peripheral arterial disease. *Journal of Vascular Surgery*, *54*(4), 1074–1080. doi:10.1016/j.jvs.2011.03.287 PMID:21784604

Hunt, J. L. (n.d.). *Radiations in Environment*. Creative Commons. San Francisco. Instrument, M. Table of Total Emissivity.

Ibarra-Castanedo, C. (2005). *Quantitative Subsurface Defect Evaluation by Pulsed Phase Thermography: Depth Retrieval with the Phase* (Doctoral dissertation). Available from Collection Mémoires et thèses électroniques, Université Laval.

Infernuso, T., Loughin, C. A., Marino, D. J., Umbaugh, S. E., & Solt, P. S. (2010). Thermal Imaging of Normal and Cranial Cruciate Ligament-Deficient Stifles in Dogs. *Veterinary Surgery*, *39*(4), 410–417. doi:10.1111/j.1532-950X.2010.00677.x PMID:20459492

International Air Transport Association. (2009). A global approach to reducing aviation emissions. First stop: carbon-neutral growth from 2020. Author.

Ishimwe, R., Abutaleb, K., & Ahmed, F. (2014). Applications of Thermal Imaging in Agriculture—A Review. *Advances in Remote Sensing*, *3*(3), 128–140. doi:10.4236/ars.2014.33011

Izhar, L. I., & Petrou, M. (2012). Thermal imaging in medicine. *Advances in Imaging and Electron Physics*, *171*, 41–114. doi:10.1016/B978-0-12-394297-5.00002-7

Jain, A. K., Ross, D., & Pankanti, S. (1999). A prototype hand geometry based verification system. *Proc. AVBPA*, *1*, 166-171.

James, C. M., Matthew, L. S., & Michael, E. M. (1998). *Crew fatigue and performance on US coast guard cutters*. Springfield, VA: National Technical Information Service.

Janssen, J. H., Ijsselsteijn, W. A., Westerink, J. H., Tacken, P., & de Vries, G. J. (2013). The tell-tale heart: Perceived emotional intensity of heartbeats. *International Journal of Synthetic Emotions*, *4*(1), 65–91. doi:10.4018/jse.2013010103

Jensen, B. (1980). *Iridology Simplified* (2nd ed.). Escondido.

Jetto, L. (1999). Stochastic modelling and 3D minimum variance recursive estimation of image sequences. *Multidimensional Systems and Signal Processing*, *10*(3), 225–254. doi:10.1023/A:1008406707503

Jiang, S., Hu, P., Mo, S., & Chen, Z. (2010). Optical modeling for a two-stage parabolic trough concentrating photovoltaic/thermal system using spectral beam splitting technology. *Sol Energ Mater Sol Cells*, (94), 1686-96.

Jing, X., Li, W., & Lan, C. (2011) Orthogonal Complex Locality Preserving Projections Based on Image Space Metric for Finger-Knuckle-Print Recognition. *Proc., International Conference on Hand-Based Biometrics (ICHB)*, *4*(1), 1-6.

Jones, B. F., & Plassmann, P. (2002). Digital infrared thermal imaging of human skin. *Engineering in Medicine and Biology Magazine, IEEE*, *21*(6), 41–48. doi:10.1109/MEMB.2002.1175137 PMID:12613210

Jumb, V., Sohani, M., & Shrivas, A. (2014). Color image segmentation using k-means clustering and Otsu's adaptive thresholding. *International Journal of Innovative Technology and Exploring Engineering*, *3*(9), 72–76.

K, N., Dat, R., & K, A. (2013). Advances in Infrared Detector Array Technology. In P. S. Pyshkin (Ed.), *Optoelectronics - Advanced Materials and Devices*. In Tech.

Kakuta, N., Yokoyama, S., & Mabuchi, K. (2002). Human thermal models for evaluating infrared images. *Engineering in Medicine and Biology Magazine, IEEE*, *21*(6), 65–72. doi:10.1109/MEMB.2002.1175140 PMID:12613213

Kapoor, P., & Prasad, S. V. A. V. (2010, February). Image processing for early diagnosis of breast cancer using infrared images. *2010 2nd International Conference on Computer and Automation Engineering (ICCAE)*, (Vol. 3, pp. 564-566). IEEE.

Kapoor, P., Prasad, D. S., & Patni, S. (2012). Automatic Analysis of Breast Thermograms for tumor detection based on Biostatistical feature extraction and ANN. *International Journal of Emerging Trends in Engineering and Development*, *2*(7), 245–255.

Kass, M., Witkin, A., & Terzopoulos, D. (1987). Snakes: Active Contour Models. *International Journal of Computer Vision*, 321–331.

Khaled A. (2012). *Technical and economic performance of parabolic trough in Jordan*. Academic Press.

Kim K., Chalidabhongse T.H., Harwood D., Davis. L. (2004). Background Modeling and Subtraction by Codebook Construction. *Proceedings of the Int. Cont. on Image Processing (ICIP '04)* (Vol. 5, pp. 3061-3064).

Kim, K., Chalidabhongse, T. H., Harwood, D., & Davis, L. (2005). Real-time Foreground-background Segmentation Using Codebook Model. *Elsevier Real-time Imaging*, *11*(3), 172–185. doi:10.1016/j.rti.2004.12.004

Kim, Y. C., Bahk, J. H., Lee, S. C., & Lee, Y. W. (2003). Infrared thermographic imaging in the assessment of successful block on lumbar sympathetic ganglion. *Yonsei Medical Journal*, *44*(1), 119–124. doi:10.3349/ymj.2003.44.1.119 PMID:12619184

King, P., Comley, P., & Sansom, C. (2014). Parabolic trough surface form mapping using photogrammetry and its validation with a large Coordinate Measuring Machine. *Energy Procedia*, *49*, 118–125. doi:10.1016/j.egypro.2014.03.013

Kiri, W., Claire, C., Seth, R., & Stefan, S. (2001). Constrained K-means Clustering with Background Knowledge.*Proc. of the 18th Int. Conf. on Machine Learning (ICML '01)* (pp. 577-584).

Kleih, J. (1991). Dish-Stirling test facility. *Solar Energy Materials*, *24*(1-4), 231–237. doi:10.1016/0165-1633(91)90063-Q

Kolarić, D., Lipić, T., Grubiaić, I., Gjenero, L., & Skala, K. (2011, September). Application of infrared thermal imaging in blade system temperature monitoring. *ELMAR, 2011 Proceedings* (pp. 309-312). IEEE.

Kotyk, T., Dey, N., Ashour, A. S., Balas-Timar, D., Chakraborty, S., Ashour, A. S., & Tavares, J. M. R. (2016). Measurement of glomerulus diameter and Bowmans space width of renal albino rats. *Computer Methods and Programs in Biomedicine*, *126*, 143–153. doi:10.1016/j.cmpb.2015.10.023 PMID:26796351

Kramer, D., & Aghdasi, F. (1999). Texture analysis techniques for the classification of micro calcifications in digitised mammograms. *IEEE AFRICON*, *1*, 395–400.

Kumar, A. (2014). Importance of Being Unique From Finger Dorsal Patterns: Exploring Minor Finger Knuckle Patterns in Verifying Human Identities. *IEEE Transactions on Information Forensics and Security.*, *9*(8), 1288–1298. doi:10.1109/TIFS.2014.2328869

Kumar, A., & Prathyusha, K. V. (2008). Personal authentication using hand vein triangulation. *Proc., SPIE Biometric Technology for Human Identification*, *6*(1), 640-644.

Kumar, A., & Ravikanth, Ch. (2009). Personal Authentication Using Finger Knuckle Surface. *IEEE Transactions on Information Forensics and Security*, *4*(1), 98–110. doi:10.1109/TIFS.2008.2011089

Kumar, A., & Santhi, V. (2011). A review on geometric invariant digital image watermarking techniques. *International Journal of Computers and Applications*, *12*(9), 31–36. doi:10.5120/1708-2275

Kumar, A., & Wang, B. (2015). Recovering and matching minutiae patterns from finger knuckle images. *Pattern Recognition*, *68*(1), 361–367. doi:10.1016/j.patrec.2015.08.013

Kumar, A., & Zhang, D. (2006). Personal recognition using hand shape and texture. *IEEE Transactions on Image Processing*, *15*(8), 2454–2461. doi:10.1109/TIP.2006.875214 PMID:16900698

Kumar, A., & Zhang, D. (2010). Improving Biometric Authentication Performance from the User Quality. *IEEE Transactions on Instrumentation and Measurement*, *59*(3), 730–735. doi:10.1109/TIM.2009.2028773

Kwon, S. H. (2004). Threshold selection based on cluster analysis. *Pattern Recognition Letters*, *25*(9), 1045–1050. doi:10.1016/j.patrec.2004.03.001

Larsen, C. (2011). *Document flash thermography* (Master of Science Thesis). Available from Utah State University DigitalCommons@USU.

Lee, F. F., Chen, F., & Liu, J. (2015). Infrared thermal imaging system on a mobile phone. *Sensors (Basel)*, *15*(5), 10166–10179. doi:10.3390/s150510166 PMID:25942639

Lee, H. (2014). The geometric-optics relation between surface slope error and reflected ray error in solar concentrators. *Solar Energy*, *101*, 299–307. doi:10.1016/j.solener.2013.12.035

Levi, K., & Weiss, Y. (2004). Learning object detection from a small number of examples: the importance of good features.*Proc. of the IEEE Conference on CVPR*, 53–60. doi:10.1109/CVPR.2004.1315144

Li, Q., Qiu, Z., Sun, D., & Wu, J. (2004). Personal Identification using knuckle print. *Sinobiometrics, 1*(1), 680-689.

Lim, G., Manzur, T., & Kar, A. (2014). Noise sources and improved performance of a mid-wave infrared uncooled silicon carbide optical photodetector. *Applied Optics*, *53*(36), 8410–8423. doi:10.1364/AO.53.008410 PMID:25608189

Liu, Tian, & Li. (2014). A new approach for inner-knuckle-print recognition. *Journal of Visual Languages and Computing, 25*(1), 33-42.

Liu, X., Dong, S., An, M., Bai, L., & Luan, J. (2015). Quantitative assessment of facial paralysis using infrared thermal imaging. In *2015 8th International Conference on Biomedical Engineering and Informatics (BMEI)* (pp. 106-110). doi:10.1109/BMEI.2015.7401482

Liu, R., Kong, Y., Jiao, B., Li, Z., Shang, H., Lu, D., & Zhang, Q. et al. (2013). A substrate-free optical readout focal plane array with a heat sink structure. *Journal of Semiconductors*, *34*(2), 024005. doi:10.1088/1674-4926/34/2/024005

Livadas, C., Walsh, B., Lapsley, D., & Strayer, T. (2006). *Using machine learning techniques to identify botnet traffic*. Second IEEE LCN workshop on network security (WNS), Tampa, FL.

Li, Z., Yang, J., Han, P., Ran, Y., & Chai, Z. (2015). Intrusion detection and cooperative tracking using PTZ network thermal imagers. *International Journal of Distributed Sensor Networks*, *2015*, 1.

Lodin & Demea. (2009). Design of an iris-based medical diagnosis system. *International Symposium on Signals, Circuits and Systems*, 1(4), 9-10.

Loose, M., Beletic, J., Garnett, J., & Xu, M. (2007). High-performance focal plane arrays based on the HAWAII-2RG/4G and the SIDECAR ASIC. In Optical Engineering+ Applications (pp. 66900C-66900C). International Society for Optics and Photonics.

López, F., Nicolau, V., Ibarra-Castanedo, C., & Maldague, X. (2014). Thermal-numerical model and computational simulation of pulsed thermography inspection of carbon fiber reinforced composites. *International Journal of Thermal Sciences*, *86*, 325–340. doi:10.1016/j.ijthermalsci.2014.07.015

Lowe, D. G. (2004). Distinctive image feature from scale-invariant keypoints. *International Journal of Computer Vision*, *60*(2), 91–110. doi:10.1023/B:VISI.0000029664.99615.94

Machin, G., Simpson, R., & Broussely, M. (2009). Calibration and validation of thermal imagers. *Quantitative InfraRed Thermography Journal*, *6*(2), 133–147. doi:10.3166/qirt.6.133-147

Madruga, F., Albendea, P., Ibarra-Castanedo, C., & López-Higuera, J. (2010). Signal to noise ratio (SNR) comparison for lock-in thermographic data processing methods in CFRP specimen. In *Proceedings of 10th International Conference on Quantitative InfraRed Thermography (QIRT 2010)*. Québec, Canada: QIRT conference open archives. doi:10.21611/qirt.2010.079

Mahmoudzadeh, E., Montazeri, M. A., Zekri, M., & Sadri, S. (2015). Extended hidden Markov model for optimized segmentation of breast thermography images. *Infrared Physics & Technology*, *72*, 19–28. doi:10.1016/j.infrared.2015.06.012

Maillet, D., Houlbert, A., Didierjean, S., Lamine, A., & Degiovanni, A. (1993). Non-destructive thermal evaluation of delaminations in a laminate: Part I – Identification by measurement of thermal contrast. *Composites Science and Technology*, *47*(2), 137–153. doi:10.1016/0266-3538(93)90043-G

Ma, L., Tan, T., Wang, Y., & Zhang, D. (2003). Personal identification based on iris texture analysis. *IEEE Transactions on Pattern Analysis and Machine Intelligence*, *25*(12), 1519–1533. doi:10.1109/TPAMI.2003.1251145

Maldague, X. (2001). *Theory and Practice of Infrared Technology for Non Destructive Testing*. New York, NY: John Wiley & Sons.

Mandel, C., & Autexier, S. (2016, May). People Tracking in Ambient Assisted Living Environments Using Low-Cost Thermal Image Cameras. In *International Conference on Smart Homes and Health Telematics* (pp. 14-26). Springer International Publishing. doi:10.1007/978-3-319-39601-9_2

Mao, Z., Ma, L., Zhao, M., & Li, Z. (2006). Feature-preserving mesh denoising based on contextual discontinuities. *Journal of Zhejiang University SCIENCE A*, *7*(9), 1603–1608. doi:10.1631/jzus.2006.A1603

Marinetti, S., Grinzato, E., Bison, P. G., Bozzi, E., Chimenti, M., Pieri, G., & Salvetti, O. (2004). Statistical analysis of IR thermographic sequences by PCA. *Infrared Physics & Technology*, *46*(1-2), 85–91. doi:10.1016/j.infrared.2004.03.012

Martinez, B., Binefa, X., & Pantic, M. (2010). Facial Component Detection in Thermal Imagery. *Proceedings of the IEEE computer Society Conference on Computer Vision and Pattern Recognition Workshops (CVPRW)*, 48-54. doi:10.1109/CVPRW.2010.5543605

März, T., Prahl, C., Ulmer, S., & Weber, S. W. C. (2011). Validation of two Optical Measurement Methods for the Qualification of the Shape Accuracy of Mirror Panels for Concentrating Solar Systems. *Journal of Solar Energy Engineering*, 133.

Masek, L. (2003). *Recognition of Human Iris Patterns for Biometric Identification*. The University of Western Australia.

McDannold, N. J., King, R. L., Jolesz, F. A., & Hynynen, K. H. (2000). Usefulness of MR Imaging-Derived Thermometry and Dosimetry in Determining the Threshold for Tissue Damage Induced by Thermal Surgery in Rabbits. *Radiology*, *216*(2), 517–523. doi:10.1148/radiology.216.2.r00au42517 PMID:10924580

Meguro, M., Taguchi, A., & Hamada, N. (2009). Data-Dependent Weighted Average Filtering For Image Sequence Enhancement. In *Proceedings of 1999 International Conference on Image Processing (ICIP'99)*. Kobe, Japan: IEEE Computer Society.

Meraoumia, Chitroub & Bouridane. (2011). Fusion of Finger-Knuckle-Print and Palm print for an Efficient Multi-biometric System of Person Recognition. In *Proc., IEEE International Conferences on Communications (ICC)*, *1*(1), 1-5.

Mercer, J. B., & Ring, E. F. (2009). Fever Screening and Infrared Thermal Imaging: Concerns and Guidelines.Thermology International, 19(3), 67-69.

Mertzios, B. G., & Tsirikolias, K. (1998). Coordinate Logic Filters and Their Applications in Image Processing and Pattern Recognition. *Circuits Systems Signal Processing*, *17*(4), 517–538. doi:10.1007/BF01201506

Meystel, A. M., & Albus, J. M. (2002). *Intelligent systems architecture, design, and control*. New York: John Wiley & Sons, Inc.

Michael, Connie & Jin. (2010). Robust Palm Print and Knuckle Print Recognition System Using a Contactless Approach. *Proc., 5th IEEE Conference on Industrial Electronics & Application (ICIEA 2010)*, *1*(1), 15-17.

Minkina, W., & Dudzik, S. (2009). *Infrared Thermography Errors and Uncertainties*. John Wiley & Sons, Ltd. doi:10.1002/9780470682234

Mohamad, H. S., & Mahmood, F. (2008). Real-time Background Modeling/Subtraction using Two- Layer Codebook Model.Proc. of the Int. Multiconf. Of Engineers and Computer Scientists (IMECS '08).

More & Pergad. (2015). On A Methodology For Detecting Diabetic Presence From Iris Image Analysis. *International Journal of Engineering Sciences & Research Technology*, *4*(7), 2277–9655.

Mukkamala, S., Janoski, G., & Sung, A. (2001). Monitoring systsem security using neural networks and support vector machines. *International workshop on hybrid intelligent systems*, 121–138.

Mukkamala, S., Sung, A. H., & Abraham, A. (2003). Intrusion detection using ensemble of soft computing paradigms. *Third international conference on intelligent systems design and applications, advances in soft computing*. Springer. doi:10.1007/978-3-540-44999-7_23

Mukkamala, S., Sung, A. H., & Abraham, A. (2004b). Modeling intrusion detection systems using linear genetic programming approach. *17th international conference on industrial and engineering applications of artificial intelligence and expert systems*, 633–642. doi:10.1007/978-3-540-24677-0_65

Mukkamala, S., Sung, A., & Abraham, A. (2004a). Designing intrusion detection systems: architectures and perspectives. The international engineering consortium (IEC) annual review of communications, 57, 1229–1241.

Mukkamala, S., Sung, A., & Abraham, A. (2007). Hybrid multi-agent framework for detection of stealthy probe. *Appl Soft Computing J*, *7*(3), 631–641. doi:10.1016/j.asoc.2005.12.002

Murphy, F. E. (2015). i4Toys: Video technology in toys for improved access to play, entertainment, and education. In *2015 IEEE International Symposium on Technology and Society (ISTAS)* (pp. 1-6). doi:10.1109/ISTAS.2015.7439423

Nandi, D., Ashour, A. S., Samanta, S., Chakraborty, S., Salem, M. A., & Dey, N. (2015). Principal component analysis in medical image processing: A study. *International Journal of Image Mining*, *1*(1), 65–86. doi:10.1504/IJIM.2015.070024

Navleet, D., & Bill, T. (2005). Histograms of Oriented Gradients for Human Detection. *Proc. IEEE Conf. Computer Vision and Pattern Recognition CVPR '05*(pp. 886-893).

Ng, E. Y. K., Kaw, G. J. L., & Chang, W. M. (2004). Analysis of IR thermal imager for mass blind fever screening.Microvascular Research, 68(2), 104-109.

Ng, E. K., Fok, S. C., Peh, Y. C., Ng, F. C., & Sim, L. S. J. (2002). Computerized detection of breast cancer with artificial intelligence and thermograms. *Journal of Medical Engineering & Technology*, *26*(4), 152–157. doi:10.1080/03091900210146941 PMID:12396330

Ng, E. Y. K., & Chen, Y. (2006). Segmentation of breast thermogram: Improved boundary detection with modified snake algorithm. *Journal of Mechanics in Medicine and Biology*, *6*(2), 123–136. doi:10.1142/S021951940600190X

Ng, E. Y. K., & Kee, E. C. (2008). Advanced integrated technique in breast cancer thermography. *Journal of Medical Engineering & Technology*, *32*(2), 103–114. doi:10.1080/03091900600562040 PMID:17852648

Ng, E.Y., Ng, F.C., & Sim, L.S. (2001). Statistical analysis of healthy and malignant breast thermography. *Journal of Medical Engineering & Technology*, *25*(6), 253–263. doi:10.1080/03091900110086642 PMID:11780767

Nigam & Gupta. (2015). Designing an accurate hand biometric based authentication system fusing finger knuckle print and palmprint. *Neurocomputing*, *15*(1), 1120-1132.

Norulzahrah, M. Z., Suzaimah, R., Kamarul, H. G., Talib, M. L., & Nor Asiakin, H. (2015). A Study on Implementing Physiology-Based Approach and Optical Flow Algorithm to Thermal Screening System for Flu Detection. *International Journal of Information and Electronics Engineering*, *5*(1), 31–34.

Nur Izzati, N. B. I., Wong, W. K., Jalil, B. M. D., Yante, B. M. N., Azlan, B. S., & Lim, H. S. (2013). Fuzzy Logic based Viola Jones Fever Detection Method in Thermal Imaging System. *Proceedings of the 2013 International Conference on Emerging Trends in Computer Science & Information Technology (ICETCSIT '13)*, 23-26.

Okada, Y., Kawamata, T., Kawashima, A., & Hori, T. (2007). Intraoperative Application Of Thermography In Extracranial-Intracranial Bypass Surgery. *Neurosurgery*, *60*(4), 362–365. PMID:17415175

Othman, Z., & Prabuwono, A. S. (2010). Preliminary study on iris recognition system: Tissues of body organs in iridology. IEEE Conference on Biomedical Engineering and Sciences, 115-119.

Ourston, D., Matzner, S., Stump, W., & Hopkins, B. (2004). Coordinated internet attacks: Responding to attack complexity. *Computers & Security*, *12*(2), 165–190. doi:10.3233/JCS-2004-12201

Papageorgiou, C. P., Oren, M., & Poggio, T. (1998). A general framework for object detection. *Proceedings of theInternational Conference on Computer Vision* (pp. 555-562). doi:10.1109/ICCV.1998.710772

Pilla, M., Klein, M., Maldague, X., & Salerno, A. (2002). New Absolute Contrast for Pulsed Thermography. In *Proceedings of 6th International Conference on Quantitative InfraRed Thermography (QIRT 2002)*, Dubrovnik, Croacia: QIRT conference open archives. doi:10.21611/qirt.2002.004

Pillans, L. A. (2013). *Performance Evaluation of an Uncooled Infrared Array Camera (Ph.D)*. London: London University.

Planck, M. (1959). *The theory of heat radiation: Authorized translation by Morton Masius*. New York: Dover Publications.

Pohl, J. (1998). Ultrasonic Inspection of Adaptive CFRP-Structures. *NDT Database & Journal, 3*(9). Retrieved from: http://www.ndt.net/article/ecndt98/aero/015/015.htm

Poljak-Blazi, M., Kolaric, D., Jaganjac, M., Zarkovic, K., Skala, K., & Zarkovic, N. (2009). Specific thermographic changes during Walker 256 carcinoma development: Differential infrared imaging of tumour, inflammation and haematoma. *Cancer Detection and Prevention, 32*(5), 431–436. doi:10.1016/j.cdp.2009.01.002 PMID:19232842

Ponomaryov, V. (2007). Real-time 2D–3D filtering using order statistics based algorithms. *Journal of Real-Time Image Processing, 1*(3), 173–194. doi:10.1007/s11554-007-0021-5

Pottlera, K., Ulmera, S., Lüpferta, E., Landmanna, M., Rögerb, M., & Prahl, C. (2014). Ensuring performance by geometric quality control and specifications for parabolic trough solar fields. *Energy Procedia, 49*, 2170–2179. doi:10.1016/j.egypro.2014.03.230

Prahl, C., Stanicki, B., Hilgert, C., Ulmer, S., & Roger, M. (2013). Airborne shape measurement of parabolic trough collector fields. *Solar Energy, 91*, 68–78. doi:10.1016/j.solener.2013.01.012

President, I. H. (2011) International Consensus and Guidelines on Medical Thermography 2011 (ICGMT 2011) 2nd International Work Group for Medical Thermography Meeting (IWGMT 2011). *Neurosurgery*, 8.

Puri, C., Olson, L., Pavlidis, I., Levine, J., & Starren, J. (2005, April). StressCam: non-contact measurement of users' emotional states through thermal imaging. In CHI'05 extended abstracts on Human factors in computing systems (pp. 1725-1728). Academic Press.

Qi, H., & Diakides, N. A. (2009). Thermal infrared imaging in early breast cancer detection, Augmented Vision Perception in Infrared. *Advances in Pattern Recognition, II*, 139–152.

Radzi, S., Ghazali, K., Mohammad, A., & Naim, F. (2011). *Procedia Computer Science Using Bimodal Gaussian Mixture Model-Based Algorithm for Background Segmentation in Thermal Fever Mass Screening*. Academic Press.

Ramlee, R. A., & Ranjit, S. (2009). Using Iris Recognition Algorithm, Detecting Cholesterol Presence. *International Conference on Information Management and Engineering*, 714(717), 3-5. doi:10.1109/ICIME.2009.61

Ramlee, R. A., Ramli, A. R., & Sulaiman, H. A. (2015). A Review on Diseases Manifestation by Ocular Diseases using Computer Aided Diagnosis (CAD). *IACSIT International Journal of Engineering and Technology, 7*(4), 1343–1348.

Restrepo, A., & Loaiza, H. (2014b). Background thermal compensation by filtering (BTCF) for infrared thermography evaluation. In *Proceedings of the XIX International Simposium of Signal and Images Processing and Artificial Vision (STSIVA, 2014)*. IEEEXplore. doi:10.1109/STSIVA.2014.7010162

Restrepo, A., & Loaiza, H. (2012). Filtrado 3D Espacio-Temporal Iterativo para la Atenuación de Ruido en Secuencias de Imágenes Térmicas para END. *Ingenium*, *6*(13), 27–34. doi:10.21774/ing.v6i13.85

Restrepo, A., & Loaiza, H. (2013). Non-uniform heating compensation on thermal images using median filtering. *DYNA*, *80*(182), 74–82.

Restrepo, A., & Loaiza, H. (2014a). New 3D finite difference method for thermal contrast enhancement in slabs pulsed thermography inspection. *Journal of Non-destructive Evaluation*, *33*(1), 62–73.

Restrepo, A., & Loaiza, H. (2014c). 3D discrete model for thermal contrast enhancement and defects depth estimation in CFRP slabs. *Ingeniería y Competitividad*, *16*(2), 143–153.

Restrepo, A., & Loaiza, H. (2015). Enhanced method for flaws depth estimation in CFRP slabs from FDTC thermal contrast sequences. *Ingeniería e Investigación*, *35*(3), 61–68. doi:10.15446/ing.investig.v35n3.50552

Restrepo, A., & Loaiza, H. (2016). Background thermal compensation by filtering for contrast enhancement in active thermography. *Journal of Non-destructive Evaluation*, *35*(1), 20. doi:10.1007/s10921-016-0336-x

Ribaric, S., & Fratric, I. (2005). A biometric identification system based on Eigen palm and Eigen finger features. *IEEE Transactions on Pattern Analysis and Machine Intelligence*, *27*(11), 1698–1709. doi:10.1109/TPAMI.2005.209 PMID:16285370

Ring, E. F. J. (2000). The discovery of infrared radiation in 1800. *Imaging Science Journal*, *48*(1), 1–8. doi:10.1080/13682199.2000.11784339

Ring, E. F. J., & Ammer, K. (2012). Infrared thermal imaging in medicine. *Physiological Measurement*, *33*(3), R33–R46. doi:10.1088/0967-3334/33/3/R33 PMID:22370242

Ring, E. F. J., Jung, A., Zuber, J., Rutowsk, P., Kalicki, B., & Bajwa, U. (2008). Detecting fever in Polish children by infrared thermography. *Proceedings of the 9th International Conference on Quantitative Infrared Thermography*,125-128. doi:10.21611/qirt.2008.03_07_17

Ring, F. (2007). Pandemic: Thermography for fever screening of airport passengers. *Thermology International*, *17*(2), 67.

Ring, F. (2010). Thermal imaging today and its relevance to diabetes. *Journal of Diabetes Science and Technology*, *4*(4), 857–862. doi:10.1177/193229681000400414 PMID:20663449

Rodríguez, F. L., Ibarra-Castanedo, C., Nicolau, V., & Maldague, X. (2014b). Optimization of pulsed thermography inspection by partial least-squares regression. *NDT & E International*, *66*, 128–138. doi:10.1016/j.ndteint.2014.06.003

Rodríguez, F. L., Nicolau, V., Ibarra-Castanedo, C., & Maldague, X. (2014a). Pulsed Thermography Signal Processing Techniques Based on the 1D Solution of the Heat Equation Applied to the Inspection of Laminated Composites. *Materials Evaluation*, *72*(1), 91–102.

Rodríguez, J. M., Sánchez, D., Martínez, G. S., Bennouna, E. G., & Ikken, B. (2016). Techno-economic assessment of thermal energy storage solutions for a 1 MWe CSP-ORC power plant. *Solar Energy*, *140*, 206–218. doi:10.1016/j.solener.2016.11.007

Rogalski, A. (2003). *Infrared detectors: status and trends* (00796727). Retrieved from http://www.sciencedirect.com/science/article/pii/S0079672702000241

Rogalski, A., & Chrzanowski, K. (2014). Infrared Devices and techniques (revision). *Metrology and Measurement Systems*, *21*(4), 565–618. doi:10.2478/mms-2014-0057

Roy Alves. (2010). *Thermal technology: a hot new trend in video surveillance*. Johannesburg: Axis Communications.

Russ, J. (1998). *The Image Processing Handbook* (3rd ed.). CRC Press, Springer, IEEE Press.

Saba, L., Dey, N., Ashour, A. S., Samanta, S., Nath, S. S., Chakraborty, S., & Suri, J. S. et al. (2016). Automated stratification of liver disease in ultrasound: An online accurate feature classification paradigm. *Computer Methods and Programs in Biomedicine*, *130*, 118–134. doi:10.1016/j.cmpb.2016.03.016 PMID:27208527

Saber, E., & Tekalp, A. M. (1998). Integration of color, edge, shape, and texture features for automatic region-based image annotation and retrieval. *Journal of Electronic Imaging*, *7*(3), 684–700. doi:10.1117/1.482605

Saleh Ali, I. M., ODonovan, T. S., Reddy, K. S., & Mallick, T. K. (2013). An optical analysis of a static 3-D solar concentrator. *Solar Energy*, *88*, 57–70. doi:10.1016/j.solener.2012.11.004

Samanta, S., Choudhury, A., Dey, N., Ashour, A. S., & Balas, V. E. (2016). *Quantum inspired evolutionary algorithm for scaling factors optimization during manifold medical information embedding. In Quantum Inspired Computational intelligence: Research and Applications*. Elsevier.

Sanchez-Reillo, R., Sanchez-Avila, C., & Gonzales-Marcos, A. (2000). A. Biometric identification through hand geometry measurements. *IEEE Transactions on Pattern Analysis and Machine Intelligence*, *22*(10), 1168–1171. doi:10.1109/34.879796

Sanderson, S., & Erbetta, J. (2000). Authentication for secure environments based on iris scanning technology. *IEEE Colloquium on Visual Biometrics*, 1-7. doi:10.1049/ic:20000468

Santhi, V., Thangavelu, A., & Arulmozhivarman, P. (2013). Adaptive visible watermarking in Hadamard domain for digital images. *International Journal of Information and Computer Security*, *5*(3), 202–223. doi:10.1504/IJICS.2013.055838

Sarvesh, M. (2005). *Analyses and synthesis of iris images* (M.S. thesis). West Virginia University, Morgantown, WV.

Saxena, R., Shah, K., Chawla, R., & Santhi, V. (2014). Biometric Watermarking for Copyright Protection of Digital Images. *International Journal of Applied Engineering Research, 9*(24), 23681–23688.

Schioppa, E. J. (2014). *The color of X-rays: Spectral X-ray computed tomography using energy sensitive pixel detectors.* Academic Press.

Segui, J. (2014). Boeing Simulates Thermal Expansion in Composites with Expanded Metal Foil for Lightning Protection of Aircraft Structures. *COMSOL News - The Multiphysics Simulation Magazine, 2014*, 4-7.

Sellami, N., Mallick, T. K., & McNeil, D. A. (2012). Optical characterisation of 3-D static solar concentrator. *Energy Conversion and Management, 64*, 579–586. doi:10.1016/j.enconman.2012.05.028

Shah, K., Dave, N., Chavan, S., Mukherjee, S., Abraham, A., & Sanyal, S. (2004). Adaptive neuro-fuzzy intrusion detection system. IEEE international conference on ITCC'04, 1, 70–74.

Sharma, U. C. (2004). *Infrared Detctors.* M. Tech Credit Seminar Report, Electronic systems Group, EE Dept, IIT.

Shen, Bai, & Zhen. (2010). Hand-based biometrics fusing palm print and finger-knuckle-print. *Proc., IEEE International Workshop on Emerging Techniques and Challenges (ETCHB), 1*(1), 123-129.

Shen, X. (2009). Intrusive Evaluation of Ambient Displays. *International Journal of Ambient Computing and Intelligence, 1*(4), 12–31. doi:10.4018/jaci.2009062202

Shuai, Y., Xia, X. L., & Tan, H. P. (2008). Radiation performance of dish solar concentrator/cavity receiver systems. *Solar Energy, 82*(1), 13–21. doi:10.1016/j.solener.2007.06.005

Siegmund, D., Handtke, D., & Kaehm, O. (2016, May). Verifying isolation in a mantrap portal via thermal imaging. In *Systems, Signals and Image Processing (IWSSIP), 2016 International Conference on* (pp. 1-4). doi:10.1109/IWSSIP.2016.7502726

Sivanandam, S., Anburajan, M., Venkatraman, B., Menaka, M., & Sharath, D. (2012). Medical thermography: A diagnostic approach for type 2 diabetes based on non-contact infrared thermal imaging. *Endocrine, 42*(2), 343–351. doi:10.1007/s12020-012-9645-8 PMID:22411072

Skala, K., Kavanah, H., Lipic, T., Sovic, I., & Grubisic, I. (2013). *5D thermal imaging system for body dissipation monitoring in motion.* Paper presented at 18th International Conference on Thermal Engineering and Thermogrammetry (THERMO), University of Technology and Economics (BME), Budapest, Hungary.

Skala, K., Lipic, T., Sovic, I., Gjenero, L., & Grubisic, I. (2011). 4D thermal imaging system for medical applications. *Periodicum Biologorum, 113*(4), 407–416.

Smith, H. (2005). Screening for breast cancer. *Journal of the American Medical Association, 293*(10), 1245–1256. doi:10.1001/jama.293.10.1245 PMID:15755947

Snell, J., Jr., & Spring, R. (2007). Infrared Thermography Advances. *Featured Articles – NDT Magazine*. Retrieved November, 2010, from http://www.marinecompositesnde.com/images/Snell,_2007.pdf

Sofiah, S., Radzi, M., Kamarul Hawari, G., Nazriyah, C. Z., & Faradila, N. (2011). Using Bimodal Gaussian Mixture Model-Based Algorithm for Background Segmentation in Thermal Fever Mass Screening. *Proceedings of the2nd World Conference on Information Technology*.

Soria, R., Lucena, A. F. P., Tomaschek, J., Fichter, T., Haasz, T., Szklo, A., & Kern, J. et al. (2016). Modelling concentrated solar power (CSP) in the Brazilian energy system: A soft-linked model coupling approach. *Energy, 1*(116), 265–280. doi:10.1016/j.energy.2016.09.080

Stauffer, C., & Grimson, W. (1999). Adaptive Background Mixture Models for Real-time Tracking.*Proc. IEEE Conf. on Comp Vision and Patt. Recog. (CVPR '99)* (pp. 246-252). doi:10.1109/CVPR.1999.784637

Suard, F., Rakotomamonjy, A., & Bensrhair, A. (2006). *Pedestrian Detection using Infrared images and Histograms of Oriented Gradients*. Academic Press.

Surabhi, V., Spinello, D., & Neesulescu, D. (2012). Infrared Fever Body Identification using Shape and Temperature Filters. *Proceedings of the IEEE Instrumentation and Measurement Technology Conference (I2MTC)*, 1556-1560. doi:10.1109/I2MTC.2012.6229315

Szeliski, R. (2010). *Computer Vision, algorithms and applications*. Springer.

Tadeu, A., & Simões, N. (2005). Fundamental solutions for transient heat transfer by conduction and convection in an unbounded, half-space, slab and layered media in the frequency domain. *Engineering Analysis with Boundary Elements, 29*(12), 1130–1142. doi:10.1016/j.enganabound.2005.06.002

Tadeu, A., & Simões, N. (2006). Three-dimensional fundamental solutions for transient heat transfer by conduction in an unbounded medium, half-space, slab and layered media. *Engineering Analysis with Boundary Elements, 30*(5), 338–349. doi:10.1016/j.enganabound.2006.01.011

Tadeu, A., & Simões, N. (2007). Conduction and convection phenomena through a slab with thermal heterogeneities. *Applied Mathematical Modelling, 31*(7), 1444–1459. doi:10.1016/j.apm.2006.04.010

Tapia, D. I., & Corchado, J. M. (2009). An ambient intelligence based multi-agent system for Alzheimer health care. *International Journal of Ambient Computing and Intelligence, 1*(1), 15–26. doi:10.4018/jaci.2009010102

Tatiraju, S., & Mehta, A. (2011). Image Segmentation using k-means clustering, EM and Normalized Cuts. In *ICIAP'11 Proceedings of the 16th International Conference on Image Analysis and Processing* (vol. 2, pp. 229-240). Genova, Italy: Springer-Verlag.

Tenney, D. R., Davis, J. G., Byron, R., & Johnston, N. (2009). NASA Composite Materials Development: Lessons Learned and Future Challenges. *NATO RTO AVT-164 Workshop on Support of Composite Systems*. Bonn, Germany: NASA Technical Reports Server (NTRS).

Tkáčová, M., Foffova, P., Živčák, J., & Hudak, R. (2010, January). The methodics of medical thermography in the diagnostics of the human body musculoskeletal system. *2010 IEEE 8th International Symposium in Applied Machine Intelligence and Informatics (SAMI),* (pp. 275-277). IEEE.

Trabelsi, I., & Bouhlel, M. S. (2017). Feature Selection for GUMI Kernel-Based SVM in Speech Emotion Recognition. In Artificial Intelligence: Concepts, Methodologies, Tools, and Applications (pp. 941-953). IGI Global.

Ulmer S., Pottler K., Lüpfert E., Röger M. (2007). Measurement techniques for the optical quality assessment of parabolic trough collector fields in commercial solar power plants. *Energy Sustainability*, 27-30.

Ulmer, S., Marz, T., Prahl, C., Reinalter, W., & Belhomme, B. (2011). Automated high resolution measurement of heliostat slope errors. *Solar Energy*, *85*(4), 681–687. doi:10.1016/j.solener.2010.01.010

Usha, K., & Ezhilarasan, M. (2013). Personal Authentication based on Angular Geometric Analysis using Finger Back Knuckle Surface. *Proceedings of 3rd International Conference on Advances in Communication Network and Computing*, *1*(1), 51-58.

Usuki, H., Ishimura, K., Hagiike, M., Okano, K., Izuishi, K., Karasawa, Y., & Maeta, H. et al. (2002). Thermographic examination for carcinoma. *Biomedical Thermology*, *21*(4), 1–7.

Valdés, M., Inamura, M., Valera, J., & Yao, L. (2006). Multidimensional filtering approaches for pre-processing thermal images. *Multidimensional Systems and Signal Processing*, *17*(4), 299–325. doi:10.1007/s11045-006-0001-0

Viola, P., & Jones, M. (2001). Rapid object detection using a boosted cascade of simple features. *Proceedings of the 2001 IEEE Computer Society Conference on Computer Vision and Pattern Recognition (CVPR '01)* (Vol. 1, pp. 511-518). doi:10.1109/CVPR.2001.990517

Vollmer, M., & Möllmann, K. P. (2010). *Infrared thermal imaging: fundamentals, research and applications*. John Wiley & Sons. doi:10.1002/9783527630868

Wang, W., Gombault, S., & Guyet, T. (2008). Towards fast detecting intrusions: using key attributes of network traffic. *The third international conference on internet monitoring and protection*. IEEE Press.

Wang, J., Liu, Y., Tian, D., & Wei, D. (2007). Internet Worm Early Detection and Response Mechanism. *Journal of China Universities of Posts and Telecommunications*, *14*(3), 79–84. doi:10.1016/S1005-8885(07)60153-8

Wang, Y., Hu, J., & Phillips, D. (2007, April). A Fingerprint Orientation Model Based on 2D Fourier Expansion (FOMFE) and Its Application to Singular-Point Detection and Fingerprint Indexing. *IEEE Transactions on Pattern Analysis and Machine Intelligence*, *29*(4), 573–585. doi:10.1109/TPAMI.2007.1003 PMID:17299215

Wang, Z., & Bovik, A. (2002). A universal image quality index. *IEEE Signal Processing Letters*, *9*(3), 81–84. doi:10.1109/97.995823

Wayman, J., Jain, J. K., Maltoni, D., & Maio, D. (2005). Biometric Systems, Technology, Design and Performance Evaluation. Springer-Verlag.

Weber, C., Ulmer, S., & Koch, H. (2014). Enhancements in high-resolution dlope deviation measurement of solar concentrator mirrors. *Energy Procedia*, *49*, 2231–2240. doi:10.1016/j.egypro.2014.03.236

Wibawa & Purnomo. (2006). Early Detection on the Condition of Pancreas Organ as the Cause of Diabetes Mellitus by Real Time Iris Image Processing. *IEEE Asia Pacific Conference on Circuits and Systems*, 1008(1010), 4-7.

Więcek, B., Zwolenik, S., & Jung, A. (1999, October). Advanced thermal, visual and radiological image processing for clinical diagnostics. In *Engineering in Medicine and Biology, 1999. 21st Annual Conference and the 1999 Annual Fall Meetring of the Biomedical Engineering Society BMES/EMBS Conference, 1999.Proceedings of the First Joint* (*Vol. 2*). IEEE. doi:10.1109/IEMBS.1999.804270

Wildes, R. P. (1997). Iris Recognition: An Emerging Biometric Technology. *Proceedings of the IEEE*, *85*(9), 1348–1363. doi:10.1109/5.628669

Wildes, R. P., & Asmuth, J. C. (1994). A System for Automated Iris Recognition. *Second IEEE Workshop on Applications of Computer Vision*, 121-128.

Winfree, W. P., & Heath, D. M. (1998, March). Thermal diffusivity imaging of aerospace materials and structures. In Aerospace/Defense Sensing and Controls (pp. 282-290). International Society for Optics and Photonics.

Wolff, Socolinsky, & Eveland. (2005). Face Recognition in the Thermal Infrared. In I. P. Bir Bhanu (Ed.), Computer Vision Beyond the Visible Spectrum (pp. 167-191). Springer-Verlag London Limited.

Wolff, E. (1976). *Anatomy of the Eye and Orbit* (7th ed.). H. K. Lewis & Co. LTD.

Wong, W. K., Hui, J. H., Desa, J. B. M., Ishak, N. I. N. B., Sulaiman, A. B., & Nor, Y. B. M. (2012). Face Detection In Thermal Imaging Using Head Curve Geometry. *Proceedings of the 2012 5th International Congress on Image and Signal Processing (CISP '12)*, 1038-1041.

Wu, S. Y., Xiao, L., Cao, Y., & Li, Y. R. (2010). A parabolic dish/AMTEC solar thermal power system and its performance evaluation. *Applied Energy*, *87*(2), 452–462. doi:10.1016/j.apenergy.2009.08.041

Xiao, J., Wei, X., Lu, Z., Yu, W., & Wu, H. (2012). A review of available methods for surface shape measurement of solar concentrator in solar thermal power applications. *Renewable & Sustainable Energy Reviews*, *16*(5), 2539–2544. doi:10.1016/j.rser.2012.01.063

Xin-bo, Z., & Jiang, L. (2009). An image segmentation algorithm based on fuzzy c-means clustering. *Digital Image Processing*, *1*, 22–26.

Yahara, T., Koga, T., Yoshida, S., Nakagawa, S., Deguchi, H., & Shirouzu, K. (2003). Relationship between microvessel density and thermographic hot areas in breast cancer. *Surgery Today*, *33*(4), 243–248. doi:10.1007/s005950300055 PMID:12707816

Yang, Sun, & Sun., (2011). Finger-Knuckle-Print Recognition using Gabor Feature and OLDA. *Proc., of the 30th Chinese Control Conference*, *4*(1), 78-84.

Yang, Y. (2007). Image Segmentation by fuzzy c-means clustering algorithm with a novel penalty term. *Computing and Informatics*, *26*, 17–31.

Yan, H., & Zhang, Y. (2011). Medical Eye Image Detection System. *ICAIC, Part V, CCIS*, *228*, 619–627.

Yared, R., Abdulrazak, B., Tessier, T., & Mabilleau, P. (2015, July). Cooking risk analysis to enhance safety of elderly people in smart kitchen. In *Proceedings of the 8th ACM International Conference on Pervasive Technologies Related to Assistive Environments* (p. 12). doi:10.1145/2769493.2769516

Yu, Zhou, & Li. (2014). Personal Identification Using Finger-Knuckle-Print Based On Local Binary Pattern, Applied. *Mechanics of Materials*, *44*(1), 703–706.

Zanganeh, G., Bader, R., Pedretti, A., Pedretti, M., & Steinfeld, A. (2012). A solar dish concentrator based on ellipsoidal polyester membrane facets. *Solar Energy*, *86*(1), 40–47. doi:10.1016/j.solener.2011.09.001

Zhang, Li, & Shen. (2011). A Novel Riesz Transforms based Coding Scheme for Finger-Knuckle-Print Recognition. *Proc., IEEE International Conference on Hand-Based Biometrics (ICHB)*, *1*(1), 1-6.

Zhang, Zhang, & Zhang. (2009). Finger-knuckle-print: a new biometric identifier. *Proc.IEEE International Conference on Image Processing*, *1*(1), 76-82. doi:10.1109/ICIP.2009.5413734

Zhang, Zhang, & Zhang. (2010). Monogenic Code: A Novel Fast Feature Coding Algorithm with Applications to Finger-Knuckle-Print Recognition. *IEEE International Workshop on Emerging Techniques and Challenges (ETCHB)*, *1*(1), 222-231. doi:10.1109/ETCHB.2010.5559286

Zhang, L., & Li, H. (2012). Encoding local image patterns using Riesz transforms With applications to palm print and finger-knuckle-print recognition. *Image and Vision Computing*, *30*(1), 1043–1105. doi:10.1016/j.imavis.2012.09.003

Zhang, L., Zhang, L., & Zhang, D. (2009). Finger-Knuckle-Print Verification Based on Band-Limited Phase-Only Correlation. *LNCS, 5702*, 141–148.

Zhang, L., Zhang, L., Zhang, D., & Zhu, H. (2011). Ensemble of local and global information for finger–knuckle-print recognition. *Pattern Recognition, 44*(1), 1990–1998. doi:10.1016/j.patcog.2010.06.007

Zhang, X. H., Guimond, Y., & Bellec, Y. (2003). Production of complex chalcogenide glass optics by molding for thermal imaging. *Journal of Non-Crystalline Solids, 326*, 519–523. doi:10.1016/S0022-3093(03)00464-2

Zhiqiang, L., Lapp, J., & Lipinski, W. (2012). Optical design of a flat-facet solar concentrator. *Solar Engineering*, (86), 1962–6.

About the Contributors

V. Santhi has received her Ph.D. in Computer Science and Engineering from VIT University, Vellore, India. She has pursued her M.Tech. in Computer Science and Engineering from Pondicherry University, Puducherry. She has received her B.E. in Computer Science and Engineering from Bharathidasan University, Trichy, India. Currently she is working as Associate Professor in the School of Computing Science and Engineering, VIT University, Vellore, India. She has authored many national and international journal papers and one book. She is currently in the process of editing two books. Also, she has published many chapters in different books published by International publishers. She is senior member of IEEE and she is holding membership in many professional bodies like CSI, ISTE, IACSIT, IEEE and IAENG. Her areas of research include Image Processing, Digital Signal Processing, Digital Watermarking, Data Compression and Computational Intelligence.

* * *

Afnan Althoupety is working as a teaching assistant with the Department of Computer Science, Taif University. She obtained her BCs from Taif University with excellent pass. She is teaching various lab courses and introductory courses to computer science. She is also serving as a volunteer for the community works such as Big Sister Little Sister conducted in her college.

Amira S. Ashour, an Asst. Professor and Vice Chair of Computer Engineering Department of in Computers and Information Technology College, Taif University, KSA. Vice Chair of Computers Science Department of in Computers and Information Technology College, Taif University, KSA during 2006 till 2015. She is a Lecturer of Electronics and Electrical Communications Engineering, Faculty of Engg., Tanta University, Egypt. She got her Ph.D. in Smart Antenna (2005) in the Electronics and Electrical Communications Engineering, Tanta University, Egypt. She had her master in "Enhancement of Electromagnetic Non-Destructive Evaluation Performance Using Advanced Signal Processing Techniques," Faculty of Engineering,

Egypt, 2000. Her research interests include Image Processing, Medical Imaging, Smart Antenna and Adaptive Antenna Arrays.

Salwa Bouadila, Engineer of Energy Engineering, received her Master of Energy and Transfer from National Institute of Applied Science and Technology of Tunisia, 2010 and her PhD of Physics from the University of Tunis, 2015. Actually, her is an Assistant Professor of Energetic Engineering in the Research and Technology Center of Energy – CRTEn. This main R&D activity is focused on the solar energy applications and latent storage research Team.

Nilanjan Dey is an Asst. Professor in the Department of Information Technology in Techno India College of Technology, Rajarhat, Kolkata, INDIA. He holds an honorary position of Visiting Scientist at Global Biomedical Technologies Inc., CA, USA and Research Scientist of Laboratory of Applied Mathematical Modeling in Human Physiology, Territorial Organization-of-gientifig and Engineering Unions, BULGARIA. He is the Editor in Chief of Int. J. of Rough Sets and Data Analysis, IGI Global,US Managing Editor of International Journal of Image Mining (IJIM), Inderscience, Regional Editor-Asia of International Journal of Intelligent Engineering Informatics (IJIEI), Inderscience and Associated Editor of International Journal of Service Science, Management, Engineering, and Technology, IGI Global. His research interests include: Medical Imaging, Soft computing, Data mining, Machine learning, Rough set, Mathematical Modeling and Computer Simulation, Modeling of Biomedical Systems, Robotics and Systems, Information Hiding, Security, Computer Aided Diagnosis, Atherosclerosis. He has 8 books and 160 international conferences and journal papers. He chaired sessions in INDIACom, 2014 Delhi, IC3T 2015, Hyderabad. He is chairing sessions in IEEE World Congress on Computational Intelligence, 2016 in Canada. He is the Program chair of AISI'15, Egypt and Publication Co-chair of CREATE'15,Inida, General Chair of RICE 2016, and Session Chair of FuzzyIEEE- 2016, Canada. He is a life member of IE, UACEE, IS.

M. Ezhilarasan was born on May 30, 1968. He has obtained Bachelor of Technology in Computer Science and Engineering, Master of Technology in Computer Science and Engineering and Ph. D. in Computer Science and Engineering from Pondicherry University in 1990, 1996 and 2007 respectively. He is Professor and Head of the Department of Information Technology, Pondicherry Engineering College, Puducherry, India. He is life member of Indian Society for Technical Education (ISTE) and member of Institute of Electrical and Electronics Engineers (IEEE) Circuits and Systems Society. He has published 71 papers in National and International Conference Proceedings, Journals and Book Chapters. His research interests include Multimedia Processing/Coding, Multi-Biometrics and Cognitive Science.

Nur Izzati Nadiah binti Ishak is currently working with Telekom Research and Development Sdn Bhd as a Research Officer. Her research interest fall under computer vision, infrared imaging and image processing.

Tariq Javid obtained PhD in Electronics Engineering from NUST, Pakistan in 2010. He obtained ME degree in Computer Systems and BE degree in Electronics from NED University of Engineering & Technology, Karachi. He is currently serving as Associate Professor at Hamdard Institute of Engineering & Technology, Hamdard University, Karachi-74600, Pakistan. Before that, he was working as Network and Systems Engineer at the Aga Khan University Hospital, Karachi, Pakistan. His research interests include embedded systems, fuzzy systems, image processing, and information security.

Usha Kazhagamani was born on October 31st 1980. She has received her B.Tech in Computer Science and Engineering and M.Tech in Distributed Computing Systems from Pondicherry University, Pondicherry, India. Currently, She is pursuing her Ph.D in Computer Science and Engineering from Pondicherry Engineering College, Pondicherry, India. Her area of interest includes Image Processing, Hand-based Biometrics and Software Engineering.

Iqbal Uddin Khan obtained MS degree from Hamdard University. He is currently serving as Assistant Professor in Hamdard Institute of Engineering & Technology, Hamdard University, Karachi-74600, Pakistan.

Ram Kumar obtained his B. Tech in Electronics and Communication Engineering from Rajasthan Institute of Engineering and Technology, Jaipur in 2010. He obtained his M. Tech in microelectronics and VLSI from National Institute of Technology, Silchar (India) in 2013. Currently he is a Research Scholar in National Institute of Technology, Silchar. His area of interest is biomedical engineering and VLSI circuits.

Heng Siong Lim is an Associate Professor for the Faculty of Engineering and Technology, Multimedia University. His research interests are in Signal Processing, Image Processing and Wireless Communications.

Humberto Loaiza-Correa is an Electrical Engineer (1990) and Master in Automation (1995) from Universidad del Valle, Colombia; and Ph.D. in Robotics (1999) from Université d'Evry -Val-d'Essonne, France. Professor and Director of the School of Electrical and Electronic Engineering of Universidad del Valle; Co-Director of the research group on Perception and Intelligent Systems (PSI). His areas of interest are Mobile Robotic, Artificial Vision, and Digital Image and Signal Processing.

Jalil bin Md Desa work under Telekom Research and Development Sdn Bhd as a Principal Researcher / Head of Network Security. His research interest fall under computer network & networking, network security, high-speed optical network, computer knowledge management & internet of things.

Ahmar Mutaza obtained MS degree from Hamdard University. He is currently serving as Lecturer in Hamdard Institute of Engineering & Technology, Hamdard University, Karachi-74600, Pakistan.

Poonguzhali N is currently a Research Scholar in the Department of Computer Science and Engineering at Pondicherry Engineering College, Puducherry, India. She received her M.C.A. degree and M.Tech., degree in Computer Science and Engineering from Pondicherry University in the year 1999 and 2009 respectively. She has a teaching experience of 10 years and has published more than fifteen research papers in reputed International Journals and Conferences. She is a member of IEEE, CSI, IETE and ISTE professional societies. Her areas of interest include Image Processing, Biometrics and Artificial Neural Networks.

Jyotsna Rani obtained her High School degree from Cotton College, Guwahati. Currently she is pursuing her B.Tech in Electronics and Communication Engineering from National Institute of Technology, Silchar (India) (expected to graduate in 2016). Her area of interest is Biomedical Engineering and Digital Image Processing.

Aqeel Rehman received MS in Information Technology from Hamdrad University, Karachi, Pakistan in 2001 and BS in Electronic Engineering from Sir Syed University of Engineering and Technology, Karachi, Pakistan in 1998. He completed his Ph.D. degree in Computer Science from National University of Computer and Emerging Sciences, Karachi, Pakistan in 2012. He is associated with Faculty of Engineering Sciences and Technology, Hamdard University, Karachi, Pakistan as the Professor and is the Chairman of Department of Computing. His current research interests include Sensor Networks, Ubiquitous Computing, Smart Agriculture, Context Modeling and Computer Networks.

Andrés-David Restrepo-Girón is an Electronics Engineer (1999), Master in Automation (2005) and Ph.D. in Engineering (2014) from Universidad del Valle, Colombia. Professor of the School of Electrical and Electronic Engineering of Universidad del Valle, and associated to the research group on Perception and

Intelligent Systems (PSI). His areas of interest are Electronic Instrumentation and Measurement, Thermography and Ultrasonics, Micro-controlled Systems, and Digital Image and Signal Processing.

Murugan Sethuraman is a Research Scholar in Alagappa University, Karaikudi, Tamilnadu, India. He received BSc in Physics from Madurai Kamaraj University, Madurai, in 1989 and MCA degree from School of Computer Science and Engineering, Alagappa University, Karaikudi, Tamilnadu, India in 1992 and MPhil (CS) from Manonmaniam Sundaranar University, Tirunelveli, Tamilnadu, India in 2002. He has 22 years of teaching and admin experience at UG &PG level in the field of Computer Science. He has published 10 papers in the National Level and 30 papers in International Level. His research interests include: Intelligence Network Security Algorithms, Malware prevention and Detection mechanism and algorithm. He has published various books and courseware in the field of Computer Science.

Safa Skouri obtained an Energy Engineering Diploma, from the National Engineering School of Monastir (ENIM) in 2008, a master Diploma in Energy Engineering from the (ENIM) in 2009 and PHD diploma entitled realized in (ENIM) with the cooperation of Research and Technologies Center of Energy, Tunisia (CRTEn).

Fazal A. Talukdar obtained his B. E (Hons) from Regional Engineering College Silchar [now, National Institute of Technology, Silchar (India)] in 1987. He obtained his M. Tech in 1993 and PhD in 2002-2003 from Indian Institute of Technology Delhi and Jadavpur University respectively. Currently, he is working as Professor in Electronics and Communication Engineering at National Institute of Technology, Silchar (India). His area of interest is image processing and VLSI circuits.

Wai Kit Wong received the Bachelor of Engineering Science Degree (Honors) in Electronics Engineering majoring in Telecommunications (2003), Master of Engineering Sciences Degree (2006) and PhD Engineering (2012) in Multimedia University, Malaysia. In December 2010, he joined the Faculty of Engineering and Technology, Multimedia University, as a senior lecturer. His research interests fall in the general areas of electronics and telecommunications. He has worked in the areas of digital signal processing, image processing, thermal imaging and surveillance technologies. His current research interests include smart home security system and image processing tools for better teleconferencing.

Index

U

V

www.ingramcontent.com/pod-product-compliance
Lightning Source LLC
LaVergne TN
LVHW061546290926
849162LV00004B/921

* 9 7 8 1 5 2 2 5 2 4 2 3 6 *